One-Parameter Semigroups

CWI Monographs

Managing Editors

J.W. de Bakker (CWI, Amsterdam)
M. Hazewinkel (CWI, Amsterdam)
J.K. Lenstra (CWI, Amsterdam)

Editorial Board

W. Albers (Maastricht)
P.C. Baayen (Amsterdam)
R.T. Boute (Nijmegen)
E.M. de Jager (Amsterdam)
M.A. Kaashoek (Amsterdam)
M.S. Keane (Delft)
J.P.C. Kleijnen (Tilburg)
H. Kwakernaak (Enschede)
J. van Leeuwen (Utrecht)
P.W.H. Lemmens (Utrecht)
M. van der Put (Groningen)
M. Rem (Eindhoven)
A.H.G. Rinnooy Kan (Rotterdam)
M.N. Spijker (Leiden)

Centrum voor Wiskunde en Informatica
Centre for Mathematics and Computer Science
P.O. Box 4079, 1009 AB Amsterdam, The Netherlands

The CWI is a research institute of the Stichting Mathematisch Centrum, which was founded on February 11, 1946, as a nonprofit institution aiming at the promotion of mathematics, computer science, and their applications. It is sponsored by the Dutch Government through the Netherlands Organization for the Advancement of Pure Research (Z.W.O.)

CWI Monograph 5

One-Parameter Semigroups

Ph. Clément
H. J. A. M. Heijmans
S. Angenent
C. J. van Duijn
B. de Pagter

1987

North-Holland
Amsterdam · New York · Oxford · Tokyo

ISBN: 0 444 70284 9

Publishers:
Elsevier Science Publishers B.V.
P.O. Box 1991
1000 BZ Amsterdam
The Netherlands

Sole distributors for the U.S.A. and Canada:
Elsevier Science Publishing Company, Inc.
52 Vanderbilt Avenue
New York, N.Y. 10017
U.S.A.

Cover: Tobias Baanders

Library of Congress Cataloging-in-Publication Data

One-parameter semigroups / Ph. Clément ... [et al.].
p. cm. -- (CWI monographs ; 5)
Bibliography: p.
Includes index.
ISBN 0-444-70284-9
1. Semigroups of operators. I. Clément, Ph. II. Series: CWI monograph ; 5.
QA329.054 1987
512'.2--dc19 87-16470
CIP

Printed in the Netherlands

Preface

This book grew out of a seminar organized by Ph. Clément and held at the Department of Mathematics and Informatics of the Technical University of Delft during the academic year 1984-1985.

The theory of semigroups of operators was initiated by E. Hille in his monograph "Functional Analysis and Semigroups" which appeared in 1948. In the years thereafter the theory was developed further by W. Feller, T. Kato, R.S. Phillips, K. Yosida and many others. The possible range of applications is enormous and includes problems in mathematical physics, probability theory and control theory.

It is not the purpose of this book to give a systematic treatment of the theory of one-parameter semigroups, but rather to illustrate the richness of this theory by presenting in one volume some of its various aspects. The book is written in such a way that all three parts can be read more or less independently. It is assumed that the reader is familiar with some of the basic principles of functional analysis.

We want to express our thanks to Odo Diekmann for his encouragement during the preparation of this volume. Furthermore we are much indebted to G. Da Prato, R. Grimmer, P. Grisvard, and in particular to J.A.P. Heesterbeek and C.B. Huijsmans for reading parts of the manuscript and correcting many errors. B. de Pagter gratefully acknowledges financial support by the Netherlands Organization for the Advancement of Pure Research (ZWO). Finally we wish to thank the Centre for Mathematics and Computer Science in Amsterdam (CWI) and its typing staff for the production of this volume by its phototypesetting system, Wim Aspers for his editorial assistance, Tobias Baanders for drawing the figures, but most particularly Carolien Swagerman for her patient and careful typing of the manuscript.

Ph. Clément
H.J.A.M. Heijmans

Amsterdam, March 1987

Author's adresses

S. Angenent
Department of Mathematics
California Institute of Technology
Pasadena
California 91125, USA

From september 1987 on:
Department of Mathematics
University of Wisconsin - Madison
Madison
Wisconsin 53706, USA

Ph. Clément
Department of Mathematics and Informatics
Technical University Delft
P.O. Box 356
2600 AJ Delft, The Netherlands

C.J. van Duijn
Department of Mathematics and Informatics
Technical University Delft
P.O. Box 356
2600 AJ Delft, The Netherlands

H.J.A.M. Heijmans
Centre for Mathematics and Computer Science
Kruislaan 413
1098 SJ Amsterdam, The Netherlands

B. de Pagter
Department of Mathematics and Informatics
Technical University Delft
P.O. Box 356
2600 AJ Delft, The Netherlands

Contents

Chapter 1

Introduction

0. Preliminaries

Let $(X, \|\cdot\|)$ be a (real or complex) Banach space. By $\mathfrak{L}(X)$ we denote the algebra of all bounded linear operators on X. A family $\{T(t)\}_{t\geqslant 0}$ in $\mathfrak{L}(X)$ is called a one-parameter semigroup of bounded linear operators in X if

$$T(0) = I \tag{1.1}$$

$$T(s+t) = T(s)T(t), \; s,t \geqslant 0. \tag{1.2}$$

Here I stands for the identity operator. If, in addition, the function $t \rightarrow T(t)$ is continuous with respect to the strong operator topology of $\mathfrak{L}(X)$, i.e. $t \rightarrow T(t)x$ is continuous on $[0, \infty)$ for every $x \in X$, then $\{T(t)\}_{t\geqslant 0}$ is called a *strongly continuous semigroup,* or also *C_0-semigroup*. This definition can be extended to nonlinear semigroups, but in this section we will restrict ourselves to the linear case. In a similar way, one can define *weakly continuous* semigroups, *uniformly continuous* semigroups, and, if X is a dual space, *weakly * continuous* semigroups. It turns out that every weakly continuous semigroup is automatically strongly continuous (see YOSIDA [1980]). Furthermore, there are infinite-dimensional Banach spaces X on which every C_0-semigroup is uniformly continuous (in particular if $X = L_\infty(S, \mu)$: see Theorem 7.36, Corollary 7.37). It is easily seen that

$$T(t) = e^{tA} = \sum_{k=0}^{\infty} \frac{1}{k!} t^k A^k, \; t \geqslant 0, \tag{1.3}$$

defines a uniformly continuous semigroup on X if A is a bounded linear operator on X. In this example, A is called the *infinitesimal generator* of $\{T(t)\}_{t\geqslant 0}$. Note that A is the right-derivative of $t \rightarrow T(t)$ at $t=0$. One of the basic facts from the theory of C_0-semigroups is the existence of a nontrivial infinitesimal generator A for every C_0-semigroup $\{T(t)\}_{t\geq 0}$. In general, A is unbounded and has a domain $\mathfrak{D}(A)$ which is a subspace of X. The precise definitions of $\mathfrak{D}(A)$ and A are:

$$\mathfrak{D}(A) = \{x \in X : \lim_{h\downarrow 0} \frac{1}{h}(T(h)x - x) \text{ exists in } X\}, \tag{1.4a}$$

$$Ax = \lim_{h\downarrow 0} \frac{1}{h}(T(h)x - x), \; x \in \mathfrak{D}(A). \tag{1.4b}$$

Then $\mathfrak{D}(A)$ is a dense linear subspace of X, and the linear operator $A : \mathfrak{D}(A) \to X$ is a closed (usually unbounded) operator. Actually, the infinitesimal generator of a strongly continuous semigroup is bounded (and everywhere defined) if and only if this semigroup is uniformly continuous. One always has that $T(t)$ leaves $\mathfrak{D}(A)$ invariant, and

$$AT(t)x = T(t)Ax, \; x \in \mathfrak{D}(A), \; t \geq 0.$$

Moreover, for every $x \in \mathfrak{D}(A)$, the function u given by $u(t) = T(t)x$, $t \geq 0$, is differentiable on $[0, \infty)$ and the unique solution to the abstract Cauchy problem

$$\frac{du}{dt}(t) = Au(t), \; u(0) = x, \; t \geq 0. \tag{1.5}$$

Thus the semigroup $\{T(t)\}_{t\geq 0}$ is completely characterized by its infinitesimal generator. One remark about notation: in this book we shall sometimes write $T_A(t)$ instead of $T(t)$, and A_T instead of A, in order to emphasize the one-one relation between semigroup and generator.

Now, let $\{T(t)\}_{t\geq 0}$ be a C_0-semigroup with generator A (we often omit the adjective infinitesimal). The type $\omega_0(A)$ of $\{T(t)\}_{t\geq 0}$ is defined by (see Section 3.7)

$$\omega_0(A) = \inf_{t>0} \frac{1}{t} \log \|T(t)\|.$$

It can be shown that $\omega_0(A) < \infty$ and that for every $\omega > \omega_0(A)$ there is a constant $M \geq 1$ such that

$$\|T(t)\| \leq Me^{\omega t}, \; t \geq 0. \tag{1.6}$$

Moreover, $\{\lambda \in \mathbb{C} : \mathrm{Re}\lambda > \omega_0(A)\}$ is contained in $\rho(A)$, the *resolvent set* of A, and

$$(\lambda I - A)^{-1}x = \int_0^\infty e^{-\lambda t} T(t)x dt, \tag{1.7}$$

for every $x \in X$ and $\lambda \in \mathbb{C}$ with $\mathrm{Re}\lambda > \omega_0(A)$. Thus the resolvent operator $R(\lambda, A) = (\lambda I - A)^{-1}$ of A equals the *Laplace transform* of the semigroup for $\mathrm{Re}\lambda > \omega_0(A)$: see also Chapter 3. Combining (1.6) and (1.7) one can easily show that

$$\|R(\lambda,A)^n\| \leqslant M(\mathrm{Re}\lambda-\omega)^{-n}, n\in\mathbb{Z}^+, \mathrm{Re}\lambda>\omega. \tag{1.8}$$

The *Hille-Yosida theorem* (see Chapter 3) states that this property characterizes generators of C_0-semigroups.

The verification of condition (1.8) for a given operator A is in practice by no means trivial since it involves an estimate for every iterate of $R(\lambda,A)$. However, there is a special but important case where this condition is workable, that is when $M=1$. Then it is sufficient to satisfy (1.8) only for $n=1$. In particular if $\omega=0$, it suffices to prove that $\|R(\lambda,A)\| \leqslant 1$. Then the corresponding semigroup $\{T(t)\}_{t\geqslant 0}$ also satifies $\|T(t)\| \leqslant 1$, $t \geqslant 0$, and is called a *contraction* C_0-semigroup. Contraction semigroups are studied in Part I.

Another important class of semigroups for which (1.8) only has to be verified for $R(\lambda,A)$ is the class of *analytic semigroups.* In this context one can also introduce the important notion of *maximal regularity.* This maximal regularity property, if present, is a very useful tool in the study of nonlinear Cauchy problems via linearization. Analytic semigroups are studied in part II.

It turns out that the asymptotic behaviour of a C_0-semigroup is intimately related to the spectral properties of its infinitesimal generator. In this respect, positivity of the semigroup, if present, can play an important role. Part III contains a thorough study of positive semigroups and their spectral properties.

1. Introduction to Part I

Let F be a nonempty closed subset of the Banach space X. The operator $L:F\to X$ is called a *contraction operator* if

$$\|Lx - Ly\| \leqslant \|x-y\|, \; x,y\in F.$$

A strongly continuous semigroup $T(t):F\to F$ is called a *contraction semigroup* if $T(t)$ is a contraction operator, for every $t\geqslant 0$. In Chapter 2 we discuss the relation between contraction semigroups and the so-called *dissipative operators* (see Definition 2.2). In Section 2.1 we present without proof the biunique correspondence between linear contraction semigroups and their infinitesimal generators. This introductory section serves as a motivation for the study of nonlinear contraction semigroups generated by dissipative operators (see Sections 2.2-2.4). In Section 2.2, the basic generation theorem due to Crandall-Liggett is given. Roughly speaking, the content of this theorem is the following. Given a (linear or nonlinear, not necessarily singlevalued) operator A with domain $\mathfrak{D}(A)$ and range X, one can associate with A a family of (linear or nonlinear) operators $\{T_A(t)\}_{t\geqslant 0}$ defined on $\overline{\mathfrak{D}(A)}$, the closure of the domain. The operators $T_A(t):\overline{\mathfrak{D}(A)}\to\overline{\mathfrak{D}(A)}$, are supposed to be the "*solution operators*" for the initial value problem

$$\begin{aligned} &\frac{du}{dt}(t)\in Au(t), \quad t>0, \\ &u(0)=x\in\overline{\mathfrak{D}(A)}. \end{aligned} \tag{1.9}$$

That is, $u(t)$ defined by $u(t)=T_A(t)x$ is supposed to be a solution of (1.9). In order to "solve" (1.9), one considers the Euler implicit scheme obtained by replacing $(du/dt)(t)$ by the differential quotient $h^{-1}(u(t+h)-u(t))$, and by evaluating Au at $t+h$. Starting at $t=0$, one obtains an approximation of the solution of (1.9) at $t=h$ by solving the equation

$$\frac{1}{h}(u(h)-x)\in Au(h), \tag{1.10}$$

or, equivalently, $x\in(I-hA)u(h)$. If one assumes that (1.10) is uniquely solvable for every $x\in X$ and every $h>0$, and if $(I-hA)^{-1}x$ is the unique solution of (1.10), then the definition

$$(I-\frac{t}{n}A)^{-n}x=(I-\frac{t}{n}A)^{-1}.....(I-\frac{t}{n}A)^{-1}x$$

makes sense for $t>0$, $n\in\mathbb{Z}^+$, and $x\in X$. Under the additional assumption that for every $h>0$, the operators $(I-hA)^{-1}:X\to X$ are contractions, in which case A is called *m-dissipative,* one finds that $\lim_{n\to\infty}(I-t/n\cdot A)^{-n}x$ exists for every $t\geqslant 0$ and $x\in\overline{\mathfrak{D}(A)}$: see Theorem 2.3. For future use we introduce the notation

$$E_h^A=(I-hA)^{-1}. \tag{1.11}$$

We call E_h^A the *Euler operator,* being the operator obtained by performing one "Euler-implicit" step. Setting

$$T_A(t)x=\lim_{n\to\infty}(I-\frac{t}{n}A)^{-n}x,$$

we find that $T_A(t):\overline{\mathfrak{D}(A)}\to\overline{\mathfrak{D}(A)}$ is a contraction operator on $\overline{\mathfrak{D}(A)}$ for every $t\geqslant 0$, and that the semigroup property is satisfied, that is:

$$T_A(0)=I|_{\overline{\mathfrak{D}(A)}},$$
$$T_A(t+s)=T_A(t)T_A(s),\ t,s\geqslant 0.$$

Moreover, the function $t\to T_A(t)x$ is Lipschitz continuous on $\mathbb{R}_+$ whenever $x\in\mathfrak{D}(A)$, and thus continuous on $\mathbb{R}_+$ if $x\in\overline{\mathfrak{D}(A)}$. So we come to the conclusion that $\{T_A(t)\}_{t\geqslant 0}$ is a strongly continuous contraction semigroup on $\overline{\mathfrak{D}(A)}$. If the function $t\to T_A(t)x$ is right-differentiable at $t=t_0$, then $T_A(t_0)x\in\mathfrak{D}(A)$, and

$$\frac{d^+}{dt}T_A(t)x|_{t=t_0}\in AT_A(t_0)x$$

(see Theorem 2.15). Thus, if $t\to T_A(t)x$ is differentiable on $\mathbb{R}_+$, then $u(t)=T_A(t)x$ is a solution of (1.9).

However, if A is nonlinear, then it may happen that the function $u(t)$ is *not* differentiable. Nevertheless, it "often" occurs that, when (1.9) is an abstract formulation of a "concrete" problem (e.g. a partial differential equation), then

u is the "right" solution to this problem, even though it is not differentiable. For an interesting illustration of this phenomenon we refer the reader to Chapter 4.

In Section 2.3 an alternative approximation method for solutions of (1.9) is considered. In the linear case, this method was introduced by Yosida. The idea is to replace A by the *Yosida approximation*

$$A_h = \frac{1}{h}((I - hA)^{-1} - I), \tag{1.12}$$

where $h > 0$. Since A_h is Lipschitz continuous on X, the initial value problem

$$\begin{aligned} &\frac{du}{dt}(t) = A_h u(t), \quad t > 0, \\ &u(0) = x \in X, \end{aligned} \tag{1.13}$$

possesses a unique differentiable solution u_h, which can be obtained by a standard iteration procedure. Now, provided that $x \in \overline{\mathfrak{D}(A)}$, $u_h(t)$ converges to $u(t) = T_A(t)x$, uniformly on bounded t-intervals: see Corollary 2.10. It can be shown that A_h is m-dissipative, and that $u_h(t) = T_{A_h}(t)x$. Even if A is singlevalued and $x \in \mathfrak{D}(A)$, it is not true in general that $\lim_{h\downarrow 0} A_h x = Ax$. However, there is convergence of A_h to A in the sense of the resolvents, to be precise

$$\lim_{h\downarrow 0} (I - \lambda A_h)^{-1} x = (I - \lambda A)^{-1} x,$$

for every $\lambda > 0$ and $x \in X$ (see Proposition 2.8).

If one has a sequence $\{A_n\}_{n \geq 1}$ of m-dissipative operators in X converging in the sense of the resolvents to some m-dissipative operator A, then the corresponding functions $t \to T_{A_n}(t)x$ converge uniformly on bounded t-intervals to $t \to T_A(t)x$, for every $x \in X$ (see Theorem 2.7). This shows that the semigroup $\{T_A(t)\}_{t \geq 0}$ depends continuously on A in a certain sense, and thus it is amply justified to call $t \to T_A(t)x$ the solution of (1.9).

These results together constitute the core of the theory of nonlinear contraction semigroups in general Banach spaces. However, a lot more can be said, if one knows in addition that X possesses specific geometrical properties. In Section 2.1 we state without proof some basic theorems concerning the Hilbert space case. In Section 2.5 we briefly mention some extentions of the results of Sections 2.2-2.4 to the more general case that $A - \omega I$ is m-dissipative for some $\omega > 0$. Nothing, however, is said about the inhomogeneous problem

$$\begin{aligned} &\frac{du}{dt}(t) \in Au(t) + f(t), \ t > 0, \\ &u(0) = x, \end{aligned} \tag{1.14}$$

where $f:[0,\infty)\to X$ is a given function. There is however one exception: if $f(t)=y$, $t\geqslant 0$, where $y\in X$, then (1.14) reduces to (1.9) if A is replaced by $A+y$. Here the operator $A+y$ is defined by

$$(A+y)x = Ax + y, \; x\in\mathfrak{D}(A).$$

Chapter 3 is devoted to the problem of generation and approximation of C_0-semigroups of linear contractions and more generally of bounded linear operators in Banach spaces. As a rule we have adopted the point of view of reducing (when possible) the general case to the "contraction" case. This "reduction" is performed in Section 3.2 and also partly in Section 3.7. There we have tried to use the results of Chapter 2, even if there are shorter ways in the linear case. In contrast with the nonlinear case, a C_0-semigroup of linear contractions (or, more generally, of bounded linear operators) is completely determined by its infinitesimal generator: see Section 3.1. Since the domain of an infinitesimal generator of a C_0-semigroup $\{T(t)\}_{t\geqslant 0}$ by definition consists of all $x\in X$ such that the function $t\to T(t)x$ is differentiable at $t=0$, and hence for every $t>0$, it follows that the initial value problem (1.9) can be solved in the usual sense. If A is a linear, m-dissipative operator, and if $\{T_A(t)\}_{t\geqslant 0}$ is the semigroup associated with A, then the function $t\to T_A(t)x$ is right-differentiable at $t=0$ if and only if $x\in\mathfrak{D}(A)$ and $Ax\in\overline{\mathfrak{D}(A)}$: see Corollary 3.10. This characterization also corresponds to the set of $x\in\mathfrak{D}(A)$ for which $\lim_{h\downarrow 0}A_hx=Ax$. Similarly, precise information about the differentiability of the solution to the inhomogeneous problem (1.14) can be obtained in the linear case. The regularity of the solution of (1.14) can be obtained by using a result of Da Prato and Grisvard concerning the sum of two "commuting" m-dissipative operators. This is explained in Section 3.6.

In Section 3.4 we consider *dual semigroups.* Considerable attention is paid to the notion of *Favard class.* The Favard class or *generalized domain* $\mathrm{Fav}(T)$ of a C_0-semigroup $\{T(t)\}_{t\geqslant 0}$ is the subspace of all $x\in X$ for which the function $t\to T(t)x$ is Lipschitz continuous on $\mathbb{R}_+$, or more precisely

$$\mathrm{Fav}(T) = \{x\in X\colon \sup_{h>0}\frac{1}{h}\|T(h)x - x\|<\infty\}.$$

It turns out that there is a nice characterization of the Favard class in terms of dual semigroups. In Section 3.5 we give a very recent perturbation result for dual semigroups.

In Section 3.7, we complexify the space X and the semigroup $\{T(t)\}_{t\geqslant 0}$, and use the theory of Banach space-valued holomorphic functions to obtain more information about the resolvent of A. More results of this kind can be found in Chapters 5 and 6.

In Chapter 4 we give an application of the contraction semigroup theory as presented in Chapter 2. The application concerns the initial value problem

$$\text{(I)}\quad \begin{aligned} &u_t = (\phi(u))_{xx}\ , && (t,x)\in\mathbb{R}^+\times\mathbb{R},\\ &u(0,x) = u_0(x)\ , && x\in\mathbb{R}, \end{aligned}$$

where ϕ is a maximal monotone graph in $\mathbb{R}^2$. In the above equation, the subscripts t and x denote differentiation with respect to these variables. We consider here the one-dimensional case only for the sake of simplicity. Nearly all results can be generalized to higher dimensions. First we reformulate Problem (I) as an abstract initial value problem in a Banach space X. We choose to work here in the space $X=L_1(\mathbb{R})$. We associate a nonlinear operator A_ϕ in $L_1(\mathbb{R})$ with the formal expression $A_\phi u=(\phi(u))_{xx}$. This is done as follows. Let $A_\phi:\mathfrak{D}(A_\phi)\subseteq L_1(\mathbb{R})\rightarrow L_1(\mathbb{R})$ be given by

$$\mathfrak{D}(A_\phi) = \{u\in L_1(\mathbb{R}) : \exists w\in L_1^{loc}(\mathbb{R}) \text{ such that } w(x)\in\phi(u(x)) \text{ a.e. on } \mathbb{R} \text{ and } w''\in L_1(\mathbb{R}) \text{ in distributional sense}\},$$

and

$$A_\phi u = w'' \text{ for every } u\in\mathfrak{D}(A_\phi).$$

In Sections 4.2 and 4.3 we show that A_ϕ is m-dissipative in $L_1(\mathbb{R})$. In Section 4.2 this is done for the case $0\in\phi(0)$ and $0\in\operatorname{int}\mathfrak{D}(\phi)$ and in Section 4.3 for the case $\mathfrak{D}(\phi)\subseteq(0,\infty)$. Note however that in these sections we rewrite and discuss the problem in terms of the inverse maximal monotone graph $\beta=\phi^{-1}$. Then it follows from the Crandall-Liggett theorem (Theorem 2.3) that the operator A_ϕ generates a semigroup of contractions $\{T_{A_\phi}(t)\}_{t\geqslant 0}$ on $\overline{\mathfrak{D}(A_\phi)}$. The function $u:[0,\infty)\rightarrow X$, defined by

$$u(t) = T_{A_\phi}(t)u_0\ ,\ u_0\in\overline{\mathfrak{D}(A_\phi)} \text{ and } t\geqslant 0,$$

is the unique mild solution of the abstract problem

$$\begin{aligned} &\frac{du}{dt}\in A_\phi(u),\\ &u(0) = u_0. \end{aligned}$$

As a function of (t,x), $u(t,x)=(T_{A_\phi}(t)u_0)(x)$, we call u the semigroup solution of Problem (I). From estimates obtained in Sections 4.2 and 4.3 we find that u satisfies a *comparison principle*: if $u_0,v_0\in\overline{\mathfrak{D}(A_\phi)}$ are such that $u_0\geqslant v_0$ a.e. on $\mathbb{R}$, then

$$(T_{A_\phi}(t)u_0)(x)\geqslant(T_{A_\phi}(t)v_0)(x)\ , \text{ a.e. on } \mathbb{R}^+\times\mathbb{R}.$$

Furthermore, u satisfies the *mass conservation property*:

$$\int_{\mathbb{R}} u(t,x)dx = \int_{\mathbb{R}} u_0(x)dx \text{ for all } t\geqslant 0.$$

In Section 4.4 we show that if ϕ is continuous and nondecreasing such that either $\phi(0)=0$ and $0\in\operatorname{int}\mathfrak{D}(\phi)$ or $\mathfrak{D}(\phi)\subseteq(0,\infty)$, and if the initial value is essentially bounded, then the semigroup solution is in fact a weak solution of Problem (I): it satisfies the differential equation in the sense of distributions. In

Section 4.5 we continue with the case $0 \in \phi(0)$ and $0 \in \operatorname{int} \mathfrak{D}(\phi)$, and consider the dependence on ϕ of solutions of Problem (I). The main problem here is to show that if $\{\phi_n\}_{n \geq 1}$ is a sequence of functions converging to ϕ and having similar properties as ϕ, then $A_{\phi_n} \to A_\phi$ in the sense of the resolvent, i.e.

$$(I - A_{\phi_n})^{-1} f \to (I - A_\phi)^{-1} f \quad \text{for every } f \in L_1(\mathbb{R}).$$

We also characterize the set $\overline{\mathfrak{D}(A_\phi)}$ which is of importance for applications. We show that

$$\overline{\mathfrak{D}(A_\phi)} = \{u \in L_1(\mathbb{R}) : \inf \mathfrak{D}(\phi) \leq u(x) \leq \sup \mathfrak{D}(\phi) \text{ a.e. on } \mathbb{R}\}.$$

Clearly if $\mathfrak{D}(\phi) = \mathbb{R}$, then $\overline{\mathfrak{D}(A_\phi)} = L_1(\mathbb{R})$.

Chapter 4 is concluded with a *regularity property* of the semigroup solution of Problem (I) for the special case where $\phi(s) = |s|^m \operatorname{sign}(s)$, $s \in \mathbb{R}$, $m > 0$ $(m \neq 1)$. In this case, a solution $u(t,x)$ of Problem (I) generates by rescaling the family of solutions

$$u_\lambda(t,x) = \lambda^{1/(m-1)} u(\lambda t, x) \quad \lambda > 0,$$

corresponding to the initial values $\lambda^{1/(m-1)} u_0(x)$. Then the associated semigroup $\{T_{A_\phi}(t)\}_{t \geq 0}$ satisfies the homogeneity property

$$T_{A_\phi}(t)(\lambda^{1/(m-1)} u_0) = \lambda^{1/(m-1)} T_{A_\phi}(\lambda t) u_0$$

for all $u_0 \in \overline{\mathfrak{D}(A_\phi)} = L_1(\mathbb{R})$. In Section 4.5 we first give two general results for a family of mappings $\{T(t)\}_{t \geq 0}$ which satisfy this homogeneity property on a suitably chosen subset C of a normed vector space X. Application of these results to the semigroup solution of Problem (I) gives

$$\limsup_{h \downarrow 0} \frac{\|u(t+h) - u(t)\|_1}{h} \leq \frac{2\|u_0\|_1}{|m-1|t},$$

and for $u_0(x) \geq 0$ a.e. on $\mathbb{R}$ (which implies $u(t,x) \geq 0$ a.e. on $\mathbb{R}^+ \times \mathbb{R}$)

$$(m-1) u_t \geq -\frac{u}{t} \quad \text{in } \mathcal{D}'((0,\infty) \times \mathbb{R}).$$

2. Introduction to part II

In part II we present some of the classical results in the theory of *analytic semigroups,* and also the more recent results of Da Prato and Grisvard on the interaction between the interpolation theory of Banach spaces and the theory of analytic semigroups.

The choice of material for Chapter 5 has been influenced by the author's prejudice towards applications of the theory to (nonlinear parabolic) PDE's. Thus a number of topics which may be interesting from a functional analytic point of view have been omitted. One such topic deals with the many different ways in which an analytic semigroup can be characterized (van Casteren [1985] gives more than ten equivalent definitions!). In view of the fine textbooks that have already appeared (DAVIES [1980], PAZY [1983], VAN CASTEREN [1985]) the omissions do not seem to be serious.

In our presentation of the theory we regard the generator A of the semigroup as the given object and then try to construct the semigroup from A. We regard the generator A as a bounded linear operator $A:E_1 \rightarrow E_0$ between two different Banach spaces E_1 and E_0, where E_1 is a dense subspace of E_0. The traditional notation for this situation would be $E_0 = X$ and $E_1 = \mathfrak{D}(A)$, and with this notation A would be called an unbounded linear operator in X with domain $\mathfrak{D}(A)$. We feel that, as far as applications to parabolic PDE's are concerned, this notation overemphasizes the role of the space X. (Indeed, when dealing with explicitly given operators, X is often just one of the many Banach (-Sobolev) spaces lying on your desk!). Therefore we have adopted a slightly different notation, which is also natural from the point of view that comes with interpolation theory.

Given the pair of spaces $E_1 \subset E_0$, we introduce in Section 5.2 a class of operators $A:E_1 \rightarrow E_0$ which we denote by $Hol(E)$ (they turn out to be the generators of holomorphic semigroups on E_0 with domain E_1). In the section on the resolvent we define $\sigma(A)$ and establish the "sectorial property" of an $A \in Hol(E)$. Then, in Section 5.4, we prove some perturbation theorems. One of these theorems says that $Hol(E)$ is an open subset of the bounded operators from E_1 to E_0 (with the norm topology). This fact turns out to be quite useful, since it allows us to do calculus on $Hol(E)$. By contrast, the set $Gen(E)$ of all $A:E_1 \rightarrow E_0$ which merely generate a C_0 semigroup does not have this property.

Calculus on $Hol(E)$ is discussed in Section 5.5. A class of holomorphic functions $\mathfrak{F}$ is defined, and by using a Dunford-integral a corresponding operator $f(A)$ is constructed in a standard way. We have chosen the class $\mathfrak{F}$ in such a manner that it contains the exponential functions $e^{t\lambda}$ for $t>0$ but is otherwise small enough to allow an easy definition of $f(A)$.

Using the calculus of Section 5.5 the next section finally shows that the $A \in Hol(E)$ are generators of holomorphic semigroups on E_0. In fact the functional calculus gives an exponential mapping exp: $Hol(E) \rightarrow \mathfrak{L}(E)$ which is holomorphic (this makes sense because we know that $Hol(E)$ is an open subset of the Banach space $\mathfrak{L}(E_1, E_0)$ and hence has an analytic structure). Again using the calculus, we derive some of the classical estimates on e^{tA} and discuss the domains of integral powers of A.

Chapter 5 is concluded with a number of examples and exercises. The first few examples deal with operators A which do or do not generate holomorphic semigroups. These examples are all multiplication operators on a weighted $L_2(\mathbb{R})$ space so that they can only illustrate the less subtle distinctions between $Hol(E)$ and the rest of $\mathfrak{L}(E_1, E_0)$.

Another exercise is devoted to *fractional powers* of an $A \in Hol(E)$. Finally, in the last exercise we mention, perhaps the most classical of all analytic semigroups, the *Gaussian semigroup.* This exercise shows that the functional calculus can sometimes be used to obtain explicit expressions for e^{tA}.

The sixth chapter is about inhomogeneous initial value problems of the form

$$\begin{aligned} &\frac{du}{dt}(t)=Au(t)+f(t), \quad 0\leqslant t\leqslant T, \\ &u(0)=x, \end{aligned} \tag{1.15}$$

where A belongs to $Hol\,(E)$. If the known function $f(t)$ has a certain amount of regularity (e.g. if $f{:}[0,T]\rightarrow E_1$ is continuous so that $Af(t)$ is well defined) then one can easily show that the solution of (1.15) exists. One can even write down the solution:

$$u(t) = e^{tA}x + \int_0^t e^{(t-s)A}f(s)ds$$

(this is the *variation-of-constants formula*). Now this expression still makes sense if f belongs to $L_1((0,T);E_0)$, but it is not clear if the corresponding $u(t)$ is in any sense a solution of (1.15). On first sight one might expect that, if f were continuous with values in E_0, then the other two terms in (1.15) might inherit this property. In other words, one might expect that $f\in C([0,T];E_0)$ would imply $u\in C([0,T];E_1)\cap C^q([0,T];E_0)$. In general this turns out to be false; however, the maximal regularity result of Section 6.5 indicates a class of triples (A,E_1,E_0) for which the mentioned expectation is indeed true. The formulation of this result (which is due to Da Prato and Grisvard) involves some concepts from the interpolation theory of Banach spaces. Therefore we have given a brief description of a small part of this theory in Sections 6.1, 6.2 and 6.3.

After mentioning a few generalities about interpolation methods in Section 6.1 we define the *continuous interpolation method.* This method is closely related to the real (θ,∞) method. Since the definition of the continuous interpolation method is not the one we need to prove the maximal regularity result, we have denoted one section to a number of equivalent definitions of this interpolation method. The techniques used in this section are quite standard in interpolation theory; since we are dealing with only one interpolation method they may serve as an illustration of the more advanced theory presented in the books of BERGH & LÖFSTRÖM [1976] or TRIEBEL [1978].

Section 6.4 introduces the problem (1.15) and considers the solution of (1.15) which is given by the variation-of-constants formula. Some of the easier classical results on the regularity of the solution $u(t)$ are proved in Lemma 6.9. Then, at last, we state and prove the maximal regularity result in Section 6.5.

The chapter ends with a number of examples and exercises. In the first few exercises the reader is asked to compute several interpolation spaces, and some attention is given to the little Hölder spaces $h^{\theta}(T)$ on the circle. Then there are two exercises on *fractional power spaces,* and their relation to the continuous interpolation spaces. The last set of exercises is meant to illustrate the maximal regularity result of Section 6.5. Some examples are given in which maximal regularity fails, and the reader is asked to think about the implications of Theorem 6.10 for the standard heat equation.

3. INTRODUCTION TO PART III

Let X be an ordered Banach space with positive cone X^+. For unknown terminology we refer the reader to Chapter 7. Let $\{T(t)\}_{t\geq 0}$ be a C_0-semigroup of bounded linear operators with generator A. Then $\{T(t)\}_{t\geq 0}$ is called *positive* if every operator $T(t)$, $t\geq 0$ is positive, i.e. $T(t)X^+ \subseteq X^+$, $t\geq 0$. In Chapter 7 we address ourselves to the following question: what kind of operators are generators of positive semigroups ?

In Section 7.2 various characterizations of generators of contractive positive semigroups are presented. These characterizations are given in terms of p-dissipativity of the generator, where p is the canonical half-norm associated with the ordered Banach space X (i.e., $p(x)$ denotes the distance of the vector $-x$ to the positive cone of X). This leads to theorems which are analogous to the Hille-Yosida-Phillips theorem.

The characterization of generators of positive C_0-semigroups without any metric conditions (such as contractivity) is in general a more difficult problem. On $C(\Omega)$, with Ω a compact Hausdorff space, the situation is particularly easy (Example 7.31): a densely defined linear operator A on $X=C(\Omega)$ is the generator of a positive C_0-semigroup if and only if $\mathfrak{R}(\lambda I-A)=X$, for $\lambda\in\mathbb{R}$ large enough, and satisfies the *positive-off-diagonal* (= POD)-property:

$$(Ax)(\omega)\geq 0 \text{ if } 0\leq x\in\mathfrak{D}(A) \text{ and } \omega\in\Omega \text{ with } x(\omega)=0.$$

Here $\mathfrak{R}(\cdot)$ denotes the range. Note that this property is a kind of abstract *maximum principle.* The POD-property can be generalized to an arbitrary Banach space: see Definition 7.18. The result stated above for $C(\Omega)$ can be carried over to ordered Banach spaces for which X^+ is normal and has nonempty interior, i.e. Banach spaces with an order unit (e.g. certain spaces of differentiable functions, or the self-adjoint part of a unitary C^*-algebra): see Theorem 7.29. It is easily seen that $R(\lambda,A)$ is a positive operator for $\lambda\in\mathbb{R}$ sufficiently large if A is the generator of a positive C_0-semigroup. Conversely, if X is an ordered Banach space with normal cone X^+ whose interior is nonempty, then a densely defined linear operator A is the generator of a C_0-semigroup if $\lambda\in\mathbb{R}$ sufficiently large implies that $\lambda\in\rho(A)$ and $R(\lambda,A)\geq 0$ (see Theorem 7.29). In these results the assumption that X^+ has nonempty interior cannot be omitted: see Example 7.32.

The *abstract Kato inequality* opens the way to extend the above results to situations where the cone does have an empty interior. This abstract Kato inequality is nothing but a generalization of the classical Kato inequality for the

Laplacian Δ on $\mathbb{R}^n$, which can be stated as follows:

for all $f \in L_1^{loc}(\mathbb{R}^n)$ with $\Delta f \in L_1^{loc}(\mathbb{R}^n)$, the inequality

$$\mathrm{Re}(\mathrm{sign}\, f \cdot \Delta f) \leqslant \Delta |f|$$

holds in the sense of distributions.

The inequality is related to the positivity of the semigroup on $L_2(\mathbb{R}^n)$ generated by Δ.

Let X be a Dedekind σ-complete Banach lattice and define for every $x \in X$ the linear *sign-mapping* $\sigma_x : X \to X$ as in Section 7.4. Then the operator A is said to satisfy the abstract Kato inequality if

$$<\sigma_x(Ax), \phi> \leqslant <|x|, A^* \phi>,$$

for every $x \in \mathcal{D}(A)$ and $0 \leqslant \phi \in \mathcal{D}(A^*)$. It is not difficult to check that every generator of a positive C_0-semigroup on a Dedekind σ-complete Banach lattice X satisfies the abstract Kato inequality (Proposition 7.39). Conversely, if X is a Dedekind σ-complete Banach lattice, and if A is the generator of a C_0-semigroup $\{T(t)\}_{t \geqslant 0}$ (and if some additional assumptions hold), then the validity of the abstract Kato inequality for A is a guarantee for the positivity of the semigroup $\{T(t)\}_{t \geqslant 0}$: see Theorem 7.44. This important result is extended to general Banach lattices in Proposition 7.46.

Chapter 7 is concluded with the *Beurling-Deny theorem,* which gives a criterion for the positivity of the semigroup generated by a form-positive self-adjoint operator on the Hilbert space $L_2(S, \mu)$.

An important issue in semigroup theory is the *asymptotic behaviour.* From the theory of linear autonomous ODE's it is well-known that the solution of the n-dimensional linear system

$$\frac{du}{dt} = Au, \; u(0) = x,$$

where A is an $n \times n$-matrix, and $x \in \mathbb{R}^n$, is given by $u(t) = T(t)x = e^{tA}x$. The asymptotic behaviour of such solutions is completely determined by the eigenvalues of A, in particular by the *dominant eigenvalue* of A, that is the eigenvalue with the largest real part. This is easily seen if one transforms A into the Jordan canonical form.

In the general case that X is an arbitrary Banach space and $\{T(t)\}_{t \geqslant 0}$ is a C_0-semigroup with infinitesimal generator A, life is not so easy. Here the spectrum of A as well as the spectrum of $T(t)$, may contain infinitely many elements, and these elements need not be eigenvalues. Moreover, the spectrum of $\{T(t)\}_{t \geqslant 0}$ is not always "faithfull" to the spectrum of A, but can be strictly larger, in contrary to the finite-dimensional case. The various relations between the spectrum of $T(t)$ and that of A are described in Section 8.1. An important role is given to the so-called *Browder essential spectrum* $\sigma_{ess}(\cdot)$ (see Appendix A.3.1). We can define the *essential type* $\omega_{ess} = \omega_{ess}(A)$ by

$$\omega_{ess}(A) = \inf_{t>0} \frac{1}{t} \log |T(t)|_\alpha, \tag{1.16}$$

where $|\cdot|_\alpha$ is the so-called *measure-of-noncompactness* (see Appendix A.3.1). It is important to note that $\omega_{ess}(A) = -\infty$ if $\{T(t)\}_{t\geq 0}$ is eventually compact. The spectral bound $s(A)$ of A is given by

$$s(A) = \sup \{\mathrm{Re}\lambda : \lambda \in \sigma(A)\}. \tag{1.17}$$

The following relation always holds (see Proposition 8.6):

$$\omega_0(A) = \max \{s(A), \omega_{ess}(A)\}. \tag{1.18}$$

If $\omega_0(A) < 0$, then the semigroup $\{T(t)\}_{t\geq 0}$ is *uniformly stable,* that is

$$\|T(t)\| \to 0 \text{ as } t \to \infty.$$

It is not difficult to show that uniform stability of a semigroup implies that $s(A) < 0$. However, the converse result is not true in general. This leads to the question for what kind of C_0-semigroups one has that $s(A) = \omega_0(A)$. In Section 9.1 it is shown that this relation holds in each of the following cases:

(i) $\{T(t)\}$ is eventually uniformly continuous, which is e.g. true if $\{T(t)\}_{t\geq 0}$ is an analytic semigroup;
(ii) $X = L_1(\Omega, \mu)$ or $L_2(\Omega, \mu)$ and $\{T(t)\}_{t\geq 0}$ is a positive semigroup;
(iii) $X = C(\Omega)$, where Ω is a compact Hausdorff space, and $\{T(t)\}_{t\geq 0}$ is a positive semigroup.

(ii) and (iii) suggest that perhaps $s(A) = \omega_0(A)$ holds for every positive semigroup on a Banach lattice X, but Example 9.3 shows that this is not true, unfortunately. It is still an open question whether the relation holds for each positive semigroup on the reflexive Banach lattice $L_p(\Omega, \mu)$, $1 < p < \infty$, $p \neq 2$.

For every $x \in X$, the *exponential growth bound* $\omega(x)$ of the orbit $\{T(t)x : t \geq 0\}$ is defined by

$$\omega(x) = \limsup_{t\to\infty} \frac{1}{t} \log \|T(t)x\|.$$

Then $\omega(x)$ is the abscissa of absolute convergence of the Laplace transform of the function $t \to T(t)x$. Then

$$\omega_0(A) = \sup \{\omega(x) : x \in X\}.$$

The number $\omega_1 = \omega_1(A)$ defined by

$$\omega_1 = \sup \{\omega(x) : x \in \mathfrak{D}(A)\},$$

is called the *exponential growth bound of solutions of the Cauchy problem* (1.5). If X is a Banach lattice, and $\{T(t)\}_{t\geq 0}$ is a positive semigroup, then $s(A) = \omega_1(A)$ (see Corollary 9.9).

In many applications it is possible to prove that solutions $u(t) = T(t)x$ of the Cauchy problem (1.5) behave in the following way:

$$e^{-\lambda t}T(t)x \to c x_0, \quad t \to \infty. \tag{1.19}$$

Here $\lambda \in \mathbb{C}$ and c is a constant depending linearly on the initial data x. In most cases λ is real. In Section 9.2 it is explained that, in order to get (1.19), it is sufficient that the following two conditions are satisfied:
(C_1) A has a *strictly dominant* algebraically simple eigenvalue λ_0,
(C_2) $\omega_{ess}(A) < \omega_0(A)$.
We call an eigenvalue λ_0 *dominant* resp. *strictly dominant* if $\mathrm{Re}\lambda \leqslant \mathrm{Re}\lambda_0$ resp. $\mathrm{Re}\lambda < \mathrm{Re}\lambda_0$ for every $\lambda \in \sigma(A)$, $\lambda \neq \lambda_0$. Actually, the convergence in (1.19) is exponentially if both (C_1) and (C_2) are satisfied (see Section 9.2). The positivity of a semigroup can be of considerable help in proving (C_1). If X is a Banach lattice with absolute value $|\cdot|$, and $\{T(t)\}_{t\geqslant 0}$ is a positive semigroup then there exist a number of powerful results characterizing the spectrum of its generator. The *peripheral spectrum* $\sigma_+(A)$ of A is defined as

$$\sigma_+(A) = \{\lambda \in \sigma(A) : \mathrm{Re}\lambda = s(A)\}$$

if $\sigma(A) \neq \varnothing$ and $\sigma_+(A) = \varnothing$ if $\sigma(A) = \varnothing$. If $\{T(t)\}_{t\geqslant 0}$ is a positive semigroup, then as a consequence of the *Pringsheim-Landau theorem*,

$$s(A) = \sup\{\lambda \in \mathbb{R} : \lambda \in \sigma(A)\},$$

and

$$R(\lambda, A)x = \int_0^\infty e^{-\lambda t}T(t)x\,dt,$$

for all $x \in X$ and $\lambda \in \mathbb{C}$ with $\mathrm{Re}\lambda > s(A)$, where the integral exists as a norm convergent improper integral. So $R(\lambda, A)$ is a positive operator for $\lambda > s(A)$, and

$$|R(\lambda, A)x| \leqslant R(\mathrm{Re}\lambda, A)|x|,$$

for $\lambda > s(A)$ and $x \in X$. In particular $s(A) \in \sigma(A)$ if $\sigma(A) \neq \varnothing$: see Theorem 7.4 and Corollary 7.5. So $\lambda_0 := s(A)$ is the dominant eigenvalue of A in that case. If, in addition, λ_0 is a pole of the resolvent $R(\lambda, A)$, then $\sigma_+(A)$ is additively cyclic, i.e. for every $\nu \in \mathbb{R}$,

$$\lambda_0 + i\nu \in \sigma_+(A) \;\Rightarrow\; \lambda_0 + ik\nu \in \sigma_+(A), \quad k \in \mathbb{Z}$$

(see Theorem 8.7 and Theorem 8.14). If, moreover, the semigroup $\{T(t)\}_{t\geqslant 0}$ is irreducible, then λ_0 is an algebraically simple eigenvalue, and the corresponding eigenvector x_0 is not only positive but also quasi-interior (see Theorem 8.17). Note that, if (C_2) is satisfied, then λ_0 is a pole of $R(\lambda, A)$.

In Section 9.3 it is indicated how perturbation theory can be used to verify condition (C_2) (here compactness properties come into play: see Proposition 9.20) and to establish the irreducibility of the semigroup $\{T(t)\}_{t\geqslant 0}$ (see Theorem 9.21).

In Chapter 10, finally, these abstract results are applied to some problems from structured population dynamics.

Part I

Contraction Semigroups

Chapter 2

Nonlinear Contraction Semigroups and Dissipative Operators

1. Hille-Yosida theorem and Trotter-Neveu-Kato theorem

In this section we will recall (without proof) two basic theorems from the theory of linear C_0-contraction semigroups on a Banach space, which will serve as a motivation for the study of nonlinear contraction semigroups. Let $(X, \|\cdot\|)$ be a real Banach space and let $C(X)$ denote the set of all linear C_0-contraction semigroups on X (see the Introduction). Given $T \in C(X)$, let A_T be its infinitesimal generator, that is the linear operator in X with domain

$$\mathfrak{D}(A_T) := \{x \in X : \lim_{h \downarrow 0} h^{-1}(T(h)x - x) \text{ exists}\},$$

and defined by

$$A_T x := \lim_{h \downarrow 0} h^{-1}(T(h)x - x), \tag{2.0}$$

for $x \in \mathfrak{D}(A_T)$. One part of the Hille-Yosida Generation Theorem says that A_T satisfies the following properties: for every $x \in \mathfrak{D}(A_T)$ and for every $h > 0$

$$\|x\| \leqslant \|x - hA_T x\|, \tag{2.1}$$

$$\mathfrak{R}(I - hA_T) = X \text{ for every } h > 0, \tag{2.2}$$

$$\mathfrak{D}(A_T) \text{ is dense in } X. \tag{2.3}$$

Here $\mathfrak{R}(I - hA_T)$ denotes the range of $I - hA_T$. A linear operator $A : \mathfrak{D}(A) \subset X \to X$ satisfying (2.1) is called *dissipative in* $(X, \|\cdot\|)$ (also: $-A$ is called *accretive*). A dissipative operator in $(X, \|\cdot\|)$ satisfying (2.2) is called *m-dissipative* ($-A$ is then called *m-accretive*). We shall denote by $D(X)$ the set of all densely defined m-dissipative operators in $(X, \|\cdot\|)$, and by $\mathcal{G} : C(X) \to D(X)$ the map which associates with $T \in C(X)$ its infinitesimal generator A_T.

The second part of the Hille-Yosida theorem asserts that $\mathcal{G}$ is a surjective map. We also consider as a part of this theorem that $\mathcal{G}$ is injective. Moreover the bijection $\mathcal{G}$ enjoys some "continuity" property. We shall say that a sequence $\{T_n\}$ in $C(X)$ converges to $T \in C(X)$ if

$$\lim_{n\to\infty} \max_{t\in[0,s]} \|T_n(t)x - T(t)x\| = 0, \tag{2.4}$$

for every $s > 0$ and for every $x \in X$. Given $A \in D(X)$, observe that for every $h > 0$, $I - hA$ is a bijection from $\mathcal{D}(A)$ onto X and that $E_h^A := (I - hA)^{-1}$ is a bounded operator in X satisfying $\|E_h^A\| \leqslant 1$. We shall say that a sequence $\{A_n\}$ in $D(X)$ converges to A in $D(X)$ if for every $x \in X$ and every $h > 0$,

$$\lim_{n\to\infty} \|E_h^{A_n} x - E_h^A x\| = 0. \tag{2.5}$$

It is a consequence of the theorem of Trotter-Neveu-Kato that $\{T_n\}$ converges to T in $C(X)$ iff $\{\mathcal{G}(T_n)\}$ converges to $\mathcal{G}(T)$ in $D(X)$. Thus there is a nice correspondence between linear C_0-contraction semigroups and their generators. An important consequence of these theorems is the solvability of the initial value problem

$$\begin{aligned} &\frac{du}{dt}(t) = Au(t), \quad t > 0, \\ &u(0) = x, \end{aligned} \tag{2.6}$$

for $A \in D(X)$. Indeed if $x \in \mathcal{D}(A)$ and T is the semigroup associated with A, then $t \to u(t,x) = T(t)x$ belongs to $C^1([0,\infty);X)$ and satisfies (2.6). Moreover if $\{A_n\}$ is a sequence in $D(X)$ converging to A, then the corresponding semigroups $\{T_n\}$ will converge to T in $C(X)$. In fact, in the proof of the surjectivity of $\mathcal{G}$, Yosida constructed a sequence $\{A_n\}$ of *bounded* operators in $D(X)$ converging to A. For $A \in D(X)$ and $h > 0$, he defined

$$A_h := h^{-1}(E_h^A - I). \tag{2.7}$$

A_h is usually called the *Yosida approximation* of A. We shall prove in Lemma 2.9 that if $C: X \to X$ is a (linear) contraction ($\|C\| \leqslant 1$) and $\lambda > 0$, then $\lambda(C - I) \in D(X)$. In particular $A_h \in D(X)$ for $h > 0$ and it will be shown (see Proposition 2.8) that A_{h_n} converges to A in $D(X)$ when $h_n \to 0$. Note that the semigroup generated by A_h is given by

$$\sum_{k=0}^{\infty} \frac{1}{k!}(tA_h)^k.$$

If $C(t): X \to X$, $t \in \mathbb{R}^+$ is a family of linear contractions in X, then $t^{-1}(C(t) - I) \in D(X)$ for $t \in (0,1]$. Suppose $t^{-1}(C(t) - I)$ converges to some A in $D(X)$ as $t \downarrow 0$. Then it has been proven by CHERNOFF [1968] that

$$\lim_{n\to\infty} \max_{t\in[0,s]} \|C(\frac{t}{n})^n x - T(t)x\| = 0, \tag{2.8}$$

for every $x \in X$ and $s > 0$, where T is the semigroup associated with A. In particular, if $C(t)=E_t^A$, $t>0$, then it follows that

$$\lim_{n\to\infty} \max_{t\in[0,s]} \|(E_{t/n}^A)^n x - T(t)x\| = 0, \tag{2.9}$$

for every $x\in X$ and $s>0$. This provides another representation of $T(t)$ by the familiar exponential formula $\lim_{n\to\infty}(I-t/n\cdot A)^{-n}x$, which will be used as a starting point in the nonlinear theory. This formula was first used by Hille for the construction of T. Finally it is interesting to note that if $T\in C(X)$, then $t^{-1}(T(t)-I)\in D(X)$ for $t>0$, and $t^{-1}(T(t)-I)$ converges to A_T in $D(X)$ as $t\downarrow 0$. (2.8) is trivially satisfied in this case, but the corresponding semigroups provide another approximation of T by semigroups having a bounded generator.

The goal of this chapter is to prove nonlinear versions of these theorems whenever they exist. It is a remarkable fact that all of the above theorems have a nonlinear analogue if X is a Hilbert space. For a detailed and nice treatment of the Hilbert space case we refer the reader to BREZIS [1973]. We first give the definition of a nonlinear contraction semigroup.

DEFINITION 2.1. Let F be a nonempty closed subset of a Banach space $(X,\|\cdot\|)$. A nonlinear contraction semigroup on F is a family of operators $T(t):F\to F$, $t\geq 0$ satisfying:

(I) $T(t+s)=T(t)T(s)$ for every $s,t\geq 0$, $T(0)=I_F$ (identity on F).

(II) $\|T(t)x-T(t)y\|\leq\|x-y\|$ for every $t\geq 0$ and for every $x,y\in F$.

(III) For every $x\in F$, $t\to T(t)x$ is continuous on $[0,\infty)$.

Observe that if $\{T(t)\}_{t\geq 0}$ satisfies (I), (II) and $\lim_{t\downarrow 0}T(t)x=x$ for every $x\in F$, then it also satisfies (III). We shall denote the class of all nonlinear contraction semigroups on F by $C(F;X)$. Motivated by the linear theory, we define the *infinitesimal generator* of T by (2.0). It may happen (even if $F=X$) that $\mathfrak{D}(A_T)=\varnothing$ (see CRANDALL and LIGGETT [1971a]). However if $\mathfrak{D}(A_T)\neq\varnothing$ ($\mathfrak{D}(A_T)$ is even dense in X if $F=X$ and X is a Hilbert space), then A_T satisfies

$$\|x-y\|\leq\|(x-y)-h(A_Tx-A_Ty)\| \tag{2.10}$$

for every $x,y\in\mathfrak{D}(A_T)$ and every $h>0$. Indeed if $A_{T,t}:=t^{-1}(T(t)-I)$ for $t>0$, we have for $x,y\in F$:

$$\begin{aligned}&\|(x-y)-h(A_{T,t}x-A_{T,t}y)\|\\ &=\|(1+t^{-1}h)(x-y)-t^{-1}h(T(t)x-T(t)y)\|\\ &\geq(1+t^{-1}h)\|x-y\|-t^{-1}h\|x-y\|=\|x-y\|.\end{aligned}$$

Then (2.10) holds by taking the limit as $t\downarrow 0$. Thus $I-hA_T$ is injective for every $h>0$ and $(I-hA_T)^{-1}:\mathfrak{R}(I-hA_T)\to X$ is a nonlinear contraction. Obviously (2.10) is a nonlinear version of the notion of dissipative operator.

Consider the following example, $X=\mathbb{R}$, $\|\cdot\|=|\cdot|$, and for $t\geq 0$:

$$T(t)x = \begin{cases} (x-t)^+ & \text{, if } x \geqslant 0, \\ -(x+t)^- & \text{, if } x<0, \end{cases}$$

where $a^+ = a \vee 0$ and $a^- = (-a)^+$. Then T is a contraction semigroup on $\mathbb{R}$ and

$$A_T x = \begin{cases} +1 & \text{, if } x<0, \\ 0 & \text{, if } x=0, \\ -1 & \text{, if } x>0. \end{cases}$$

A_T satisfies (2.10) but $\mathcal{R}(I - A_T) \neq \mathbb{R}$. However, if we define the graph $A \subset \mathbb{R} \times \mathbb{R}$ by setting $[x,y] \in A$ iff $y=1$ for $x<0$, $y=-1$ for $x>0$, and $y \in [-1,1]$ for $x=0$, then $\mathcal{R}(I-hA)=\mathbb{R}$ for every $h>0$ and $(I-hA)^{-1}$ is the graph of a function from $\mathbb{R} \to \mathbb{R}$ satisfying

$$|(I-hA)^{-1}x - (I-hA)^{-1}y| \leqslant |x-y|,$$

for every $h>0$, $x,y \in \mathbb{R}$. This is a nonlinear version of an "m-dissipative operator". Moreover, if we define as above $A_{T,t}$ by

$$A_{T,t} = t^{-1}(T(t) - I),\ t>0,$$

we have $(I - hA_{T,t})^{-1}: \mathbb{R} \to \mathbb{R}$ is a contraction and

$$\lim_{t \downarrow 0} (I - hA_{T,t})^{-1}x = (I-hA)^{-1}x,$$

for every $x \in \mathbb{R}$, and every $h>0$.

This example and the theory of nonlinear contraction semigroups in Hilbert spaces motivate the following definitions.

DEFINITION 2.2. Let $(X, \|\cdot\|)$ be a Banach space and let $A \subset X \times X$ be a graph. Then A is called *dissipative* ($-A$ *accretive*) if $(I-hA)^{-1}$ is the graph of a nonlinear contraction for every $h>0$. Clearly the domain of $(I-hA)^{-1}$ is $\mathcal{R}(I-hA)$. A dissipative graph in $X \times X$ is called *m-dissipative* ($-A$ is called *m-accretive*) if $(I-hA)^{-1}$ is everywhere defined in X for every $h>0$.

In Definition 2.2 and in what follows, we shall identify an operator A from $\mathcal{D}(A) \subset X \to 2^X$ with its graph in $X \times X$, which we also denote by A, and conversely we shall identify a graph A in $X \times X$ with a multivalued operator $A: \mathcal{D}(A) \subset X \to 2^X$. Thus for $A \subset X \times X$,

$$\mathcal{D}(A) := \{u \in X : [u,v] \in A\},$$

$$\mathcal{R}(A) := \{v \in X : [u,v] \in A\}.$$

$\mathcal{D}(A)$ is called the domain of A and $\mathcal{R}(A)$ the range of A.

$$A^{-1} := \{[u,v] \in X \times X : [v,u] \in A\}.$$

For $A, B \subset X \times X$,

$$A \pm B := \{[u,v] \in X \times X : u \in \mathfrak{D}(A) \cap \mathfrak{D}(B) \text{ and } v = v_1 \pm v_2$$
$$\text{where } [u,v_1] \in A \text{ and } [u,v_2] \in B\}.$$
$$\lambda A := \{[u,v] \in X \times X : u \in \mathfrak{D}(A) \text{ and } v = \lambda u \text{ with } [u,v] \in A\},$$

for $\lambda \in \mathbb{R}$. $I := \{[u,u] \in X \times X : u \in X\}$. We shall also use the notation $y \in Ax$ for $[x,y] \in A \subset X \times X$. We shall denote by $D(F;X)$ the set of all m-dissipative operators A in X such that $\overline{\mathfrak{D}(A)} = F$ where F is a nonempty closed subset of X. We shall say that a sequence $\{T_n\}$ in $C(F;X)$ converges to $T \in C(F;X)$ if

$$\lim_{n \to \infty} \max_{t \in [0,s]} \|T_n(t)x - T(t)x\| = 0,$$

for every $x \in F$ and every $s > 0$, and that $\{A_n\}$ in $D(F;X)$ converges to $A \in D(F;X)$ if

$$\lim_{n \to \infty} \|(I - hA_n)^{-1}x - (I - hA)^{-1}x\| = 0,$$

for every $x \in X$ and every $h > 0$. With these definitions, we can state the nonlinear version of the Hille-Yosida theorem for a Hilbert space H. Let $T \in C(K;H)$, where K is a nonempty closed convex subset of H. Then there is a unique $A_T \in D(K;H)$ such that $A_{T,h}$ defined by $h^{-1}(T(h) - I)$ for $h > 0$, converges to A_T in $D(K;H)$, as $h \downarrow 0$. We shall denote by $\mathcal{G}$ the map $T \to A_T$. Moreover $\mathcal{G}$ is a bijection and $\{T_n\}$ converges to T in $C(K;H)$ iff $\{\mathcal{G}(T_n)\}$ converges to $\mathcal{G}(T)$ in $D(K;H)$. This nonlinear version of the "Hille-Yosida-Trotter-Neveu-Kato" theorem surely motivates the use of multivalued operators. We mention that $A \subset H \times H$ is dissipative iff $-A$ is monotone in H, and A is m-dissipative iff $-A$ is maximal monotone in H. See Brezis [1973] and Appendix 1.

The situation in a general Banach space is not so nice. As we already mentioned, if $T \in C(X;X)$ then $\mathfrak{D}(A_T)$ may be empty. Moreover it is shown in Crandall and Liggett [1971a], that even if X is finite-dimensional, with dim $X \geqslant 3$, and if $T \in C(X;X)$, then $A_{T,t}$ above defined does not need to converge to some $A \in D(X;X)$ as $t \downarrow 0$. It is interesting to note that if $T \in C(F;X)$, then the contraction semigroups with infinitesimal generator $A_{T,t} = t^{-1}(T(t) - I)$ for $t > 0$, converge to T in $C(F;F)$ as $t \downarrow 0$ (see Kobayashi [1974]). It follows that even if a sequence of contraction semigroups $T_n \in C(X;X)$ converges to a contraction semigroup T in $C(X;X)$, then their infinitesimal generators, if they exist, do not need to converge to some element of $D(X;X)$. Thus for a general Banach space there is no counterpart of the map $\mathcal{G}: C(X) \to D(X)$ in the nonlinear case. In what follows, we shall see that an analogue of $\mathcal{F} := \mathcal{G}^{-1}: D(X) \to C(X)$ does exist in the nonlinear case for a general Banach space, with the same "continuity property". However this map $\mathcal{F}$ is not injective in general (see Crandall and Liggett [1971b]). Thus two different m-dissipative operators may generate the same semigroup. Moreover it is still an open problem to know whether the map $\mathcal{F}: D(X;X) \to C(X;X)$ is surjective; see Crandall [1986].

2. The Crandall-Liggett theorem

The goal of this section is to prove the following result.

Theorem 2.3 (Crandall-Liggett). *Let A be m-dissipative in X, let $t \geqslant 0$, and $x \in \overline{\mathfrak{D}(A)}$. Then the sequence $\{(I - t/n \cdot A)^{-n} x\}_{n \geqslant 1}$ is a Cauchy sequence in X. The limit, which we denote by $T_A(t)x$, belongs to $\overline{\mathfrak{D}(A)}$, and $T_A(t): \overline{\mathfrak{D}(A)} \to \overline{\mathfrak{D}(A)}$ is a nonlinear contraction semigroup on $\overline{\mathfrak{D}(A)}$. Moreover if $x \in \mathfrak{D}(A)$, we have:*

$$\| T_A(t)x - (I - \frac{t}{n} A)^{-n} x \| \leqslant |Ax| t \sqrt{1/n}, \quad t \geqslant 0, n \geqslant 1, \tag{2.11}$$

and

$$\| T_A(t)x - T_A(s)x \| \leqslant |Ax| \cdot |t - s|, \quad t,s \geqslant 0, \tag{2.12}$$

where

$$|Ax| := \sup_{h>0} h^{-1} \| (I - hA)^{-1} x - x \|. \tag{2.13}$$

Before going to the proof of this theorem a first remark concerning (2.13) is in order. One has to show that $|Ax| < \infty$ for $x \in \mathfrak{D}(A)$. Since A is m-dissipative in X, E_h^A defined by

$$E_h^A := (I - hA)^{-1}, \text{ for } h > 0, \tag{2.14}$$

is an overall contraction in X with range in $\mathfrak{D}(A)$. Even more, E_h^A is a bijection from X onto $\mathfrak{D}(A)$. As in the linear case, we define the Yosida approximation of A by

$$A_h := h^{-1}(E_h^A - I), \text{ for } h > 0. \tag{2.15}$$

Thus we get that $|Ax| = \sup_{h>0} \|A_h x\|$. In order to prove that $|Ax| < \infty$, we first observe that for every $x \in X$ and every $h > 0$, $[E_h^A x, A_h x] \in A$. Indeed from $(I - hA)E_h^A \supseteq I$, we get $E_h^A - hAE_h^A \supseteq I$ and thus $AE_h^A \supseteq h^{-1}(E_h^A - I) = A_h$. Moreover since A is dissipative, we have for every $[x_1, y_1], [x_2, y_2] \in A$ and $h > 0$:

$$\|x_1 - x_2\| \leqslant \|(x_1 - x_2) - h(y_1 - y_2)\|.$$

By choosing $x_1 = E_h^A x$, $x_2 = x$ and $y_1 = A_h x$, with $x \in \mathfrak{D}(A)$, we obtain

$$\|E_h^A x - x\| \leqslant \|(E_h^A x - x) - h(A_h x - y)\| = \|hy\| = h\|y\|,$$

for every $y \in X$ such that $[x,y] \in A$. Thus for $x \in \mathfrak{D}(A)$, we have:

$$|Ax| \leqslant \inf\{\|y\| : [x,y] \in A\} < \infty. \tag{2.16}$$

It also follows from (2.13) and (2.16) that for $x \in \mathfrak{D}(A)$,

$$\|E_h^A x - x\| = h\|A_h x\| \leqslant h|Ax| \to 0 \text{ as } h \downarrow 0.$$

Moreover, one easily proves (see (2.37)) that

$$\lim_{h \downarrow 0} \|E_h^A x - x\| = 0 \text{ if and only if } x \in \overline{\mathfrak{D}(A)}. \tag{2.17}$$

If $x \in X$ satisfies $|Ax| < \infty$ then $\lim_{h \downarrow 0} \| E_h^A x - x \| = 0$ and thus $x \in \overline{\mathfrak{D}(A)}$. It may happen that $|Ax| < \infty$ and $x \notin \mathfrak{D}(A)$. The class of $x \in \overline{\mathfrak{D}(A)}$ such that $|Ax| < \infty$ is called the *generalized domain* of A (see CRANDALL [1973]). It is easily verified that (2.11) and (2.12) hold if x belongs to the generalized domain of A.

Theorem 2.3 will be a consequence of the following

FUNDAMENTAL LEMMA 2.4. *Let A be dissipative in X,*

$$K, \eta > 0,$$

$$t_0 = 0 < t_1 \cdots < t_i < \cdots < t_m,$$

$$s_0 = 0 < s_1 \cdots < s_j < \cdots < s_n,$$

$$\gamma_i = t_i - t_{i-1},\ i = 1, \ldots, m,$$

$$\delta_j = s_j - s_{j-1},\ j = 1, \ldots, n,$$

$$x_i \in \mathfrak{D}(A),\ y_j \in \mathfrak{D}(A),\ i = 1, \ldots, m;\ j = 1, \ldots, n,$$

$$f_i, g_j \in X,\ i = 1, \ldots, m\ ;\ j = 1, \ldots, n,$$

$$x_0, y_0 \in X,$$

where x_i, y_i, f_i, g_j satisfy

$$\frac{x_i - x_{i-1}}{t_i - t_{i-1}} \in Ax_i + f_i,\ i = 1, \ldots, m, \tag{2.18}$$

$$\frac{y_j - y_{j-1}}{s_j - s_{j-1}} \in Ay_j + g_j,\ j = 1, \ldots, n, \tag{2.19}$$

$$\left\| \frac{x_i - x_{i-1}}{t_i - t_{i-1}} \right\| \leqslant K,\ i = 1, \ldots, m, \tag{2.20}$$

$$\left\| \frac{y_j - y_{j-1}}{s_j - s_{j-1}} \right\| \leqslant K,\ j = 1, \ldots, n, \tag{2.21}$$

$$\|f_i\|, \|g_j\| \leqslant \eta,\ i = 1, \ldots, m\ ;\ j = 1, \ldots, n. \tag{2.22}$$

Then

$$\|x_m - y_n\| \leqslant \|x_0 - y_0\| + 2\eta \cdot \min(t_m, s_n) + K \sqrt{(t_m - s_n)^2 + \min(t_m, s_n) \cdot \max_{\substack{i=1,\ldots,m \\ j=1,\ldots,n}} (\gamma_i + \delta_j)}. \tag{2.23}$$

Before giving the proof of the Fundamental Lemma we draw some consequences which will imply Theorem 2.3. We have

PROPOSITION 2.5. *Let A be m-dissipative in X. For $x \in \overline{\mathfrak{D}(A)}$, $\mu, \nu > 0$, $m, n \in \mathbb{Z}^+$, we have:*

$$\|(E_\mu^A)^m x - (E_\nu^A)^n x\| \leqslant |Ax| \sqrt{(m\mu - n\nu)^2 + \min(m\mu, n\nu)\cdot(\mu+\nu)}. \quad (2.24)$$

PROOF. We apply the fundamental lemma with $x_0 = y_0 = x \in \mathfrak{D}(A)$, $f_i = g_j = 0$, $i = 1, \ldots, m$, $j = 1, \ldots, n$. Set $t_i = i\mu$, $s_j = j\nu$ ($\gamma_i = \mu$, $\delta_j = \nu$). Since A is m-dissipative, i.e. $\mathfrak{R}(I - hA) = X$, we get that $x_i = E_\mu^A x_{i-1}$, $y_j = E_\nu^A y_{j-1}$, $i = 1, \ldots, m$, $j = 1, \ldots, n$. We have

$$\frac{x_i - x_{i-1}}{t_i - t_{i-1}} \in A x_i, \; i = 1, \ldots, m,$$

$$\frac{y_j - y_{j-1}}{s_j - s_{j-1}} \in A y_j, \; j = 1, \ldots, n,$$

and

$$\left\|\frac{x_i - x_{i-1}}{t_i - t_{i-1}}\right\| = \left\|\frac{x_i - x_{i-1}}{\mu}\right\| = \frac{1}{\mu}\|(E_\mu^A)^i x - (E_\mu^A)^{i-1} x\|$$

$$\leqslant \frac{1}{\mu}\|E_\mu^A x - x\|,$$

since E_μ^A are contractions. Moreover

$$\frac{1}{\mu}\|E_\mu^A x - x\| \leqslant |Ax|.$$

Similar estimates hold for y_j, thus we can choose $K = |Ax|$ and $\eta = 0$. Moreover $\|x_0 - y_0\| = 0$. Thus (2.18)-(2.22) hold since $m\mu = t_m$ and $n\nu = s_n$. □

By setting $\mu = t/m$ and $\nu = t/n$ in (2.24), we obtain for $x \in \mathfrak{D}(A)$:

$$\|(E_{t/m}^A)^m x - (E_{t/n}^A)^n x\| \leqslant |Ax| t \sqrt{1/m + 1/n}.$$

Hence $\{(E_{t/m}^A)^m x\}_{m \geqslant 1}$ is a Cauchy sequence in X. Since $\|(E_{t/m}^A)^m\|_{Lip} \leqslant 1$, it follows that $\{E_{t/m}^A x\}_{m \geqslant 1}$ is a Cauchy sequence if $x \in \overline{\mathfrak{D}(A)}$. Since X is complete, $\lim_{m\to\infty} (E_{t/m}^A)^m x$ exists for $x \in \overline{\mathfrak{D}(A)}$, and we denote the limit by $T_A(t)x$. Since $(E_{t/m}^A)^m x$ belongs to $\mathfrak{D}(A)$ for $m \geqslant 1$, we may conclude that $T_A(t)x \in \overline{\mathfrak{D}(A)}$. Moreover $\|T_A(t)\|_{Lip} \leqslant 1$. Thus $T_A(t): \overline{\mathfrak{D}(A)} \to \overline{\mathfrak{D}(A)}$ is a contraction. Clearly

$$\|T_A(t)x - (E_{t/n}^A)^n x\| \leqslant |Ax| t \sqrt{1/n},$$

which proves (2.11). For $x \in \mathfrak{D}(A)$ and $t, s > 0$, we get from (2.24)

$$\|(E_{t/m}^A)^m x - (E_{s/n}^A)^n x\| \leqslant |Ax| \sqrt{(t-s)^2 + \min(t,s)\cdot\left(\frac{t}{m} + \frac{s}{n}\right)}.$$

Thus

$$\|T_A(t)x - T_A(s)x\| \leqslant |Ax| \cdot |t - s|,$$

which proves (2.12). By using the triangle inequality and the fact that $T_A(t)$ are contractions, one shows that $t \to T_A(t)x$ is continuous on $[0, \infty)$. In order to complete the proof of Theorem 2.3 we need to prove the semigroup property $T_A(t+s)=T_A(t)T_A(s)$ for every $t,s\geqslant 0$. By using the continuity of $t\to T_A(t)x$ for $x\in\overline{\mathfrak{D}(A)}$, it is sufficient to prove that $T_A(t+s)=T_A(t)T_A(s)$ for t,s rational and positive. This is a consequence of $T_A(mt)=T_A(t)^m$ for every $m\in\mathbb{Z}^+$ and $t>0$. Indeed if this holds, we have for $p,q,r,s\in\mathbb{Z}^+$, and $x\in\overline{\mathfrak{D}(A)}$,

$$\begin{aligned}T_A(\frac{p}{q}+\frac{r}{s})x &= T_A(\frac{ps+rq}{qs})x = T_A(\frac{1}{qs})^{ps+qr}x =\\ &= T_A(\frac{1}{qs})^{ps}T_A(\frac{1}{qs})^{qr}x = T_A(\frac{1}{qs})^{ps}T_A(\frac{r}{s})x =\\ &= T_A(\frac{p}{q})T(\frac{r}{s})x.\end{aligned}$$

Finally we prove that $T_A(mt)=T_A(t)^m$ for every $t>0$ and every $m\in\mathbb{Z}^+$. For $m=1$, it is trivial. We proceed by induction.

$$T_A(mt)x = \lim_{n\to\infty}(E^A_{mt/n})^n x.$$

Choose $n=km$, $k\in\mathbb{Z}^+$. Then

$$\begin{aligned}T_A(mt)x &= \lim_{k\to\infty}(E^A_{mt/km})^{km}x =\\ &= \lim_{k\to\infty}(E^A_{t/k})^{km}x = \lim_{k\to\infty}(E^A_{t/k})^k(E^A_{t/k})^{k(m-1)}x =\\ &= \lim_{k\to\infty}(E^A_{t/k})^k(E^A_{t(m-1)/k(m-1)})^{k(m-1)}x.\end{aligned}$$

Note that

$$\lim_{k\to\infty}(E^A_{t(m-1)/k(m-1)})^{k(m-1)}x = T_A((m-1)t)x.$$

Moreover if $x_k\in\overline{\mathfrak{D}(A)}$ and $\lim_{k\to\infty}x_k=y$, then $\lim_{k\to\infty}(E^A_{t/k})^k x_k=T_A(t)y$. Use

$$\begin{aligned}\|T_A(t)y-(E^A_{t/k})^k x_k\| &\leqslant \|T_A(t)y-(E^A_{t/k})^k y\| + \|(E^A_{t/k})^k y-(E^A_{t/k})^k x_k\|\\ &\leqslant \|T_A(t)y-(E^A_{t/k})^k y\| + \|y-x_k\|\to 0\end{aligned}$$

as $k\to\infty$. Hence

$$\begin{aligned}T_A(mt)x &= \lim_{k\to\infty}(E^A_{t/k})^k(E^A_{t(m-1)/k(m-1)})^{k(m-1)}x\\ &= T_A(t)T_A((m-1)t)x.\end{aligned}$$

We have $T_A((m-1)t)x=T_A(t)^{m-1}x$ by induction, and therefore $T_A(mt)x= T_A(t)T_A(t)^{m-1}x=T_A(t)^m x$. □

Proof of the fundamental lemma 2.4.

Since A is dissipative, it follows from (2.18) and (2.19) that

$$\begin{aligned}\|x_i-y_j\| \leqslant &\frac{\gamma_i}{\gamma_i+\delta_j}\|x_i-y_{j-1}\| + \frac{\delta_i}{\gamma_i+\delta_j}\|x_{i-1}-y_j\|\\ &+\frac{\gamma_i\delta_i}{\gamma_i+\delta_j}\|f_i-g_j\|.\end{aligned} \tag{2.25}$$

Indeed, since A is dissipative, we have for every $h>0$,

$$\|x_i - y_j\| \leqslant \|(x_i - y_j) - h[(\frac{x_i - x_{i-1}}{\gamma_i} - f_i) - (\frac{y_j - y_{i-1}}{\delta_j} - g_j)]\|.$$

Choose $h = (\gamma_i\delta_j)/(\gamma_i + \delta_j)$, then

$$(x_i - y_j) - \frac{\gamma_i\delta_j}{\gamma_i + \delta_j}[(\frac{x_i - x_{i-1}}{\gamma_i} - f_i) - (\frac{y_j - y_{j-1}}{\delta_j} - g_j)]$$

$$= \frac{\gamma_i}{\gamma_i + \delta_j}(x_i - y_{j-1}) + \frac{\delta_j}{\gamma_i + \delta_j}(x_{i-1} - y_j) + \frac{\gamma_i\delta_j}{\gamma_i + \delta_j}(f_i - g_j).$$

Then use the triangle inequality. Define $a_{i,j} := \|x_i - y_j\|$ for $i = 1, \ldots, m$, $j = 1, \ldots, n$.

$$R := \{(i,j) : i = 0, 1, \ldots, m\, ;\, j = 0, 1, \ldots, n\},$$

$$\partial R := \{(i,j) \in R : i = 0 \text{ or } j = 0\},$$

$$R^0 := R \setminus \partial R.$$

We have

$$a_{i,j} \leqslant \frac{\gamma_i}{\gamma_i + \delta_j} a_{i,j-1} + \frac{\delta_j}{\gamma_i + \delta_j} a_{i-1,j} + 2\frac{\gamma_i\delta_j}{\gamma_i + \delta_j}\eta \tag{2.26}$$

on R^0. For $j = 0$, $a_{i,0} = \|x_i - y_0\| \leqslant \|x_i - x_0\| + \|x_0 - y_0\|$. Now

$$\|x_i - x_0\| \leqslant \sum_{k=1}^{i} \|x_k - x_{k-1}\| \leqslant K \sum_{k=1}^{i} (t_k - t_{k-1}) = Kt_i,$$

by using (2.20). Thus $a_{i,0} \leqslant a_{0,0} + Kt_i$, for $i = 1,..,m$. Similarly $a_{0,j} \leqslant a_{0,0} + Ks_j$, for $j = 1, \ldots, n$. We have

$$a_{i,j} \leqslant a_{0,0} + K|t_i - s_j| \text{ on } \partial R. \tag{2.27}$$

Consider the equation:

$$\begin{cases} b_{i,j} = \dfrac{\gamma_i}{\gamma_i + \delta_j} b_{i,j-1} + \dfrac{\delta_j}{\gamma_i + \delta_j} b_{i-1,j} + \dfrac{\gamma_i\delta_j}{\gamma_i + \delta_j} h_{i,j} \text{ on } R^0, \\ b_{i,j} = e_{i,j} \text{ on } \partial R, \end{cases} \tag{2.28}$$

where $h : R^0 \to \mathbb{R}$ and $e : \partial R \to \mathbb{R}$ are given. Then:

(a) for each (e,h), (2.28) possesses a unique solution which we denote by $b(e,h)$.

(b) $b(e,h) = b(e,0) + b(0,h)$ and $e \to b(e,0)$, $h \to b(0,h)$ are linear.

(c) $e \geqslant 0$ and $h \geqslant 0$ imply $b(e,h) \geqslant 0$.

(d) If $a(e,h)$ satisfies (2.28), where the equalities are replaced by inequalities ($\leqslant$), then $a(e,h) \leqslant b(e,h)$. Thus

(e) $a_{i,j}(a_{0,0} + K|t - s|, 2\eta) \leqslant b_{i,j}(a_{0,0} + K|t - s|, 2\eta) = a_{0,0} b_{i,j}(\mathbf{1}, 0) + K b_{i,j}(|t - s|, 0) + 2\eta b_{i,j}(0, \mathbf{1})$, where $\mathbf{1}(i,j) = 1$ and $|t - s|(i,j) = |t_i - s_j|$.

(f) $b_{i,j}(\mathbf{1}, 0) = 1$, since $\dfrac{\gamma_i}{\gamma_i + \delta_j} + \dfrac{\delta_j}{\gamma_i + \delta_j} = 1$.

(g) $b_{i,j}(t,\mathbf{1})=t_i$, since $t_i=\frac{\gamma_i}{\gamma_i+\delta_j}t_i+\frac{\delta_j}{\gamma_i+\delta_j}t_{i-1}+\frac{\gamma_i\delta_j}{\gamma_i+\delta_j}$. Similarly we find that $b_{i,j}(s,\mathbf{1})=s_j$.

(h) $b_{i,j}(0,\mathbf{1})\leqslant\min(t_i,s_j)$. Indeed $b_{i,j}(0,\mathbf{1})\leqslant b_{i,j}(t,\mathbf{1})=t_i$, $b_{i,j}(0,\mathbf{1})\leqslant b_{i,j}(s,\mathbf{1})=s_j$.

(i) $b^2_{i,j}(e,0)\leqslant b_{i,j}(e^2,0)$.

$b^2_{i,j}=[\frac{\gamma_i}{\gamma_i+\delta_j}b_{i,j-1}+\frac{\delta_j}{\gamma_i+\delta_j}b_{i-1,j}]^2\leqslant\frac{\gamma_i}{\gamma_i+\delta_j}b^2_{i,j-1}+\frac{\delta_j}{\gamma_i+\delta_j}b^2_{i-1,j}$, since $\frac{\gamma_i}{\gamma_i+\delta_j},\frac{\delta_j}{\gamma_i+\delta_j}\geqslant 0$ and $\frac{\gamma_i}{\gamma_i+\delta_j}+\frac{\delta_j}{\gamma_i+\delta_j}=1$.

Thus $b^2(e,0)$ satisfies the inequality with e^2 on ∂R and $h=0$. (i) follows then from (d).

(j) $(t_i-s_j)^2=b_{i,j}((t-s)^2,-\tilde{h})$ with $\tilde{h}_{i,j}=\gamma_i+\delta_j$.

$$\frac{\gamma_i}{\gamma_i+\delta_j}(t_i-s_{j-1})^2+\frac{\delta_j}{\gamma_i+\delta_j}(t_{i-1}-s_j)^2=$$
$$\frac{\gamma_i}{\gamma_i+\delta_j}[(t_i-s_j)+\delta_j]^2+\frac{\delta_j}{\gamma_i+\delta_j}[(t_i-s_j)-\gamma_i]^2=$$
$$\frac{\gamma_i}{\gamma_i+\delta_j}(t_i-s_j)^2+\frac{\delta_j}{\gamma_i+\delta_j}(t_i-s_j)^2+2\frac{\gamma_i\delta_j}{\gamma_i+\delta_j}(t_i-s_j)-2\frac{\gamma_i\delta_j}{\gamma_i+\delta_j}(t_i-s_j)$$
$$+\frac{\delta_j^2\gamma_i+\gamma_i^2\delta_j}{\gamma_i+\delta_j}=(t_i-s_j)^2+\frac{\gamma_i\delta_j}{\gamma_i+\delta_j}(\gamma_i+\delta_j).$$

(k) $b_{i,j}((t-s)^2,0)\leqslant(t_i-s_j)^2+\max_{1\leqslant i\leqslant m,1\leqslant j\leqslant n}(\gamma_i+\delta_j)b_{i,j}(0,\mathbf{1})$.

This follows from $b_{i,j}((t-s)^2,0)=b_{i,j}((t-s)^2,-\tilde{h})+b_{i,j}(0,\tilde{h})$, and (j) and (h).

Finally we have

$$b_{i,j}(|t-s|,0)\leqslant\sqrt{(t_i-s_j)^2+\min(t_i,s_j)\cdot\max_{\substack{1\leqslant i\leqslant m\\1\leqslant j\leqslant n}}(\gamma_i+\delta_j)},$$

and

$$a_{i,j}(a_{0,0}+K|t-s|,2\eta)\leqslant b_{i,j}(a_{0,0}+K|t-s|,2\eta)$$
$$\leqslant a_{0,0}+Kb_{i,j}(|t-s|,0)+2\eta b_{i,j}(0,\mathbf{1}),$$

which proves (2.23). □

In Theorem 2.3, we have constructed the semigroup T_A by considering uniform partitions of the interval $[0,t]$, $t_0=0, t_1=t/n,\ldots,t_i=it/n,\ldots,t_n=t$, and by solving the backward difference scheme of the initial value problem $du/dt\in Au$, $u(0)=x$. In the next theorem, we show that if one takes nonuniform partitions of $[0,t]$, one gets the same limit as the step size tends to zero.

THEOREM 2.6. *Let A be an m-dissipative operator in X, and $x\in\overline{\mathfrak{D}(A)}$. Let $P_n:=\{s_0^n=0<s_1^n<\cdots<s^n_{N(n)}=\bar{t}\}$ be a sequence of partitions of $[0,\bar{t}]$ such that $\lim_{n\to\infty}h(n)=0$, where $h(n):=\max_{1\leqslant j\leqslant N(n)}(s_j^n-s_{j-1}^n)$. Let $y_j^n\in\mathfrak{D}(A)$ satisfy*

$$\begin{cases} \dfrac{y_j^n - y_{j-1}^n}{s_j^n - s_{j-1}^n} \in Ay_j^n, \ j=1,\ldots,N(n), \\ y_0^n = x. \end{cases} \tag{2.29}$$

Define

$$u^n(t) := \begin{cases} x, & \text{for } t=0, \\ y_j^n, & \text{for } t \in (s_{j-1}^n, s_j^n)], j=1,\ldots,N(n). \end{cases}$$

Then

$$\lim_{n\to\infty} \sup_{t\in[0,\bar{t}]} \|u^n(t) - T_A(t)x\| = 0.$$

More precisely, we have

$$\|u^n(t) - T_A(t)x\| \leqslant 2\|x - \tilde{x}\| + |A\tilde{x}| \cdot (\sqrt{th(n)} + h(n)), \tag{2.30}$$

for all $\tilde{x} \in \mathfrak{D}(A)$.

PROOF. It is sufficient to prove (2.30). If $\tilde{y}_j^n$ satisfies (2.29) with x replaced by $\tilde{x}$, then it is not difficult to see that $\|\tilde{y}_j^n - y_j^n\| \leqslant \|\tilde{x} - x\|$, and thus $\|\tilde{u}^n(t) - u(t)\| \leqslant \|\tilde{x} - x\|$, where $\tilde{u}^n(t)$ is defined as $u^n(t)$ by replacing x by $\tilde{x}$ and y_j^n by $\tilde{y}_j^n$. We obtain

$$\begin{aligned} \|u^n(t) - T_A(t)x\| &\leqslant \|u^n(t) - \tilde{u}^n(t)\| + \|\tilde{u}^n(t) - T_A(t)\tilde{x}\| \\ &\quad + \|T_A(t)\tilde{x} - T_A(t)x\| \\ &\leqslant 2\|x - \tilde{x}\| + \|\tilde{u}^n(t) - T_A(t)\tilde{x}\|. \end{aligned}$$

Thus it suffices to prove (2.30) with $x = \tilde{x} \in \mathfrak{D}(A)$. Then

$$\begin{aligned} \sup_{t\in[0,\bar{t}]} \|u^n(t) - T_A(t)x\| &= \max_{1\leqslant j\leqslant N(n)} \sup_{t\in(s_{j-1}^n, s_j^n]} \|y_j^n - T_A(t)x\| \\ &\leqslant \max_{1\leqslant j\leqslant N(n)} [\|y_j^n - T_A(s_j^n)x\| + \sup_{t\in(s_{j-1}^n, s_j^n]} \|T_A(t)x - T_A(s_j^n)x\|] \\ &\leqslant \max_{1\leqslant j\leqslant N(n)} \|y_j^n - T_A(s_j^n)x\| + |Ax| \cdot h(n), \end{aligned}$$

by using (2.12) and the definition of $h(n)$.

$$\|y_j^n - T_A(s_j^n)x\| \leqslant \|y_j^n - (E^A_{s_j/m})^m x\| + |Ax| \bar{t} \sqrt{1/m},$$

by using (2.11). In order to estimate $\|y_j^n - (E^A_{s_j/m})^m x\|$ we can apply the Fundamental Lemma 2.4. First observe that

$$\left\|\frac{y_j^n - y_{j-1}^n}{s_j^n - s_{j-1}^n}\right\| = \frac{1}{s_j^n - s_{j-1}^n} \|E^A_{s_j^n - s_{j-1}^n} y_{j-1}^n - y_{j-1}^n\| = \|A_{s_j^n - s_{j-1}^n} y_{j-1}^n\|.$$

From (2.13) and (2.15) we have

$$\left\|\frac{y_j^n - y_{j-1}^n}{s_j^n - s_{j-1}^n}\right\| \leqslant |Ay_{j-1}^n|,$$

and by (2.16),

$$|Ay_{j-1}^n| \leqslant \inf\{\|y\| : [y_{j-1}^n, y] \in A\}.$$

In particular

$$y = \frac{y_{j-1}^n - y_{j-2}^n}{s_{j-1}^n - s_{j-2}^n} \in Ay_{j-1}^n.$$

Thus

$$\left\|\frac{y_j^n - y_{j-1}^n}{s_j^n - s_{j-1}^n}\right\| \leqslant \left\|\frac{y_{j-1}^n - y_{j-2}^n}{s_{j-1}^n - s_{j-2}^n}\right\|.$$

By induction we obtain

$$\left\|\frac{y_j^n - y_{j-1}^n}{s_j^n - s_{j-1}^n}\right\| \leqslant \left\|\frac{y_1^n - y_0^n}{s_1^n - s_0^n}\right\| = \|A_{s_1^n - s_0^n} x\| \leqslant |Ax|.$$

Thus in Lemma 2.4 we can choose $K = |Ax|$ and $\eta = 0$. It follows that

$$\|y_j^n - E_{s_j/m}^m x\| \leqslant |Ax| \sqrt{s_j^n \cdot (s_j^n / m + h(n))}.$$

Finally we have

$$\|y_j^n - T_A(s_j^n)x\| \leqslant \left(\sqrt{s_j^n (s_j^n/m + h(n))} + \bar{t}\sqrt{1/m}\right) \cdot |Ax|,$$

for every $m \geqslant 1$. By taking the limit as $m \to \infty$,

$$\|y_j^n - T_A(s_j^n)x\| \leqslant \sqrt{s_j^n \cdot h(n)} \cdot |Ax| \leqslant \sqrt{\bar{t}h(n)} \cdot |Ax|. \quad \square$$

REMARK. In Theorems 2.3 and 2.5 we have used the fact that A is m-dissipative to insure the existence of x_i and y_j satisfying (2.18), (2.19) and (2.29). It would be sufficient to assume for example a range condition

$$\bigcap_{h>0} \mathfrak{R}(I - hA) \supseteq \overline{\mathfrak{D}(A)}$$

for this purpose. Then one would also obtain a nonlinear contraction semigroup defined on $\overline{\mathfrak{D}(A)}$, under this weaker assumption. For a discussion of the weakening of assumption $\mathfrak{R}(I - hA) = X$, we refer the reader to CRANDALL [1986], and REICH [1986]. Finally we note that solutions of the initial value problem $du/dt \in Au$, $u(0) = x$ obtained by limits of backward difference schemes are called *mild solutions* in CRANDALL [1986]. We shall not introduce this interesting notion here.

3. APPROXIMATION OF CONTRACTION SEMIGROUPS

In this section we prove a nonlinear version of one half of the Trotter-Neveu-Kato theorem, and Chernoff's formula.

THEOREM 2.7. *Let A and A_n $(n = 1, 2, \cdots)$ be m-dissipative operators in X and $x \in \overline{\mathfrak{D}(A)}$, $x_n \in \overline{\mathfrak{D}(A_n)}$, $n = 1, 2, \cdots$.*

(i) *If for some $h>0$ and for every $y\in X$*

$$\lim_{n\to\infty}\|E_h^{A_n}y - E_h^A y\| = 0, \tag{2.31}$$

then (2.31) *holds for every $h>0$ and every $y\in X$.*

(ii) *If* (2.31) *holds for some $h>0$ and every $y\in X$, and if $\lim_{n\to\infty}\|x_n - x\|=0$, then*

$$\lim_{n\to\infty}\max_{t\in[0,s]}\|T_{A_n}(t)x_n - T_A(t)x\| = 0, \tag{2.32}$$

for every $s>0$.

PROOF.
(i) Assume that (2.31) holds for $h_0>0$. Let $h>0$ and $y\in X$. Set $x:=(I-hA)^{-1}y$ and $x_n=(I-A_n)^{-1}y$, for $n\in\mathbb{Z}^+$. We have to prove $\lim_{n\to\infty}x_n=x$. We shall prove it for $h>0$ satisfying $|1-h_0/h|<1$, i.e. for $h\in(h_0/2,\infty)$. Then by induction this will be true for every $h>0$. Let the element z_n be defined by $z_n:=(I-h_0A_n)^{-1}\,[(1-h_0/h)x+(h_0/h)y]$. Then $\lim_{n\to\infty}z_n=(I-h_0A)^{-1}\,[(1-h_0/h)x+(h_0/h)y]$, by assumption. We claim that

$$(I-h_0A)^{-1}[(1-h_0/h)x+(h_0/h)y]=x.$$

Indeed, $x\in hAx+y$, or equivalently $(h_0/h)x\in h_0Ax+(h_0/h)y$. Therefore $x\in h_0Ax+(1-h_0/h)x+(h_0/h)y$, that is

$$x=(I+h_0A)^{-1}[(1-h_0/h)x+(h_0/h)y].$$

Similarly

$$x_n=(I+h_0A_n)^{-1}[(1-h_0/h)x_n+(h_0/h)y].$$

Therefore $\|x_n-z_n\|\leqslant|1-h_0/h|\cdot\|x-x_n\|$. From the inequality $\|x_n-x\|\leqslant\|x_n-z_n\|+\|z_n-x\|$ we obtain that

$$\|x_n-x\|\leqslant(1-|1-h_0/h|)^{-1}\|z_n-x\|.$$

Since $\lim_{n\to\infty}z_n=x$, we have $\lim_{n\to\infty}x_n=x$.

(ii) We shall first prove the theorem under the additional assumption

$$x_n\in\mathcal{D}(A_n)\text{ and }\sup_{n\geqslant1}|A_nx_n|<\infty. \tag{2.33}$$

Set $M=\max(|Ax|,\sup_{n\geqslant1}|A_nx_n|)$. For $t\in[0,s]$, $s>0$ and $m\in\mathbb{Z}^+$, we have:

$$\begin{aligned}\|T_{A_n}(t)x_n - T_A(t)x\|&\leqslant\|T_{A_n}(t)x_n-(E_{t/m}^{A_n})^m x_n\| + \\ &\quad\|(E_{t/m}^{A_n})^m x_n-(E_{t/m}^{A})^m x\| + \|(E_{t/m}^{A})^m x - T_A(t)x\| \\ &\leqslant 2Mt\sqrt{1/m} + \|(E_{t/m}^{A_n})^m x_n-(E_{t/m}^{A})^m x\|,\end{aligned}$$

by using (2.11) and (2.33). We claim that

$$\lim_{n\to\infty}\|(E_{t/m}^{A_n})^m x_n-(E_{t/m}^{A})^m x\| = 0. \tag{2.34}$$

$$\begin{aligned}\|(E_{t/m}^{A_n})^m x_n - (E_{t/m}^{A})^m x\| &\leq \|(E_{t/m}^{A_n})^m x_n - (E_{t/m}^{A_n})^m x\| \\ &\quad + \|(E_{t/m}^{A_n})^m x - (E_{t/m}^{A})^m x\| \\ &\leq \|x_n - x\| + \|(E_{t/m}^{A_n})^m x - (E_{t/m}^{A})^m x\|.\end{aligned}$$

Since $\lim_{n\to\infty}\|x_n - x\| = 0$, it suffices to prove that

$$\lim_{n\to\infty} \|(E_{t/m}^{A_n})^m x - (E_{t/m}^{A})^m x\| = 0. \tag{2.35}$$

We proceed by induction. Clearly

$$\lim_{n\to\infty} \|(E_{t/m}^{A_n})^j x - (E_{t/m}^{A})^j x\| = 0$$

for $j=1$, by using (i). Assume the result true for $j-1$. We have

$$\begin{aligned}&\|(E_{t/m}^{A_n})^j x - (E_{t/m}^{A})^j x\| \leq \\ &\|E_{t/m}^{A_n}(E_{t/m}^{A_n})^{j-1} x - E_{t/m}^{A_n}(E_{t/m}^{A})^{j-1} x\| \\ &+ \|E_{t/m}^{A_n}(E_{t/m}^{A})^{j-1} x - E_{t/m}^{A}(E_{t/m}^{A})^{j-1} x\| \\ &\leq \|(E_{t/m}^{A_n})^{j-1} x - (E_{t/m}^{A})^{j-1} x\| + \|E_{t/m}^{A_n} z - E_{t/m}^{A} z\|,\end{aligned}$$

by using $\|E_{t/m}^{A_n}\|_{Lip} \leq 1$, and by setting $z = (E_{t/m}^{A})^{j-1} x$. By (i) and the induction hypothesis, we obtain (2.35) and (2.34). Then

$$\limsup_{n\to\infty} \|T_{A_n}(t)x_n - T_A(t)x\| \leq 2Mt\sqrt{1/m},$$

for all $m \geq 1$, hence

$$\lim_{n\to\infty} \|T_{A_n}(t)x_n - T_A(t)x\| = 0.$$

Next observe that

$$\begin{aligned}&|\|T_{A_n}(t_1)x_n - T_A(t_1)x\| - \|T_{A_n}(t_2)x_n - T_A(t_2)x\|| \\ &\leq \|T_{A_n}(t_1)x_n - T_{A_n}(t_2)x_n - T_A(t_1)x + T_A(t_2)x\| \\ &\leq \|T_{A_n}(t_1)x_n - T_{A_n}(t_2)x_n\| + \|T_A(t_1)x - T_A(t_2)x\| \\ &\leq 2M|t_1 - t_2|,\end{aligned}$$

for $0 \leq t_1, t_2 \leq s$, by using (2.12) and (2.33). Then (2.32) follows. Finally we remove the additional assumption (2.33). For $m \geq 1$, set

$$x^m := E_{1/m}^{A} x \text{ and } x_n^m := E_{1/m}^{A_n} x_n.$$

For every $m \geq 1, \lim_{n\to\infty}\|x_n^m - x^m\| = 0$, by (i), thus there is a constant $M_m > 0$ such that $\|x_n^m\| \leq M_m$. Note that $(A_n)_{1/m} x_n = A_n E_{1/m}^{A_n} x_n$, thus by (2.16):

$$|A_n x_n^m| \leq \|(A_n)_{1/m} x_n\| = m\|E_{1/m}^{A_n} x_n - x_n\| \leq m(M_m + \|x_n\|) \leq M_m{}',$$

for all $n \in \mathbb{Z}^+$, for some $M_m{}' > 0$. Thus the sequence $\{x_n^m\}_{n\geq 1}$ satisfies the condition (2.33) for all $m \geq 1$. We have

$$\max_{t\in[0,s]}\|T_{A_n}(t)x_n - T_A(t)x\|$$
$$\leqslant \max_{t\in[0,s]}\{\|T_{A_n}(t)x_n - T_{A_n}(t)x_n^m\| + \|T_{A_n}(t)x_n^m - T_A(t)x^m\|$$
$$+ \|T_A(t)x^m - T_A(t)x\|\}$$
$$\leqslant \|x_n - x_n^m\| + \|x - x^m\| + \max_{t\in[0,s]}\|T_{A_n}(t)x_n^m - T_A(t)x^m\|.$$

Hence

$$\limsup_{n\to\infty}\max_{t\in[0,s]}\|T_{A_n}(t)x_n - T_A(t)x\| \leqslant 2\|x - x^m\|, \tag{2.36}$$

for every $m \geqslant 1$. Next we observe that for every m-dissipative operator B in X,

$$\lim_{h\downarrow 0}\|E_h^B y - y\| = 0, \text{ for } y \in \overline{\mathfrak{D}(B)}. \tag{2.37}$$

Indeed

$$\|E_h^B y - y\| \leqslant \|E_h^B y - E_h^B \tilde{y}\| + \|E_h^B \tilde{y} - \tilde{y}\| + \|\tilde{y} - y\|$$
$$\leqslant 2\|y - \tilde{y}\| + h|A\tilde{y}|,$$

for every $\tilde{y} \in \mathfrak{D}(B)$. This proves (2.37) and

$$\lim_{m\to\infty}\|x - x^m\| = 0.$$

By taking the limit as $m\to\infty$ of (2.36), we obtain (2.32). □

As a first application of Theorem 2.7, we shall prove that the semigroups associated with the Yosida approximation of an m-dissipative operator A converge to the semigroup associated with A. This is a direct consequence of

PROPOSITION 2.8. *Let A be m-dissipative and let A_h be the Yosida approximation of A for $h>0$. Then*
(i) *A_h is m-dissipative,*
(ii) *$\lim_{h\downarrow 0}\|E_\lambda^{A_h}x - E_\lambda^A x\| = 0$, for every $\lambda>0$ and every $x\in X$.*

PROOF. (i) Let $A_h = h^{-1}(E_h^A - I)$, with $\|E_h^A\|_{Lip} \leqslant 1$. Now the following lemma implies (i).

LEMMA 2.9. *Let $C: X\to X$ be a contraction ($\|C\|_{Lip}\leqslant 1$). Then $\lambda(C-I)$ is m-dissipative if $\lambda \geqslant 0$.*

PROOF OF LEMMA 2.9. For $x,y\in X$ and $h>0$, we have

$$\|(x-y) - h\lambda[(Cx - Cy) - (x-y)]\|$$
$$= \|(1+h\lambda)(x-y) - h\lambda(Cx - Cy)\|$$
$$\geqslant (1+h\lambda)\|x-y\| - h\lambda\|x-y\| = \|x-y\|.$$

This implies that $\lambda(C-I)$ is dissipative. Next we consider the equation

$$x - h\lambda(Cx - x) = y$$

for $y \in X$. This is equivalent to

$$x = h\lambda(1 + h\lambda)^{-1}[Cx + (h\lambda)^{-1}y].$$

Note that $\|S\|_{Lip} < 1$, where $S(x) := h\lambda(1+h\lambda)^{-1}[Cx + (h\lambda)^{-1}y]$. By the Picard-Banach fixed point theorem S possesses a fixed point. □

We continue the proof of Proposition 2.8.
(ii) Set $y := E_\lambda^A x$ and $y_h := E_\lambda^{A_h} x$. Then $y - \lambda Ay \ni x$ and $y_h - \lambda A_h y_h = x$. Thus $y - \lambda A_h y = x + \lambda z - \lambda A_h y$, where $z \in Ay$. Since A_h is dissipative, we have

$$\|y_h - y\| \leqslant \|\lambda z - \lambda A_h y\| \leqslant \lambda(\|z\| + |Ay|) = M$$

independent of h. Hence $\|y_h\| \leqslant \|y\| + M$ and

$$\|A_h y_h\| = \lambda^{-1}\|y_h - x\| \leqslant \lambda^{-1}(\|y\| + M + \|x\|).$$

Moreover, since A is dissipative and

$$E_h^A y_h - \lambda A E_h^A y_h \ni E_\lambda^A y_h - y_h + x,$$

$$y - \lambda Ay \ni x,$$

we have

$$\|y - E_h^A y_h\| \leqslant \|E_h^A y_h - y_h\|.$$

Hence

$$\begin{aligned}\|y - y_h\| &\leqslant \|y - E_h^A y_h\| + \|E_h^A y_h - y_h\| \\ &\leqslant 2\|E_h^A y_h - y_h\| = 2h\|A_h y_h\| \\ &\leqslant 2h\lambda^{-1}(\|y\| + M + \|x\|)\end{aligned}$$

tends to zero as $h \downarrow 0$. □

Corollary 2.10. *Let A be m-dissipative in X. Then for every $x \in \overline{\mathfrak{D}(A)}$ and for every $s > 0$*

$$\lim_{h\downarrow 0} \max_{t\in[0,s]} \|T_{A_h}(t)x - T_A(t)x\| = 0, \tag{2.38}$$

where A_h denotes the Yosida approximation of A.

Since A_h is Lipschitz continuous, one can associate a semigroup with A_h via the initial value problem

$$\begin{aligned}&\frac{du}{dt} = A_h u,\ t \geqslant 0,\\ &u(0) = x \in X,\end{aligned} \tag{2.39}$$

by setting $T_h(t):=u(t,x)$. One may expect that $T_h=T_{A_h}$. This is a consequence of

PROPOSITION 2.11. *Let $A:X\to X$ be m-dissipative and Lipschitz continuous. Then $u:[0,\infty)\to X$ defined by $u(t):=T_A(t)x$, $x\in X$ is continuously differentiable on $[0,\infty)$ and satisfies*

$$\frac{du}{dt}(t)=Au(t),\, t\geqslant 0,\, u(0)=x.$$

PROOF. For $t>0$ and $n\geqslant 1$ we have

$$E^A_{t/n}x-x=(t/n)AE^A_{t/n}x,$$

$$(E^A_{t/n})^2x-E^A_{t/n}x=(t/n)A(E^A_{t/n})^2x,$$

$$\vdots$$

$$(E^A_{t/n})^n x-(E^A_{t/n})^{n-1}x=(t/n)A(E^A_{t/n})^n x.$$

By adding these equalities we obtain

$$(E^A_{t/n})^n x-x=(t/n)\sum_{j=1}^{n}A(E^A_{t/n})^j x.$$

Since by Theorem 2.3, $\lim_{n\to\infty}(E^A_{t/n})^n x=T_A(t)x$, we may conclude that

$$\lim_{n\to\infty}\frac{t}{n}\sum_{j=1}^{n}A(E^A_{t/n})^j x$$

exists. Moreover $t\to Au(t)=AT_A(t)x$ is continuous, hence

$$\lim_{n\to\infty}\left(\frac{t}{n}\right)\sum_{j=1}^{n}Au\left(\frac{jt}{n}\right)=\int_0^t Au(z)dz.$$

It is easily seen that

$$\left\|\left(\frac{t}{n}\right)\sum_{j=1}^{n}Au\left(\frac{jt}{n}\right)-\left(\frac{t}{n}\right)\sum_{j=1}^{n}A(E^A_{t/n})^j x\right\|$$

$$\leqslant\frac{t}{n}\|A\|_{Lip}\sum_{j=1}^{n}\left\|u\left(\frac{jt}{n}\right)-(E^A_{t/n})^j x\right\|.$$

By using (2.11) we have

$$\|u(jt/n)-(E^A_{jt/jn})^j x\|\leqslant|Ax|\frac{jt}{n}\sqrt{1/j}$$

for $j=1,\ldots,n$. (Note that since A is continuous, $|Ax|=\|Ax\|$). Then

$$\|(\frac{t}{n})\sum_{j=1}^{n}Au(\frac{jt}{n})-(\frac{t}{n})\sum_{j=1}^{n}A(E_{t/n}^{A})^{j}x\|$$

$$=(\frac{t}{n})\|A\|_{Lip}|Ax|t\Theta(\sqrt{n})\to 0 \text{ as } n\to\infty.$$

We obtain

$$u(t)-x=\int_0^t Au(z)dz, \text{ for every } t>0.$$

Hence $u\in C^1([0,\infty);X)$ and u is the unique solution to the initial value problem

$$\frac{du}{dt}(t)=Au(t),\quad t\geq 0,\ u(0)=x. \quad \Box$$

As a second application of Theorem 2.7 we shall consider a nonlinear version of *Chernoff's formula.* We need the following estimate.

PROPOSITION 2.12 (CHERNOFF'S ESTIMATE). *Let $C:X\to X$ be a contraction, $\lambda>0$, $x\in X$. Then, if we denote by $u_\lambda(t;x)$ the unique solution of the initial value problem*

$$\frac{du}{dt}=\lambda^{-1}(Cu-u),\ t\geq 0,$$
$$u(0)=x,$$

we have

$$\|u_\lambda(t;x)-C^n x\|\leq\sqrt{(n-t/\lambda)^2+t/\lambda}\cdot\|x-Cx\|, \tag{2.40}$$

for every $n\geq 1$ and $t>0$.

PROOF. Observe that $u_\lambda(t;x)=u_1(t/\lambda;x)$. Thus it is sufficient to prove (2.40) for $\lambda=1$. We denote $u_1(t;x)$ by $u(t)$. We have

$$u(t)=e^{-t}x+\int_0^t e^{-(t-s)}Cu(s)ds.$$

Note that $C^n x=e^{-t}C^n x+\int_0^t e^{-(t-s)}C^n x ds$. Thus

$$\|u(t)-C^n x\|\leq e^{-t}\|x-C^n x\|+\int_0^t e^{-(t-s)}\|Cu(s)-C^n x\|ds$$

$$\leq e^{-t}\|x-C^n x\|+\int_0^t e^{-(t-s)}\|u(s)-C^{n-1}x\|ds,$$

(with $C^0=I$) for $n\geq 1$. If $x=Cx$, (2.40) is satisfied. Otherwise we define $\phi_n(t):=\|u(t)-C^n x\|\cdot\|x-Cx\|^{-1}$ for $n=0,1,2,\cdots$. Note that

$$\|x - C^n x\| \leqslant \|x - Cx\| + \|Cx - C^2 x\| + \cdots + \|C^{n-1}x - C^n x\|$$
$$\leqslant n\|x - Cx\|.$$

If follows that ϕ_n satisfies

$$\phi_n(t) \leqslant e^{-t} n + \int_0^t e^{-(t-s)} \phi_{n-1}(s)ds, \; n \geqslant 1, \; t \geqslant 0. \tag{2.41}$$

Next we have:

$$\phi_0(t) \leqslant t \; , \; t \geqslant 0. \tag{2.42}$$

Indeed, $C - I$ is m-dissipative by Lemma 2.9 and Lipschitz continuous. Thus $u(t) = T_{C-I}(t)x$ by Proposition 2.11, hence by using (2.12) we obtain

$$\|u(t) - x\| = \|u(t) - u(0)\| \leqslant |(C - I)x| t = \|(C - I)x\| t.$$

It follows that (2.42) holds. In fact (2.42) can be proven in a direct way. We give a sketch of the proof. Denote by S_{C-I} the semigroup associated to $C - I$ by setting $S_{C-I}(t)x = u(t;x)$, where $u(t;x)$ is the unique solution to

$$\frac{du}{dt} = (C - I), \; u(0) = x.$$

One verifies that

$$\|S_{C-I}(t)x - S_{C-I}(t)y\| \leqslant \|x - y\|,$$

for every $t > 0$ and $x,y \in X$. Then

$$\|u(t + h;x) - u(t;x)\| \leqslant \|u(h;x) - u(0;x)\|,$$

for every $t > 0$ and $h > 0$. Divide by h and let h go to zero: we obtain

$$\|\frac{du}{dt}(t)\| \leqslant \|\frac{d^+ u}{dt}(0)\| = \|(C - I)x\|.$$

Then

$$\|u(t) - x\| = \|u(t) - u(0)\| = \|\int_0^t \frac{du}{ds}(s)ds\|$$
$$\leqslant \int_0^t \|\frac{du}{ds}(s)\| ds \leqslant t\|(C - I)x\|,$$

which proves (2.42).
We claim that $\Phi_n(t) := \phi_n^2(t)$ satisfies

$$\Phi_n(t) \leqslant e^{-t} n^2 + \int_0^t e^{-(t-s)} \Phi_{n-1}(s)ds \; , \; t \geqslant 0 \, , \; n \geqslant 1,$$
$$\Phi_0(t) \leqslant t^2 \, , \; t \geqslant 0. \tag{2.43}$$

It is sufficient to prove the first inequality. One way to do it is to define on the space $\mathbb{R}\times C[0,t]$ the inner product

$$<(\alpha,f),(\beta,g)>:=e^{-t}\alpha\beta+\int_0^t e^{-(t-s)}f(s)g(s)ds,$$

with $(\alpha,f),(\beta,g)\in\mathbb{R}\times C[0,t]$. Note that $<(1,\mathbf{1}),(1,\mathbf{1})>=1$. Thus

$$\begin{aligned}(e^{-t}\alpha+\int_0^t e^{-(t-s)}f(s)ds)^2 &= <(1,\mathbf{1}),(\alpha,f)>^2\\ &\leqslant <(1,\mathbf{1}),(1,\mathbf{1})>\cdot<(\alpha,f),(\alpha,f)>\\ &= e^{-t}\alpha^2+\int_0^t e^{-(t-s)}f^2(s)ds,\end{aligned}$$

by using the Cauchy-Schwarz inequality. Then (2.43) follows from (2.41). Next we denote by $X_n(t)$ the solution of

$$\begin{cases} X_n(t)=e^{-t}n^2+\int_0^t e^{-(t-s)}X_{n-1}(s)ds, & n\geqslant 1,\ t\geqslant 0\\ X_0(t)=t^2+t, & t\geqslant 0. \end{cases} \tag{2.44}$$

It is easy to verify that $\Phi_n(t)\leqslant X_n(t)$, $t\geqslant 0$, $n\geqslant 0$. Finally (2.44) is equivalent to

$$\begin{cases} \frac{d}{dt}X_n(t)+X_n(t)=X_{n-1}(t), & n\geqslant 1,\ t>0\\ X_n(0)=n^2, & n\geqslant 1,\\ X_0(t)=t^2+t, & t\geqslant 0. \end{cases} \tag{2.45}$$

Observe that $X_n(t)=(t-n)^2+t$, $t\geqslant 0, n\geqslant 0$. Thus $\Phi_n(t)\leqslant\sqrt{(n-t)^2+t}$. □

An easy consequence of Chernoff's formula is the following approximation result.

PROPOSITION 2.13. *Let A be m-dissipative in X and let T_A denote the semigroup associated with A. Then for every $x\in X$ and every $s>0$,*

$$\lim_{n\to\infty}\max_{t\in[0,s]}\|T_{A_n}(t)x-T_A(t)x\|=0,$$

where $A_n:=n(T_A(1/n)-I)$.

REMARK. As mentioned in Section 2.1, this result holds for every nonlinear contraction semigroup T on a closed set of X.

We are now in a position to prove the nonlinear version of Chernoff's formula. Let A be an m-dissipative operator in X. Let $F(\lambda)$, $\lambda>0$ be a family of contractions everywhere defined on X. Then for every $\lambda>0$, the operator $A_{(\lambda)}:=\lambda^{-1}(F(\lambda)-I)$ is m-dissipative by Lemma 2.9, and Lipschitz continuous. Thus $E_h^{A_{(\lambda)}}$ is well-defined for $h>0$. We have

THEOREM 2.14. *Assume that*

$$\lim_{\lambda\downarrow 0}\|E_h^{A_{(\lambda)}}x - E_h^A x\| = 0, \tag{2.46}$$

for some $h>0$ *and every* $x\in X$, *then*

$$\lim_{n\to\infty}\max_{t\in[0,s]}\|F(\frac{t}{n})^n x - T_A(t)x\| = 0,$$

for every $s>0$ *and every* $x\in\overline{\mathfrak{D}(A)}$.

PROOF. Let $x\in\overline{\mathfrak{D}(A)}$, $s>0$, and $t\in[0,s]$. For $h>0$ define $y:=E_h^A x$ and $y_\lambda:=E_h^{A_{(\lambda)}}x$. Then

$$\|F(\frac{t}{n})^n x - T_A(t)x\| \leqslant \|F(\frac{t}{n})^n x - F(\frac{t}{n})^n y_\lambda\|$$

$$+ \|F(\frac{t}{n})^n y_\lambda - T_A(t)y\| + \|T_A(t)y - T_A(t)x\|$$

$$\leqslant \|x-y_\lambda\| + \|x-y\| + \|F(\frac{t}{n})^n y_\lambda - T_A(t)y\|.$$

Choose $\lambda=t/n$, then

$$\|F(\frac{t}{n})^n y_{t/n} - T_A(t)y\| \leqslant \|F(\frac{t}{n})^n y_{t/n} - T_{A_{(t/n)}}(t)y_{t/n}\|$$

$$+ \|T_{A_{(t/n)}}(t)y_{t/n} - T_A(t)y\|.$$

From Theorem 2.7, we obtain

$$\lim_{n\to\infty}\max_{t\in[0,s]}\|T_{A_{(t/n)}}(t)y_{t/n} - T_A(t)y\| = 0, \tag{2.47}$$

by noting that $\lim_{n\to\infty}\|y_{t/n}-y\|=0$ holds by using (2.46) and Theorem 2.7-(i). On the other hand, we have

$$\|F(\frac{t}{n})^n y_{t/n} - T_{A_{(t/n)}}(t/n)y_{t/n}\|$$

$$\leqslant \sqrt{(n-t/(t/n))^2 + t/(t/n)}\cdot\|y_{t/n} - F(\frac{t}{n})y_{t/n}\|$$

$$= \sqrt{n}\,\|y_{t/n} - F(\frac{t}{n})y_{t/n}\|$$

by using (2.40) and Proposition 2.11. There is an $M(h)>0$ independent of n and $t\in[0,s]$ such that

$$\|y_{t/n} - F(\frac{t}{n})y_{t/n}\| \leqslant (\frac{t}{n})M(h). \tag{2.48}$$

Indeed, $E_h^{A_{(t/n)}}x - hA_{(t/n)}E_h^{A_{(t/n)}}x = x$, thus $A_{(t/n)}y_{t/n} = h^{-1}(y_{t/n} - x)$ and

$$\|A_{(t/n)}y_{t/n}\| \leq h^{-1}(\|y_{t/n}\| + \|x\|) \leq h^{-1}M =: M(h),$$

where M is independent of t and n. Then

$$\|y_{t/n} - F(\frac{t}{n})y_{t/n}\| = (t/n)\|A_{(t/n)}y_{t/n}\| \leq (\frac{t}{n})M(h),$$

which proves (2.48). It follows that

$$\|F(\frac{t}{n})^n y_{t/n} - T_{A_{(t/n)}}(t/n)y_{t/n}\| \leq M(h)t\frac{1}{\sqrt{n}}. \tag{2.49}$$

From (2.47) and (2.49), we get

$$\lim_{n\to\infty} \max_{t\in[0,s]} \|F(t/n)^n y_{t/n} - T_A(t)y\| = 0,$$

$$\limsup_{n\to\infty}\{\max_{t\in[0,s]} \|F(t/n)^n x - T_A(t)x\|\} \leq 2\|x - y\|.$$

Finally, since $x \in \overline{\mathfrak{D}(A)}$, $\lim_{h\downarrow 0} y = \lim_{h\downarrow 0} E_h^A x = x$, we have

$$\limsup_{n\to\infty}\{\max_{t\in[0,s]} \|F(t/n)^n x - T_A(t)x\|\} = 0. \quad \square$$

Application. If X is a Hilbert space and A is m-dissipative and single-valued, one verifies that

$$F(\lambda) := (I + \tfrac{1}{2}\lambda A)(I - \tfrac{1}{2}\lambda A)^{-1}$$

satisfies the assumption of the theorem. This proves the convergence of the scheme of *Crank-Nicholson* in the Hilbert space case: see Kato [1978].

4. Initial value problems and m-dissipative operators

In Section 2.2 we have constructed a nonlinear contraction semigroup T_A associated with an m-dissipative operator A in X. In Section 2.3 we have proved that this semigroup is the "limit" of nonlinear contraction semigroups T_{A_h} generated by the Yosida approximation A_h, and that $t \to u_h(t,x) = T_{A_h}(t)x$ is the unique solution to the initial value problem

$$\frac{du}{dt} = A_h u, \quad t \geq 0,$$
$$u(0) = x.$$

A natural question to ask now is whether $t \to u(t;x) = T_A(t)x$ is a solution to

$$\frac{du}{dt} \in Au, \quad t > 0,$$
$$u(0) = x.$$

In the linear case, when $\overline{\mathfrak{D}(A)} = X$, $t \to T_A(t)x$ is right differentiable at $t = t_0 \geq 0$ if and only if $T_A(t_0)x \in \mathfrak{D}(A)$. In the nonlinear case for a general Banach space only one implication is true, namely:

THEOREM 2.15. *Let A be an m-dissipative operator in X, let T_A be its associated semigroup, and $x \in \overline{\mathfrak{D}(A)}$. If $t \to T_A(t)x$ is right-differentiable at $t_0 \geq 0$, then*
(i) $T_A(t_0)x \in \mathfrak{D}(A)$,
(ii) $\dfrac{d^+}{dt} T(t)x|_{t=t_0} \in AT_A(t_0)x$.

In order to prove Theorem 2.15, we shall need to differentiate the function $t \to \|u(t)\|$. Therefore we need to differentiate the norm.

For $x,y \in X$, the function $t \to \|x+ty\|$ is convex, thus $t \to t^{-1}(\|x+ty\|-\|x\|)$ is nondecreasing for $t \neq 0$, and the right and left limits exist at $t=0$. We shall use the following notations. For $x,y \in X$ and $\lambda \in \mathbb{R} \setminus \{0\}$:

$$[x,y]_\lambda := \lambda^{-1}(\|x+\lambda y\| - \|x\|), \tag{2.50}$$

$$[x,y]_+ := \inf_{\lambda>0} [x,y]_\lambda = \lim_{\lambda \downarrow 0} [x,y]_\lambda, \tag{2.51}$$

$$[x,y]_- := \sup_{\lambda<0} [x,y]_\lambda = \lim_{\lambda \uparrow 0} [x,y]_\lambda. \tag{2.52}$$

We have the following proposition (see CRANDALL [1986]).

PROPOSITION 2.16. *Let $x,y,z \in X$ and $\alpha,\beta \in \mathbb{R}$. Then*
(0) $[x,y]_- = -[x,-y]_+$,
(1) $[\cdot,\cdot]_+ : X \times X \to \mathbb{R}$ *is upper-semicontinuous,*
(2) $[\alpha x, \beta y]_+ = |\beta| \cdot [x,y]_+$, *if* $\alpha\beta > 0$,
(3) $[x, \alpha x + y]_+ = \alpha\|x\| + [x,y]_+$,
(4) $|[x,y]_+| \leq \|y\|$ *and* $[0,y]_+ = \|y\|$,
(5) $-[x,-y]_+ \leq [x,y]_+$,
(6) $[x,y+z]_+ \leq [x,y]_+ + [x,z]_+$,
(7) $|[x,y]_+ - [x,z]_+| \leq \|y-z\|$,
(8) *if* $u:(t_1,t_2] \to X$ *is left-differentiable at* t_2, *then*

$$\frac{d^-}{dt}\|u(t)\||_{t=t_2} = [u(t_2), \frac{d^- u}{dt}(t_2)]_-;$$

if $u:[t_1,t_2) \to X$ *is right-differentiable at* t_1 *then*

$$\frac{d^+}{dt}\|u(t)\|_{t=t_1} = [u(t_1), \frac{d^+}{dt}u(t_1)]_+,$$

(9) $[x,y]_+ \geq 0$ *iff* $\|x\| \leq \|x+\lambda y\|$ *for every* $\lambda > 0$.

PROOF. We shall need and prove (0),(5),(6),(8) and (9).
(0) follows from $[x,-y]_\lambda = -[x,y]_{-\lambda}$.
(5) follows from $[x,y]_- \leq [x,y]_+$ and (0).
(6) follows from the inequality

$$[x,y+z]_\lambda \leq [x,y]_{2\lambda} + [x,z]_{2\lambda}, \ \lambda > 0.$$

(8) $|\|u(t_1)+h\frac{d^+}{dt}u(t_1)+o(h)\|-\|u(t_1)-h\frac{d^+}{dt}u(t_1)\||=o(h)$, for $h>0$. Thus

$$h^{-1}[\|u(t_1+h)-u(t_1)\|]=h^{-1}[\|u(t_1)+h\frac{d^+}{dt}u(t_1)\|-\|u(t_1)\|+o(h)]$$

$$=[u(t_1),\frac{d^+}{dt}u(t_1)]_h+h^{-1}o(h),$$

for $h>0$. Hence $\frac{d^+}{dt}\|u(t)\||_{t=t_1}$ exists and is equal to $[u(t_1),\frac{d^+}{dt}u(t_1)]_+$.

Similarly for $\frac{d^-}{dt}\|u(t)\||_{t=t_2}$.

(9) is trivial. □

As an exercise the reader may compute $[\cdot,\cdot]_+$ and $[\cdot,\cdot]_-$ for the L_p-norms, $1\leqslant p\leqslant\infty$; see also Appendix 1. For the proof of Theorem 2.15 we need two more lemmas.

LEMMA 2.17. *Let A be an m-dissipative operator in X. If $x,y\in X$ satisfy*

$$[x-u,v-y]_+\geqslant 0 \text{ for all } [u,v]\in A, \tag{2.53}$$

then $x=E_1^A(x-y)\in\mathfrak{D}(A)$ and $y=A_1(x-y)\in Ax$.

PROOF. $E_1^A(x-y)\in\mathfrak{D}(A)$ and $A_1(x-y)\in AE_1^A(x-y)x$. Choose $u=E_1^A(x-y)$ and $v=A_1(x-y)$ in (2.53). Then

$$[x-E_1^A(x-y),A_1(x-y)-y]_1\geqslant[x-E_1^A(x-y),A_1(x-y)-y]_+\geqslant 0.$$

Thus

$$\|x-E_1^A(x-y)+A_1(x-y)-y\|\geqslant\|x-E_1^A(x-y)\|.$$

We have $x-y=E_1^A(x-y)-A_1(x-y)$, hence $\|x-E_1^A(x-y)\|=0$, $x=E_1^A(x-y)$, and $y=A_1(x-y)$, since $I=E_1^A-A_1$. □

LEMMA 2.18. *Let A be an m-dissipative operator in X. Then for every $x\in X$, and every $\lambda,\mu>0$:*

$$(A_\lambda)_\mu x=A_{\lambda+\mu}x, \tag{2.54}$$

$$\lim_{\mu\downarrow 0}(A_\lambda)_\mu x=A_\lambda x. \tag{2.55}$$

PROOF.
(i) Note that if $[\xi,\eta]\in A$ and $\nu>0$, then $\xi=E_\nu^A(\xi-\nu\eta)$ and $\eta=A_\nu(\xi-\nu\eta)$. Set $y:=(A_\lambda)_\mu x$. Then $y=\mu^{-1}(z-x)$ where z satisfies the equation $z-\mu A_\lambda z=x$, hence $y=A_\lambda z$. Moreover $A_\lambda z=\lambda^{-1}(w-z)$ where w satisfies $w-\lambda Aw\ni z$. So we obtain that $x=z-\mu y=w-\lambda A_\lambda z-\mu y=w-(\lambda+\mu)y$. Since $w\in\mathfrak{D}(A)$ we have $y=A_\lambda z\in AE_\lambda^A z=Aw$. Thus $y=A_{\lambda+\mu}(w-(\lambda+\mu)y)=A_{\lambda+\mu}x$.
(ii) $(A_\lambda)_\mu x=A_\lambda E_\mu^{A_\lambda}x$. A_λ is Lipschitz continuous and

$$\|E_\mu^{A_\lambda}x - x\| = \mu\|(A_\lambda)_\mu x\| \leqslant \mu|A_\lambda x| = \mu\|A_\lambda x\|,$$

by (2.16). Thus $\lim_{\mu\downarrow 0}(A_\lambda)_\mu x = A_\lambda x$. □

PROOF OF THEOREM 2.15. For $h>0$, let A_h be the Yosida approximation of A, and let T_{A_h} be the corresponding semigroup. We know that $t \to u_h(t) := T_{A_h}(t)x_0$ belongs to $C^1([0,\infty);X)$ for every $x_0 \in X$, and satisfies $du_h/dt = A_h u_h$, $t \geqslant 0$, and $u_h(0) = x_0$. Since A_h is dissipative (Proposition 2.8), we have

$$\|(u_h(t) - x) - \lambda(A_h u_h(t) - A_h x)\| \geqslant \|u_h(t) - x\|,$$

for every $t \geqslant 0$, $\lambda > 0$ and $x \in X$. Thus $[u_h(t) - x, A_h x - A_h u_h(t)]_\lambda \geqslant 0$ for every $\lambda > 0$. Moreover $A_h u_h(t) = (du_h/dt)(t)$ for every $t > 0$, hence

$$[u_h(t) - x\,, A_h x - \frac{d^- u_h}{dt}(t)]_\lambda \geqslant 0,$$

for every $\lambda > 0$, $t > 0$, $x \in X$. By taking the infimum over $\lambda > 0$, we have

$$[u_h(t) - x\,, A_h x - \frac{d^- u_h}{dt}(t)]_+ \geqslant 0,$$

for every $t > 0$ and $x \in X$. By using Proposition 2.16-(6) we obtain:

$$-[u_h(t) - x, -\frac{d^- u_h}{dt}(t)]_+ \leqslant [u_h(t) - x\,, A_h x]_+,$$

for $t > 0$ and $x \in X$. By using (2.16-(0)) and (2.16-(8)) and the fact that u_h is differentiable, we have

$$\frac{d^+}{dt}\|u_h(t) - x\| \leqslant [u_h(t) - x\,, A_h x]_+\,, t > 0,$$

for every $x \in X$. And also

$$\frac{d^+}{dt}\|u_h(t) - x\| \leqslant [u_h(t) - x\,, A_h x]_\lambda\,, \lambda > 0.$$

By integrating over $[s,t]$ with $0 < s < t$, we have

$$\|u_h(t) - x\| \leqslant \|u_h(s) - x\| + \int_s^t [u_h(\tau) - x\,, A_h x)]_\lambda d\tau,$$

for $0 \leqslant s < t$, $h > 0$, $x \in X$, $\lambda > 0$. Replace x by $E_1^{A_h}\xi$ with $\xi \in X$. Then by (2.54)

$$A_h x = A_h E_1^{A_h}\xi = (A_h)_1 \xi = (A_1)_h \xi.$$

By taking the limit as $h \downarrow 0$ we obtain

$$\|u(t) - E_1^A \xi\| \leqslant \|u(s) - E_1^A \xi\| + \int_s^t [u(\tau) - E_1^A(\xi), A_1 \xi]_\lambda d\tau,$$

for every $\xi \in X$, $0 \leqslant s < t$, by using (2.55). If $t \to T_A(t)x$ is right-differentiable at t_0, we get from

$$\|u(t_0+h)-E_1^A\xi\|-\|u(t_0)-E_1^A\xi\|\leqslant\int_{t_0}^{t_0+h}[u(\tau)-E_1^A\xi,A_1\xi]_\lambda d\tau,$$

with $h>0$, that

$$\frac{d^+}{dt}\|u(t_0)-E_1^A\xi\||_{t=t_0}\leqslant[u(t_0)-E_1^A\xi,A_1\xi]_\lambda,$$

for every $\lambda>0$ and $\xi\in X$. From (2.16-(8)), we have

$$[u(t_0)-E_1^A\xi,\frac{d^+}{dt}u(t_0)]_+\leqslant[u(t_0)-E_1^A\xi,A_1\xi]_+,$$

for every $\xi\in X$. By using (2.16-(6)) we get that

$$[u(t_0)-E_1^A\xi,A_1\xi-\frac{d^+}{dt}u(t_0)]_+\geqslant 0,$$

for every $\xi\in X$. Next for $[u,v]\in A$, set $\xi:=u-v$. Then $E_1^A\xi=u$ and $A_1\xi=v$, hence

$$[u(t_0)-u,v-\frac{d^+}{dt}u(t_0)]_+\geqslant 0$$

for every $[u,v]\in A$. Applying Lemma 2.17, one obtains

$$u(t_0)\in\mathfrak{D}(A)\text{ and }\frac{d^+}{dt}u(t_0)\in Ax.$$

This completes the proof of the theorem. □

REMARK. If $x\in\mathfrak{D}(A)$ then $t\to T_A(t)x$ is Lipschitz continuous as follows from (2.12). Thus, if X satisfies the *Radon-Nikodym property* (in particular if X is reflexive) then $t\to T_A(t)x$ is differentiable a.e. on $[0,\infty)$. It follows that

$$\frac{d^+u}{dt}(t)\in Au(t)$$

a.e. on $[0,\infty)$, where $u(t):=T_A(t)x$. If X is a Hilbert space, then

$$\frac{d^+u}{dt}(t)\in Au(t)$$

everywhere on $[0,\infty)$ (see BREZIS [1973]). Moreover

$$\frac{d^+u}{dt}(t)=A^0u(t),$$

where A^0x denotes the element of Ax with minimal norm. However, in many interesting cases, X does not satisfy the Radon-Nikodym property (for instance if $X=L_1(\Omega)$, see Chapter 4) and the differentiability of u does not follow from the general theory.

5. Quasi-contractive semigroups

In this short section, we restate without proof, the basic theorems of the previous three sections in the case where the operator $A \subset X \times X$ is not necessarily m-dissipative but $A - \omega I$ is m-dissipative, for some $\omega > 0$. We could call such A a quasi-m-dissipative operator in X. From now on ω is a fixed positive real number. If $A - \omega I$ is m-dissipative, then it is easy to verify that $\mathcal{R}(I - hA) = X$ for $0 < h < \omega^{-1}$. Moreover, $(I - hA)^{-1}$ is the graph of an operator for $h \in (0, \omega^{-1})$ and $\mathcal{R}(I - hA)^{-1} = \mathcal{D}(A)$. We have:

Theorem 2.19. *Let $A - \omega I$ be m-dissipative. Then for every $t \geq 0$ and $x \in \overline{\mathcal{D}(A)}$, the sequence $\{(I - t/n \cdot A)^{-n} x\}_{n > t\omega}$ is a Cauchy sequence in X. The limit, which we denote by $T_A(t)x$ belongs to $\overline{\mathcal{D}(A)}$. Furthermore, $T_A(t): \overline{\mathcal{D}(A)} \to \overline{\mathcal{D}(A)}$ satisfies*

$$T_A(t+s)x = T_A(t)T_A(s)x, \quad t, s \geq 0, x \in \overline{\mathcal{D}(A)}, \tag{2.56}$$

$$T_A(0) = I|_{\overline{\mathcal{D}(A)}}, \tag{2.57}$$

$$\|T_A(t)x - T_A(t)y\| \leq e^{\omega t}\|x - y\|, \quad t \geq 0, \ x, y \in \overline{\mathcal{D}(A)}, \tag{2.58}$$

$$t \to T_A(t)x \text{ is continuous on } [0, \infty) \text{ for every } x \in \overline{\mathcal{D}(A)}. \tag{2.59}$$

Moreover, the sequence $\{(I - t/n \cdot A)^{-n} x\}$ converges uniformly on bounded t-intervals.

Remark. A semigroup $\{T(t)\}_{t \geq 0}$ satisfying (2.58) and (2.59) is sometimes called a *quasi-contractive semigroup*.

The semigroup T_A defined in Theorem 2.19 is "stable" with respect to A in the following sense.

Theorem 2.20. *Let $A - \omega I$ and $A_n - \omega I$, $n = 1, 2, \ldots$, be m-dissipative for some $\omega > 0$.*

(i) *If for some $h_0 \in (0, \omega^{-1})$ and for every $y \in X$,*

$$\lim_{n \to \infty} \|(I - h_0 A_n)^{-1} y - (I - h_0 A)^{-1} y\| = 0 \tag{2.60}$$

holds, then (2.60) holds for every $h \in (0, \omega^{-1})$ and every $y \in X$.

(ii) *If for some $h_0 \in (0, \omega^{-1})$ and for every $y \in X$, (2.60) holds, and*

$$\lim_{n \to \infty} \|x_n - x\| = 0 \text{ with } x \in \overline{\mathcal{D}(A)}, x_n \in \overline{\mathcal{D}(A_n)},$$

then for every $s > 0$,

$$\lim_{n \to \infty} \max_{t \in [0, s]} \|T_{A_n}(t)x_n - T_A(t)x\| = 0.$$

Let $A - \omega I$ be m-dissipative in X. Then A_h defined by (2.7) for $h \in (0, \omega^{-1})$ is Lipschitz continuous, $A_h - \omega I$ is m-dissipative and

$$\lim_{h \downarrow 0} \|(I - \lambda A_h)^{-1} y\| = 0$$

for every $y \in X$ and every $\lambda \in (0, \omega^{-1})$. Then, by Theorem 2.20,

$$\lim_{h \downarrow 0} \max_{t \in [0,s]} \|T_{A_h}(t)x - T_A(t)x\| = 0$$

for every $x \in \overline{\mathcal{D}(A)}$. The analogue of Theorem 2.15 holds:

THEOREM 2.21. *Let $A - \omega I$ be m-dissipative for some $\omega > 0$. Let T_A be the semigroup defined in Theorem 2.19. If $x \in \overline{\mathcal{D}(A)}$ and $t \to T_A(t)x$ is right-differentiable at $t_0 \geqslant 0$, then*
(i) $T_A(t_0)x \in \mathcal{D}(A)$
(ii) $\dfrac{d^+}{dt} T_A(t)x|_{t=t_0} \in AT(t_0)x$.

The last theorem of this section is:

THEOREM 2.22. *Let $\omega > 0$ and let $F(\lambda)$: $X \to X$ satisfy:*

$$\|F(\lambda)x - F(\lambda)y\| \leqslant e^{\omega\lambda}\|x - y\|, \ x,y \in X, \ \lambda \in (0, \lambda_0). \tag{2.61}$$

Then $A_{(\lambda)} := \lambda^{-1}(F(\lambda) - I)$, $\lambda \in (0, \lambda_0)$, is Lipschitz continuous and the operator $A_{(\lambda)} - \lambda^{-1}(e^{\omega\lambda} - 1)I$ is m-dissipative. If moreover,

$$\lim_{\lambda \downarrow 0} \|(I - h_0 A_{(\lambda)})^{-1}y - (I - h_0 A)^{-1}y\| = 0$$

for every $y \in X$, for some $h_0 \in (0, \lambda_0(e^{\omega\lambda_0} - 1)^{-1})$ and some operator A with $A - \omega I$ m-dissipative, then

$$\lim_{n \to \infty} \max_{t \in [0,s]} \|F(\frac{t}{n})^n x - T_A(t)x\| = 0,$$

for every $s > 0$ and every $x \in \overline{\mathcal{D}(A)}$, where T_A is the semigroup defined in Theorem 2.19.

EXERCISE. Find and prove the analogue of (2.11) and (2.12) when $A - \omega I$ is m-dissipative.

Chapter 3

Generation of Linear Semigroups

1. C_0-SEMIGROUPS OF LINEAR CONTRACTIONS

The goal of this section is to prove the theorems of Hille-Yosida and of Trotter-Neveu-Kato announced in Section 2.1. We shall still use the notations $C(X)$ and $D(X)$ introduced in Section 2.1. First we have:

THEOREM 3.1. *Let $T(t):X\to X$, $t\geqslant 0$, be a C_0-semigroup of (linear) contractions. Then A_T, the infinitesimal generator of T, defined by* (2.0) *satisfies:*
(i) *A_T is m-dissipative and densely defined,*
(ii) *$(\omega I-A_T)^{-1}y=\int_0^{\infty}e^{-\omega t}T(t)y\,dt$, for every $\omega>0$ and every $y\in X$.*

PROOF. Let $T\in C(X)$. For $h>0$, we define

$$A_{T,h}:=h^{-1}(T(h)-I). \tag{3.1}$$

$A_{T,h}$ is bounded, $\|A_{T,h}\|\leqslant 2h^{-1}$ and $A_{T,h}$ is m-dissipative by Lemma 2.9. Moreover $x\in\mathfrak{D}(A_T)$ iff $\lim_{h\downarrow 0}A_{T,h}x$ exists. For $t>0$ and $x\in X$, we define

$$R_{T,t}x:=t^{-1}\int_0^t T(s)x\,ds. \tag{3.2}$$

Clearly, $R_{T,t}$ is bounded in X, $\|R_{T,t}\|\leqslant 1$, and $\lim_{t\downarrow 0}R_{T,t}x=x$, for every $x\in X$. We shall use the following lemma:

LEMMA 3.2.
(i) *For $h,t>0$, we have*

$$A_{T,h}R_{T,t}=R_{T,t}A_{T,h}, \tag{3.3}$$

$$A_{T,h}R_{T,t} = A_{T,t}R_{T,h}. \tag{3.4}$$

(ii) *For $t>0$, $x \in X$, $R_{T,t}x \in \mathfrak{D}(A_T)$, and*

$$A_T R_{T,t}x = A_{T,t}x. \tag{3.5}$$

For $h>0$, $x \in \mathfrak{D}(A_T)$, we have

$$A_{T,h}x = R_{T,h}A_Tx. \tag{3.6}$$

PROOF.
(3.3): For $x \in X$,

$$A_{T,h}R_{T,t}x = h^{-1}(T(h)-I)t^{-1}\int_0^t T(\tau)x d\tau$$

$$= t^{-1}\int_0^t T(\tau)h^{-1}(T(h)-I)x d\tau = R_{T,t}A_{t,h}x.$$

(3.4):

$$htA_{T,h}R_{T,t}x = (T(h)-I)\int_0^t T(\tau)x d\tau =$$

$$\int_0^t (T(\tau+h)x - T(\tau)x)d\tau = \int_h^{t+h} T(\tau)x d\tau - \int_0^t T(\tau)x d\tau =$$

$$\int_t^{t+h} T(\tau)x d\tau - \int_0^h T(\tau)x d\tau = \int_0^h (T(\tau+t)x - T(\tau)x)d\tau =$$

$$\int_0^h (T(t)-I)T(\tau)x d\tau = (T(t)-I)\int_0^h T(\tau)x d\tau = htA_{T,t}R_{T,h}x.$$

(3.5): From (3.4) we obtain

$$\lim_{h\downarrow 0} A_{T,h}R_{T,t}x = A_{T,t}(\lim_{h\downarrow 0} R_{T,h}x) = A_{T,t}x.$$

(3.6): Similar. □

Since $A_{T,h}$ is dissipative and $A_Tx = \lim_{h\downarrow 0}A_{T,h}x$ for $x \in \mathfrak{D}(A_T)$, we have

$$\|x\| \leqslant \lim_{h\downarrow 0}\|x - \lambda A_{T,h}x\| = \|x - \lambda A_Tx\|,$$

for every $\lambda>0$ and $x \in \mathfrak{D}(A_T)$. Hence A_T is dissipative.

For $x \in X$, $R_{T,t}x \in \mathfrak{D}(A_T)$ by Lemma 3.2, and $\lim_{t\downarrow 0}R_{T,t}x = x$, hence $\mathfrak{D}(A_T)$ is dense in X. Next we prove that A_T is closed. Let $[x_n,y_n] \in A_T$ be such that $\lim_{n\to\infty}[x_n,y_n] = [x,y]$, with $[x,y] \in X \times X$. By using (3.6), we obtain:

$$A_{T,h}x_n = R_{T,h}A_Tx_n = R_{T,h}y_n,$$

hence

$$A_{T,h}x = R_{T,h}y.$$

Then $\lim_{h\downarrow 0} A_{T,h}x = y$ and $[x,y]\in A_T$. It remains to prove that for every $\omega>0$, and $y\in X$,

$$x_\omega := \int_0^\infty e^{-\omega t}T(t)y\,dt$$

belongs to $\mathfrak{D}(A_T)$ and $(\omega I - A_T)x_\omega = y$. For $y\in X$, define

$$v(t) = \int_0^t T(\tau)y\,d\tau.$$

Then clearly $v\in C^1([0,\infty);X)$, $\frac{dv}{dt}(t)=T(t)y$, and by (3.5),

$$A_T v(t) = A_T t R_{T,t}y = tA_{T,t}y = T(t)y - y.$$

Hence

$$\frac{dv}{dt}(t) = A_T v(t) + y, \quad t\geqslant 0.$$

By setting $T_\omega(t) := e^{-\omega t}T(t)$, $\omega>0$, $t\geqslant 0$, and $v_\omega(t) := \int_0^t T_\omega(\tau)y\,d\tau$, we obtain

$$\frac{dv_\omega}{dt}(t) = AT_\omega v_\omega(t) + y,$$

hence

$$\omega v_\omega(t) + \frac{dv_\omega}{dt}(t) = A_T v_\omega(t) + y, \quad t\geqslant 0, \tag{3.7}$$

by using that $A_{T_\omega} = A_T - \omega I$. Then $x_\omega = \lim v_\omega(t)$ exists and

$$\lim_{t\to\infty}\frac{dv_\omega}{dt}(t) = \lim_{t\to\infty} e^{-\omega t}T(t)y = 0.$$

Since A_T is closed, we obtain $x_\omega \in \mathfrak{D}(A_T)$ and $(\omega I - A_T)x_\omega = y$, from (3.7). □

In the next proposition, we prove that, if $T\in C(X)$ and A_T is the infinitesimal generator of T, then $t\to u(t) := T(t)x$ is the unique solution to the initial value problem:

$$\frac{du}{dt} = A_T u, \quad t>0,$$
$$u(0) = x,$$

for $x\in\mathfrak{D}(A_T)$.

PROPOSITION 3.3.

(i) *Let* $T\in C(X)$ *and* A_T *be its infinitesimal generator. If* $u(t) := T(t)x$ *for* $x\in\mathfrak{D}(A_T)$, $t\geqslant 0$, *then* $u\in C^1([0,\infty);X)$, $u(t)\in\mathfrak{D}(A_T)$ *for* $t\geqslant 0$, *and*

$\dfrac{du}{dt}(t)=A_T u(t)$ *for* $t>0$. *Moreover*

$$A_T T(t)x = T(t)A_T x, \quad t \geqslant 0. \tag{3.8}$$

(ii) *Let* A *be dissipative and let* $u \in C([0,\infty);X) \cap C^1((0,\infty);X)$ *satisfy*

$$\begin{cases} u(t) \in \mathcal{D}(A) & ,t>0, \\ \dfrac{du}{dt}(t) = Au(t) & ,t>0, \\ u(0) = x & ,\text{for some } x \in X. \end{cases}$$

Then $\|u(t)\| \leqslant \|x\|$ *for* $t>0$. *In particular, if* $x=0$, *then* $u(t)=0$ *for all* $t>0$.

PROOF.
(i) For $h>0$, $t \geqslant 0$, $x \in \mathcal{D}(A_T)$, we have

$$h^{-1}[u(t+h) - u(t)] = h^{-1}[T(t+h)x - T(t)x] = T(t)A_{T,h}x.$$

Since $x \in \mathcal{D}(A_T)$ and $T(t)$ is continuous, we have

$$\frac{d^+}{dt}u(t) = T(t)A_T x.$$

Similarly, for $t>0$, $h \in (0,t)$ and $x \in \mathcal{D}(A_T)$,

$$h^{-1}[u(t) - u(t-h)] = h^{-1}[T(t-h)T(h)x - T(t-h)x]$$
$$= T(t-h)A_{T,h}x.$$

Clearly

$$\lim_{h \downarrow 0} T(t-h)A_{T,h}x = T(t)A_T x,$$

hence

$$\frac{d^-}{dt}u(t) = T(t)A_T x = \frac{d^+}{dt}u(t), \quad t>0.$$

Moreover, for $h>0$, $t \geqslant 0$ we have

$$T(t)A_{T,h} = A_{T,h}T(t),$$

hence $T(t)x \in \mathcal{D}(A_T)$ for $x \in \mathcal{D}(A_T)$, and

$$A_T T(t)x = T(t)A_T x,$$

which proves (3.8). Finally, since $t \to T(t)A_T x$ belongs to $C([0,\infty);X)$, the function $t \to u'(t) = T(t)A_T x$ is also continuous on $[0,\infty)$, and $u \in C^1([0,\infty);X)$.
(ii) Since A is dissipative and $u(t) \in \mathcal{D}(A), t>0$,

$$[u(t), -Au(t)]_+ \geqslant 0 \quad \text{for } t>0,$$

by Proposition 2.16-(9). Hence

$$\frac{d^-}{dt}\|u(t)\| = [u(t), \frac{du}{dt}]_- = -[u(t), -\frac{du}{dt}(t)]_+$$
$$= -[u(t), -Au(t)]_+ \leqslant 0,$$

by Proposition 2.16 again. It follows that $t \to \|u(t)\|$ is nonincreasing on $[0,\infty)$. Since $t \to \|u(t)\|$ is continuous on $[0,\infty)$, we obtain $\|u(t)\| \leqslant \|u(0)\| = \|x\|$. □

As in Section 2.1 we shall denote by $\mathcal{G}: C(X) \to D(X)$ the mapping defined by $\mathcal{G}(T) = A_T$. We have:

THEOREM 3.4. *$\mathcal{G}: C(X) \to D(X)$ is a bijection. For every $A \in D(X)$, $\mathcal{G}^{-1}(A) = T_A$, where T_A is the semigroup associated with A in Theorem 2.3. Moreover, the sequence $\{T_n\} \subset C(X)$ converges to T in $C(X)$ iff the sequence $\{\mathcal{G}(T_n)\} \subset D(X)$ converges to $\mathcal{G}(T)$ in $D(X)$.*

PROOF. "Injectivity of $\mathcal{G}$." Let $T_1, T_2 \in C(X)$ be such that $A_{T_1} = A_{T_2} = A \in D(X)$. By Proposition 3.3, $T_1(t)x = T_2(t)x$ for $t \geqslant 0$ and $x \in \mathcal{D}(A)$. Since $\overline{\mathcal{D}(A)} = X$ and $T_1(t), T_2(t)$ are continuous, they agree on X.
"Surjectivity of $\mathcal{G}$ and $\mathcal{G}(T_A) = A$". Let $A \in D(X)$ and let T_A denote the associated semigroup defined in Theorem 2.3. Then $T_A \in C(X)$. (Note that $T_A(t)$ are linear operators defined on $\overline{\mathcal{D}(A)} = X$.) We have to prove that $\mathcal{G}(T_A) = A$. Set $u(t) = T_A(t)x$, for $t \geqslant 0$ and $x \in \mathcal{D}(B)$ where $B = \mathcal{G}(T_A)$. By definition $(d^+u/dt)(0)$ exists and, moreover $(d^+u/dt)(0) = Bx$. By Theorem 2.15, $x \in \mathcal{D}(A)$ and $(d^+u/dt)(0) = Ax$. Hence $B \subseteq A$. But B is m-dissipative and A is dissipative, hence by Proposition 3.8-(i), $B = A$.
"Continuity Property of $\mathcal{G}$ and $\mathcal{G}^{-1}$" If $A_n \to A$ in $D(X)$, then $\mathcal{G}^{-1}(A_n) = T_{A_n} \to T_A = \mathcal{G}^{-1}(A)$ in $C(X)$ by Theorem 2.7. If $T_n \to T$ in $C(X)$, that is if (2.4) holds, we have by Theorem 3.1-(ii):

$$\|(\omega I - A_{T_n})^{-1}y - (\omega I - A_T)^{-1}y\| \leqslant \int_0^\infty e^{-\omega t}\|T_n(t)y - T(t)y\|dt,$$

for every $y \in X$ and $\omega > 0$. Given $\epsilon > 0$, there is an $s > 0$ such that

$$\int_s^\infty e^{-\omega t}\|T_n(t)y - T(t)y\|dt \leqslant \int_s^\infty 2e^{-\omega t}\|y\|dt < \epsilon/2.$$

By (2.4), there is an $N \in \mathbb{Z}^+$ such that for $n \geqslant N$,

$$\int_0^s e^{-\omega t}\|T_n(t)y - T(t)y\|dt < \epsilon/2.$$

Thus

$$\lim_{n\to\infty} \|(\omega I - A_{T_n})^{-1}y - (\omega I - A_T)^{-1}y\| = 0.$$

Finally observe that for $A \in D(X)$, $E_h^A = h^{-1}(h^{-1}I - A)^{-1}$ for $h > 0$. □

2. C_0-SEMIGROUPS OF BOUNDED LINEAR OPERATORS

In this section we shall extend the theorems of Section 3.1 to the class of C_0-semigroups of bounded linear operators.

Let $(X,\|\cdot\|)$ be a real normed space. A one-parameter semigroup of linear operators on $(X,\|\cdot\|)$ is a family of linear operators $T(t):X\to X$, $t\geqslant 0$, satisfying

$$\begin{cases} T(t+s)=T(t)T(s),\ t,s\geqslant 0, \\ T(0)=I. \end{cases} \tag{3.9}$$

The following holds: either

$$\sup_{t\in[0,s]}\ \sup_{\substack{x\in X\\ \|x\|\leqslant 1}} \|T(t)x\| = \infty, \text{ for every } s>0, \tag{3.10}$$

or

$$\sup_{t\in[0,s]}\ \sup_{\substack{x\in X\\ \|x\|\leqslant 1}} \|T(t)x\| < \infty, \text{ for every } s>0. \tag{3.11}$$

Indeed if (3.10) does not hold, there are $s_0>0$ and $M_0>0$ such that $\|T(t)\|\leqslant M_0$, for $t\in[0,s_0]$. Clearly $M_0\geqslant 1$ and for $n\geqslant 1$:

$$\sup_{t\in[0,ns_0]}\ \sup_{\substack{x\in X\\ \|x\|\leqslant 1}} \|T(t)x\| = \max_{1\leqslant k\leqslant n}\ \sup_{t\in[0,s_0]} \|T(kt)\|\leqslant M^n.$$

We shall be only interested in semigroups satisfying (3.11). For those semigroups, it is easy to prove that there are $M\geqslant 1$ and $\omega\in\mathbb{R}$ such that

$$\|T(t)\|\leqslant Me^{\omega t},\quad t\geqslant 0. \tag{3.12}$$

A semigroup satisfying (3.12) has a unique continuous extension to the completion of $(X,\|\cdot\|)$ if X is not complete, and moreover the extension is also a semigroup satisfying (3.12). Therefore we shall assume that $(X,\|\cdot\|)$ is a real Banach space.

Let $\{T(t)\}_{t\geqslant 0}$ be a one-parameter semigroup of linear operators in a real Banach space $(X,\|\cdot\|)$ satisfying (3.12). If $x\in X$ is such that $t\to T(t)x$ is right-continuous at 0, then $t\to T(t)x\in C([0,\infty);X)$. Therefore

$$X_0 := \{x\in X : \lim_{t\downarrow 0} T(t)x = x\} \tag{3.13}$$

is an invariant subspace of $T(t)$, for all $t>0$. It is also closed in X. Therefore $T(t)|_{X_0}:X_0\to X_0$ is a C_0-semigroup of bounded linear operators in X_0, satisfying (3.12). It may happen that $X_0=\{0\}$. We shall assume that $X_0=X$. Finally we observe that if $\{T(t)\}_{t\geqslant 0}:X\to X$ is a C_0-semigroup of *bounded* linear operators in a Banach space X, then it easily follows from the uniform boundedness theorem that T satisfies (3.11) and thus (3.12). Therefore, if $\{T(t)\}_{t\geqslant 0}$ is a one-parameter semigroup of linear bounded operators on a Banach space X satisfying (3.10), then $X_0\neq X$, where X_0 is defined in (3.13). We shall denote by $S(M,\omega)$ the set of all C_0-semigroups of bounded linear operators in a real

Banach space $(X,\|\cdot\|)$, satisfying (3.12). As in Section 3.1, we denote by A_T the infinitesimal generator of $T\in S(M;\omega)$. We have the following generalization of Theorem 3.1.

THEOREM 3.5. *Let $T(t):X\to X$, $t\geqslant 0$ be a C_0-semigroup of linear bounded operators on $(X,\|\cdot\|)$ satisfying (3.12). Then A_T, the infinitesimal generator of T, defined by (2.0), satisfies:*

(i) $\overline{\mathfrak{D}(A_T)}=X$.

(ii) *For every $\lambda>\omega$, $\lambda I-A_T:\mathfrak{D}(A_T)\to X$ is a bijection and $(\lambda I-A_T)^{-1}:X\to X$ is bounded.*

(iii) *For every $\lambda>\omega$, and every $n\in\mathbb{Z}^+$:*

$$\|(\lambda I - A_T)^{-n}\|\leqslant M(\lambda-\omega)^{-n}. \tag{3.14}$$

(iv) *For every $\lambda>\omega$ and $y\in X$,*

$$(\lambda I - A_T)^{-1}y = \int_0^\infty e^{-\lambda t}T(t)y\,dt. \tag{3.15}$$

PROOF. First observe that, if T satisfies (3.12), then $T_\omega(t):=e^{-\omega t}T(t)$ satisfies $\|T_\omega(t)\|\leqslant M$, $t\geqslant 0$ and $A_{T_\omega}=A_T-\omega I$. For $x\in X$, define

$$|||x||| = \sup_{t\geqslant 0}\|T_\omega(t)x\|. \tag{3.16}$$

Then $|||\cdot|||$ is a norm on X and

$$\|x\|\leqslant |||x|||\leqslant M\|x\|,\quad x\in X. \tag{3.17}$$

Thus $|||\cdot|||$ is an equivalent norm on X, and

$$|||T_\omega(t)x||| = \sup_{s\geqslant 0}\|T_\omega(s)T_\omega(t)x\| = \sup_{s\geqslant t}\|T_\omega(s)x\|\leqslant |||x|||.$$

Hence T_ω is a C_0-semigroup of contractions in $(X,|||\cdot|||)$. Clearly $\mathfrak{D}(A_{T_\omega})$ and A_{T_ω} defined by (2.0) in $(X,|||\cdot|||)$ are the same in $(X,\|\cdot\|)$. Hence $\overline{\mathfrak{D}(A_{T_\omega})}=X$ as well as $\overline{\mathfrak{D}(A_T)}=X$, by using Theorem 3.1-(i). This proves (i). By Theorem 3.1-(i) again,

$$\mu I - A_{T_\omega}:\mathfrak{D}(A_{T_\omega})=\mathfrak{D}(A_T)\to X$$

is a bijection for $\mu>0$, and $(\mu I-A_{T_\omega})^{-1}$ is bounded. Moreover for every $x\in X$ and $n\in\mathbb{Z}^+$,

$$|||(\mu I - A_{T_\omega})^{-n}x|||\leqslant\mu^{-n}|||x|||.$$

Since $A_{T_\omega}=A_T-\omega I$, we have

$$\begin{aligned}\|(\lambda I - A_T)^{-n}x\|&\leqslant |||(\lambda I - A_T)^{-n}x|||\leqslant(\lambda-\omega)^{-n}|||x|||\\&\leqslant M(\lambda-\omega)^{-n}\|x\|,\end{aligned}$$

which proves (3.14). Now (3.15) directly follows from Theorem 3.1-(ii). □

We shall denote by $G(M;\omega)$ the class of all linear operators A in X, densely defined, for which $\lambda I - A : \mathfrak{D}(A) \to X$ is a bijection for $\lambda > \omega$, $(\lambda I - A)^{-1} : X \to X$ is bounded and

$$\|(\lambda I - A)^{-n}\| \leqslant M(\lambda - \omega)^{-n} \text{ for } \lambda > \omega, n \in \mathbb{Z}^+.$$

By Theorem 3.5, we can define a map $\mathcal{G}_{M,\omega} : S(M;\omega) \to G(M;\omega)$ by setting $\mathcal{G}_{M,\omega}(T) = A_T$ with $T \in S(M;\omega)$. We have the following generalization of Theorem 3.4.

THEOREM 3.6. *For every $M \geqslant 1, \omega \in \mathbb{R}$,*
(i) *$\mathcal{G}_{M,\omega} : S(M;\omega) \to G(M;\omega)$ is a bijection.*
(ii) *For every $A \in G(M;\omega)$ and $h > 0$ such that $\omega h < 1$, $I - hA : \mathfrak{D}(A) \to X$ is a bijection and $\lim_{n\to\infty}(I - tA/n)^{-n}x$ exists for all $x \in X$, $t \geqslant 0$, the convergence being uniform on bounded t-intervals. If*

$$T_A(t)x := \lim_{n\to\infty}(I - \frac{t}{n}A)^{-n}x,$$

then $T_A \in S(M;\omega)$.
(iii) *For every $A \in G(M;\omega)$, $\mathcal{G}^{-1}_{M,\omega}(A) = T_A$. Furthermore $t \to T_A(t)x$, $x \in \mathfrak{D}(A)$, is the unique solution in $C([0,\infty);X) \cap C^1((0,\infty);X)$ to the initial value problem:*

$$\begin{cases} u(t) \in \mathfrak{D}(A) & ,t > 0, \\ \dfrac{du}{dt}(t) = Au(t) & ,t > 0, \\ u(0) = x. \end{cases} \tag{3.18}$$

(iv) *If $\{T_n\}$ is a sequence in $S(M;\omega)$ and $T \in S(M;\omega)$, then the formula*

$$\lim_{n\to\infty} \max_{t\in[0,s]} \|T_n(t)x - T(t)x\| = 0, \tag{3.19}$$

for every $s > 0$ and every $x \in X$ holds if and only if for every $x \in X$ and $h > 0$ satisfying $h\omega < 1$, we have

$$\lim_{n\to\infty} (I - hA_{T_n})^{-1}x = (I - hA_T)^{-1}x. \tag{3.20}$$

PROOF. We first prove (i), (ii), (iii) for $\omega = 0$, $M > 1$.
"Injectivity of $\mathcal{G}_{M,0}$." Let $T_1, T_2 \in S(M, 0)$ be such that $A_{T_1} = A_{T_2} = A \in G(M, 0)$. Let $x \in \mathfrak{D}(A)$ and $u_i(t) = T_i(t)x$, $t \geqslant 0$, $i = 1,2$. Define

$$|||y|||_i = \sup_{t\geqslant 0}\|T_i(t)y\|, \text{ for } y \in X,\ i = 1,2.$$

Since T_1 is a contraction semigroup with respect to the norm $|||\cdot|||_1$, the operator A is dissipative with respect to $|||\cdot|||_1$ by Theorem 3.1, and (3.18) possesses at most one solution for $x \in \mathfrak{D}(A)$ by Proposition 3.3. Since $A = A_{T_1}$, we get that u_1 is the solution of (3.18). Similarly, for u_2, and thus

$T_1(t)x = T_2(t)x$, $t \geqslant 0, x \in \mathfrak{D}(A)$. Since $T_1,(t), T_2(t)$ are continuous and $\overline{\mathfrak{D}(A)} = X$, we get $T_1 = T_2$.
"Surjectivity of $\mathcal{G}_{M,0}$." Let $A \in G(M;0)$. Then

$$\|(I - hA)^{-n}\| \leqslant M \text{ for every } n \in \mathbb{Z}^+, \text{ and } h > 0.$$

For $x \in X$ and $\mu > 0$ we define

$$\|x\|_\mu := \sup_{n \geqslant 0} \|(I - \mu A)^{-n} x\|. \tag{3.21}$$

Clearly $\|\cdot\|_\mu$ is a norm on X,

$$\|x\| \leqslant \|x\|_\mu \leqslant M\|x\|, \quad x \in X, \tag{3.22}$$

and

$$\|(I - \mu A)^{-1} x\|_\mu \leqslant \|x\|_\mu, \quad x \in X. \tag{3.23}$$

Moreover

$$\|(I - \lambda A)^{-1} x\|_\mu \leqslant \|x\|_\mu, \; x \in X, \; \lambda > \mu > 0. \tag{3.24}$$

Indeed, for $0 < \mu < \lambda$, we have

$$(I - \lambda A)^{-1} = (I - \mu A)^{-1}[(\frac{\mu}{\lambda})I + (1 - \frac{\mu}{\lambda})(I - \lambda A)^{-1}], \tag{3.25}$$

since

$$\begin{aligned}
&(I - \mu A)^{-1}[(\frac{\mu}{\lambda})I + (1 - \frac{\mu}{\lambda})(I - \lambda A)^{-1}] \\
&= (I - \mu A)^{-1}[(\frac{\mu}{\lambda})(I - \lambda A) + (1 - \frac{\mu}{\lambda})I](I - \lambda A)^{-1} \\
&= (I - \mu A)^{-1}(I - \mu A)(I - \lambda A)^{-1},
\end{aligned}$$

by noting that

$$\mathfrak{R}[(I - \lambda A)^{-1}] \subseteq \mathfrak{D}(A).$$

Hence

$$\begin{aligned}
\|(I - \lambda A)^{-1} x\|_\mu &= \|(I - \mu A)^{-1}[\frac{\mu}{\lambda} I + (1 - \frac{\mu}{\lambda})(I - \lambda A)^{-1}]x\|_\mu \\
&\leqslant \frac{\mu}{\lambda}\|x\|_\mu + (1 - \frac{\mu}{\lambda})\|(I - \lambda A)^{-1} x\|_\mu,
\end{aligned}$$

which proves (3.24). From (3.24) we have

$$\|(I - \lambda A)^{-n} x\|_\mu \leqslant \|x\|_\mu, \quad n \in \mathbb{Z}^+,$$

and from (3.22),

$$\|(I - \lambda A)^{-n} x\| \leqslant \|x\|_\mu, \text{ for } \lambda > \mu.$$

Hence $\|x\|_\lambda \leqslant \|x\|_\mu$ for $\lambda > \mu$. Next we set

$$|x| := \sup_{\mu > 0} \|x\|_\mu = \lim_{\mu \downarrow 0} \|x\|_\mu, \text{ for } x \in X. \tag{3.26}$$

Now $|\cdot|$ is a norm on X satisfying

$$\|x\| \leqslant |x| \leqslant M\|x\|, \quad x \in X, \tag{3.27}$$

and from (3.24), we obtain

$$|(I - \lambda A)^{-1}x| \leqslant |x|, \lambda > 0, x \in X. \tag{3.28}$$

Hence A is m-dissipative in $(X, |\cdot|)$ and densely defined. Moreover

$$\|T_A(t)x\| \leqslant |T_A(t)x| \leqslant |x| \leqslant M\|x\|, x \in X, t \geqslant 0,$$

and $T_A \in S(M;0)$. Then $\mathcal{G}_{M,0}$ is surjective by Theorem 3.4, and (ii) - (iii) follow from Proposition 3.3 and Theorem 3.4.

If $\omega \neq 0$ and $T \in S(M,\omega)$, then T_ω defined by

$$T_\omega(t) = e^{-\omega t}T(t), \quad t \geqslant 0,$$

satisfies $T_\omega \in S(M,0)$, $A_{T_\omega} = A_T - \omega I$, and $A_{T_\omega} \in G(M,0)$. It easily follows that $\mathcal{G}_{M,\omega}$ is injective. Conversely, if $A \in G(M,\omega)$, then $A - \omega I \in G(M,0)$ and $\mathcal{G}_{M,\omega}$ is also surjective by Theorem 3.4. It remains to prove that if $A \in G(M,\omega)$, then

$$\lim_{n \to \infty} (I - \frac{t}{n}A)^{-n}x = e^{\omega t}T_{A-\omega I}(t)x, t \geqslant 0, x \in X,$$

uniformly on bounded t-intervals. We have

$$(I - \frac{t}{n}A)^{-n}x = [(1 - \frac{\omega t}{n})I - \frac{t}{n}(A - \omega I)]^{-n}x$$

$$= (1 - \frac{\omega t}{n})^{-n}[I - \frac{t_n}{n}(A - \omega I)]^{-n}x,$$

with $t_n = (1 - \omega t/n)^{-1}t$. Now $B := A - \omega I$ is m-dissipative and densely defined with respect to the norm $|\cdot|$ defined in (3.26). From (2.24) we have for $x \in \mathcal{D}(B)$:

$$\|(I - \frac{t}{n}B)^{-n}x - (I - \frac{t_n}{n}B)^{-n}x\| \leqslant M\|Bx\| \cdot \sqrt{(t - t_n)^2 + \min(t, t_n) \cdot \frac{t + t_n}{n}}.$$

Hence

$$\lim_{n \to \infty} \max_{t \in [0,s]} \|(I - \frac{t}{n}B)^{-n}x - (I - \frac{t_n}{n}B)^{-n}x\| = 0,$$

for every $s > 0$. It follows that

$$\lim_{n \to \infty} \max_{t \in [0,s]} \|(I - \frac{t}{n}(A - \omega I))^{-n}x - e^{\omega t}T_{A-\omega I}(t)x\| = 0,$$

for every $s > 0$, and every $x \in \mathcal{D}(A)$, hence for every $x \in \overline{\mathcal{D}(A)} = X$. This completes the proof of (i), (ii), (iii).

(iv) The "only if" part of the proof is an easy consequence of (3.15). The "if" part and some generalization of it will be proven in Theorem 3.32. □

3. DISSIPATIVE OPERATORS

In this section we collect some results about dissipative operators in a real Banach space $(X, \|\cdot\|)$.

PROPOSITION 3.7. *Let $A:\mathfrak{D}(A)\subset X\to X$ be dissipative. Then:*

(i) *If A is closable, then the closure of $\overline{A}$ is also dissipative.*

(ii) *If $\overline{\mathfrak{D}(A)}\supseteq\mathfrak{R}(A)$, in particular if A is densely defined, then A is closable.*

(iii) *A is closed if and only if $\mathfrak{R}(I-hA)$ is closed for every $h>0$ (or equivalently, for some $h_0>0$).*

(iv) *$\mathfrak{R}(I-h_0A)=X$ for some $h_0>0$ if and only if $\mathfrak{R}(I-hA)=X$ for every $h>0$, that is, A is m-dissipative.*

PROOF.

(i) follows directly from the definition of dissipativity.

(ii) Let $x_n\in\mathfrak{D}(A)$, $y\in X$ be such that $\lim_{n\to\infty}x_n=0$ and $\lim_{n\to\infty}Ax_n=y$. We have to prove $y=0$. Let $z\in\mathfrak{D}(A)$, then for $h>0$, we have

$$\|x_n+hz\|\leqslant\|x_n+hz-hA(x_n+hz)\|,$$

and

$$\|hz\|\leqslant\|h(z-y)-h^2Az\|,$$

by taking the limit. Hence $\|z\|\leqslant\|z-y-hAz\|$ and $\|z\|\leqslant\|z-y\|$ by letting $h\downarrow 0$. Since $y\in\overline{\mathfrak{R}(A)}\subseteq\overline{\mathfrak{D}(A)}$, there is a sequence $\{z_n\}\subset\mathfrak{D}(A)$ such that $\lim_{n\to\infty}z_n=y$. Hence $\|y\|\leqslant 0$ and $y=0$.

(iii) A is closed $\Rightarrow$ $\mathfrak{R}(I-hA)$ is closed for every $h>0$.

Let $h>0$ and let $x_n\in\mathfrak{D}(A)$, $f_n=x_n-hAx_n$ be such that $\lim_{n\to\infty}f_n=f$. Since A is dissipative, $\|x_n-x_m\|\leqslant\|f_n-f_m\|$ and $\{x_n\}$ is a Cauchy sequence with limit $x\in X$. Now $Ax_n=h^{-1}(x_n-f_n)$ converges to $h^{-1}(x-f)$. Since A is closed, $x\in\mathfrak{D}(A)$ and $Ax=h^{-1}(x-f)$. Then $f\in\mathfrak{R}(I-hA)$.

Clearly $\mathfrak{R}(I-hA)=X$, for every $h>0$ implies that $\mathfrak{R}(I-h_0A)=X$ for some $h_0>0$.

$\mathfrak{R}(I-h_0A)=X\Rightarrow A$ is closed.

Let $[x_n,y_n]\in A$ be such that $\lim_{n\to\infty}[x_n,y_n]=[x,y]$. Set $f_n=x_n-h_0Ax_n$. Then f_n converges to $f=x-h_0y$. Since $\mathfrak{R}(I-h_0A)$ is closed, there is $z\in\mathfrak{D}(A)$ such that $f=z-h_0Az$. But A is dissipative, thus $\|x-z\|\leqslant\|f-f\|=0$, hence $x=z$ and $Ax=y$.

(iv) Assume $\mathfrak{R}(I-h_0A)=X$, for some $h_0>0$. Then $\mathfrak{R}(I-hA)=X$ for $h\in(h_0/2,\infty)$. Indeed, let $f\in X$ and $h\in(h_0/2,\infty)$. Then $x-hAx=f$ is equivalent to $x-h_0Ax=h_0/h\cdot f+(1-h_0/h)x$, or

$$x=(I-h_0A)^{-1}\left(\frac{h_0}{h}f+(1-\frac{h_0}{h})x\right):=F(x).$$

Since $1-h_0/h\in(-1,1)$ and $\|(I-h_0A)^{-1}\|\leqslant 1$, we get that $F:X\to X$ is Lipschitz continuous with $\|F\|_{Lip}<1$. By the Banach fixed point theorem, F possesses a unique fixed point X. Hence $f\in\mathfrak{R}(I-hA)$ and $\mathfrak{R}(I-hA)=X$, for $h>h_0/2$. By induction $\mathfrak{R}(I-hA)=X$, for $h>0$. □

PROPOSITION 3.8. *Let* $A:\mathfrak{D}(A)\subseteq X\to X$ *be m-dissipative. Then*
(i) *A is "maximal" dissipative, i.e. if B is dissipative and* $A\subseteq B$, *then* $A=B$.
(ii) *If X is reflexive, then* $\overline{\mathfrak{D}(A)}=X$.
(iii) *If* $X_0:=\overline{\mathfrak{D}(A)}$ *and if* A_0 *is the "part of A in* X_0*" i.e.*

$$\begin{cases}\mathfrak{D}(A_0) := \{x\in\mathfrak{D}(A): Ax\in X_0\},\\ A_0x := Ax \text{ for } x\in\mathfrak{D}(A_0),\end{cases}$$

then A_0 *is m-dissipative and densely defined in* $(X_0,\|\cdot\|)$.

PROOF.
(i) $A\subseteq B$ in the sense of graphs in $X\times X$ implies $(I-A)^{-1}\subseteq(I-B)^{-1}$. Now $\mathfrak{D}((I-A)^{-1})=\mathfrak{R}(I-A)=X$, hence $\mathfrak{D}((I-B)^{-1})=X$. Moreover $(I-B)^{-1}$ is the graph of a single valued operator in X, since B is dissipative. Hence $(I-A)^{-1}=(I-B)^{-1}$. It follows that $I-A=I-B$ and $A=B$.
(ii) let $x^*\in X^*$ be such that $<x,x^*>=0$ for every $x\in\mathfrak{D}(A)$. We have to show that $x^*=0$. Since $E_h^Ax\in\mathfrak{D}(A)$, for $h>0$, we have

$$<A_hx,x^*> = h^{-1}<E_h^Ax - x\,,x^*> = 0,$$

for every $x\in\mathfrak{D}(A)$. For $x\in\mathfrak{D}(A)$, $\lim_{h\downarrow 0}E_h^Ax=x$ and $\|A_hx\|\leqslant\|Ax\|$, by (2.13), (2.16). Since X is reflexive, there exists a sequence $h_n\downarrow 0$ and $z\in X$ such that

$$w-\lim_{n\to\infty} A_{h_n}x = z,$$

or

$$w-\lim_{n\to\infty} AE_{h_n}^Ax = z.$$

A is closed by Proposition 3.7-(iii) and thus A is also weakly closed. Hence $z=Ax$. Then

$$<Ax,x^*> = \lim_{n\to\infty} <A_{h_n}x,x^*> = 0,$$

and we get that $<x-Ax,x^*>=0$ for every $x\in\mathfrak{D}(A)$. Since A is m-dissipative, $\mathfrak{R}(I-A)=X$ and $x^*=0$.
(iii) Clearly A_0 is dissipative. For $h>0$ and $f\in X_0$, there is an $x\in\mathfrak{D}(A)$ such that $x-hAx=f$, since A is m-dissipative. Then $Ax=h^{-1}(x-f)\in X_0$, hence $x\in\mathfrak{D}(A_0)$ and A_0 is m-dissipative in X_0. Moreover $x=E_h^Af$ converges to f as $h\downarrow 0$ since $f\in\overline{\mathfrak{D}(A)}$. Hence $\mathfrak{D}(A_0)$ is dense in X_0. □

COROLLARY 3.9. *Let A be m-dissipative (linear) in X and let* T_A *be the* C_0-*semigroup of contractions on* $\overline{\mathfrak{D}(A)}$ *defined in Theorem 2.3. Then the infinitesimal generator of* T_A *is the part of A in* $\overline{\mathfrak{D}(A)}$.

PROOF. Let B denote the infinitesimal generator of T_A in $\overline{\mathfrak{D}(A)}$. By Theorem 2.15, $Bx=Ax$ for every $x\in\mathfrak{D}(B)$. Since $Bx\in\overline{\mathfrak{D}(A)}$, $Bx=A_0x$, for every

$x\in\mathcal{D}(B)$, where A_0 denotes the part of A in $\overline{\mathcal{D}(A)}$. Hence $B\subseteq A_0$ in $\overline{\mathcal{D}(A)}$. Since B is m-dissipative and A_0 is dissipative in $\mathcal{D}(A)$, $B=A_0$ by Proposition 3.8-(i). □

Let $A:\mathcal{D}(A)\subset X\to X$ be dissipative and densely defined. Then A is closable and the closure of A is dissipative by Proposition 3.7. Let A^* denote the adjoint of A in $(X^*,\|\cdot\|)$. Note that A^* is also the adjoint of the closure of A in X. We have:

PROPOSITION 3.10. *Let $A:\mathcal{D}(A)\subset X\to X$ be dissipative and densely defined. Then the following assertions are equivalent.*

(i) *A^* is dissipative in $(X^*,\|\cdot\|)$.*
(ii) *$\mathcal{R}(I-hA)$ is dense in X for every $h>0$.*
(iii) *$\mathcal{R}(I-h_0A)$ is dense in X for some $h_0>0$.*
(iv) *The closure of A is m-dissipative.*
(v) *A^* is m-dissipative in $(X^*,\|\cdot\|)$.*

PROOF. (i)⇒(ii). We have to prove that $\langle x-hAx,x^*\rangle=0$, for every $x\in\mathcal{D}(A)$ and some $h>0$, implies $x^*=0$. Clearly $x^*\in\mathcal{D}(A^*)$ and $\langle x,x^*-hA^*x^*\rangle=0$ for every $x\in\mathcal{D}(A)$. Since $\overline{\mathcal{D}(A)}=X$, $x^*-hA^*x^*=0$, but A^* is dissipative, hence $x^*=0$.

(ii)⇒(iii) is clear.

(iii)⇒(iv). Let $\tilde{A}$ denote the closure of A. Then $\tilde{A}$ is closed and $\mathcal{R}(I-h_0\tilde{A})$ is closed by Proposition 3.7-(iii). Moreover $\mathcal{R}(I-h_0\tilde{A})\supseteq\overline{\mathcal{R}(I-h_0A)}=X$, hence $\tilde{A}$ is m-dissipative, by Proposition 3.7-(iv).

(iv)⇒(v). We have $A^*=\tilde{A}^*$. We first prove that A^* is dissipative. Since $\overline{\mathcal{D}(\tilde{A})}=X$, we have that

$$\|x^*\|=\sup_{\substack{x\in\mathcal{D}(\tilde{A})\\ \|x\|\leqslant 1}}|\langle x,x^*\rangle|$$

for every $x^*\in X^*$. In particular for $x^*\in\mathcal{D}(A^*)$, $h>0$,

$$\|x^*-hA^*x^*\|=\sup_{\substack{x\in\mathcal{D}(\tilde{A})\\ \|x\|\leqslant 1}}|\langle x,x^*-hA^*x^*\rangle|=$$

$$\sup_{\substack{x\in\mathcal{D}(\tilde{A})\\ \|x\|\leqslant 1}}|\langle x-h\tilde{A}x,x^*\rangle|\geqslant$$

$$\sup_{\substack{x\in\mathcal{D}(\tilde{A})\\ \|x-h\tilde{A}x\|\leqslant 1}}|\langle x-h\tilde{A}x,x^*\rangle|,$$

by using the fact that $\tilde{A}$ is dissipative. Since $\mathcal{R}(I-h\tilde{A})=X$, we have

$$\sup_{\substack{x\in\mathcal{D}(\tilde{A})\\ \|x-h\tilde{A}x\|\leqslant 1}}|\langle x-h\tilde{A}x,x^*\rangle|=\sup_{\substack{y\in X\\ \|y\|\leqslant 1}}|\langle y,x^*\rangle|=\|x^*\|.$$

Then A^* is dissipative. Next we prove that $\mathfrak{R}(I-hA^*)=X^*$. Let $f^*\in X^*$ and set $x^*=[(I-h\tilde{A})^{-1}]^* f^*$, for $h>0$. Then for $f\in\mathfrak{D}(\tilde{A})$ we have:

$$|<f-h\tilde{A}f,x^*>|=|<f-h\tilde{A}f,[(I-h\tilde{A})^{-1}]^*f^*>|=$$

$$|<(I-h\tilde{A})^{-1}(f-h\tilde{A}f),f^*>|=|<f,f^*>|\leqslant\|f^*\|\,\|f\|.$$

Thus $x^*\in\mathfrak{D}(A^*)$ and

$$<f,(I-hA^*)x^*> = <(I-h\tilde{A})f,x^*> = <f,f^*>.$$

Since $\mathfrak{D}(\tilde{A})$ is dense in X, $(I-h\tilde{A})^*x^*=f^*$ and $\mathfrak{R}(I-h\tilde{A})=X^*$.
(v)$\Rightarrow$(i) is clear. □

PROPOSITION 3.11. *Let $A:\mathfrak{D}(A)\subset X\to X$ be m-dissipative and densely defined. Then*
(i) *A^* is m-dissipative in $(X^*,\|\cdot\|)$.*
(ii) *$E_h^{A^*}=(E_h^A)^*$ and $(A^*)_h=(A_h)^*$, $h>0$.*
(iii) *$\mathfrak{D}(A^*)$ is weak * dense in X^*.*
(iv) *$\mathfrak{D}(A^*)=\{x^*\in X^*:\sup_{h>0}\|A_h^*x^*\|<\infty\}$.*

PROOF.
(i) Since A is m-dissipative, $\mathfrak{R}(I-hA)=X$ and A^* is m-dissipative by Proposition 3.10.
(ii) $E_h^{A^*}=(E_h^A)^*$ follows from the last part of the proof of Proposition 3.10. Then

$$(A^*)_h=h^{-1}(E_h^{A^*}-I)=(h^{-1}(E_h^A-I))^*=(A_h)^*.$$

(iii) Let $x^*\in X^*$, then $E_h^{A^*}x^*\in\mathfrak{D}(A^*)$ for $h>0$, and for every $x\in X$,

$$\lim_{h\downarrow 0}<x,E_h^{A^*}x^*> = \lim_{h\downarrow 0}<E_h^Ax,x^*> = <x,x^*>.$$

Hence $\mathfrak{D}(A^*)$ is weak * dense in X^*.
(iv) From (2.13), (2.16) it follows that

$$\mathfrak{D}(A^*)\subseteq\{x^*\in X^*:\ \sup_{h\downarrow 0}\|A_h^*x^*\|_*<\infty\}.$$

Assume that there is an $M>0$ such that $\|(A^*)_hx^*\|\leqslant M$ for every $h>0$. We have to prove that $x^*\in\mathfrak{D}(A^*)$. For every $x\in\mathfrak{D}(A)$,

$$|<Ax,x^*>|=\lim_{h\downarrow 0}|<A_hx,x^*>|$$

$$=\lim_{h\downarrow 0}|<x,(A_h)^*x^*>|=\lim_{h\downarrow 0}|<x,(A^*)_hx^*>|$$

$$\leqslant\sup_{h>0}|<x,(A^*)_hx^*>|\leqslant M\|x\|.$$

Then $x^*\in\mathfrak{D}(A^*)$. □

REMARKS.

a) If X is reflexive and $A \in C(X)$, then A^* is m-dissipative by Proposition 3.11 and $\mathfrak{D}(A^*)$ is dense in X^* by Proposition 3.8-(ii). Then $A^* \in C(X^*)$. Moreover $A = A^{**}$ and

$$\mathfrak{D}(A) = \{x \in X : \sup_{h>0} \|A_h x\| < \infty\} = \{x \in X : |Ax| < \infty\}.$$

b) If X is a Hilbert space and A is closed, densely defined, dissipative, symmetric, but not self-adjoint, then A^* is not dissipative. Indeed, if A^* were dissipative, then A would be m-dissipative by Proposition 3.11-(i). Moreover $A \subseteq A^*$ and A^* dissipative would imply that $A = A^*$ by Proposition 3.8-(i). But A is not self-adjoint. In other words a symmetric dissipative operator in a Hilbert space is self-adjoint if and only if A is m-dissipative if and only if A^* is dissipative.

EXAMPLE. Let $X = L_2[0,1]$, $\mathfrak{D}(A) = \{u \in W_{2,2}(0,1):\ u(0) = u(1) = u'(0) = u'(1) = 0\}$, and $Au = u''$ for $u \in \mathfrak{D}(A)$. Then A is closed, densely defined, dissipative and symmetric but not m-dissipative.

We conclude this section by considering a slight but interesting generalization of the notion of dissipativity introduced in ARENDT, CHERNOFF and KATO [1982]. Let X be a real vector space. A function $p: X \to \mathbb{R}$ is called sublinear if

$$\begin{cases} p(x+y) \leqslant p(x) + p(y) & , x, y \in X, \\ p(\lambda x) = \lambda p(x) & , \lambda \geqslant 0, x \in X. \end{cases}$$

Then clearly $p(x) + p(-x) \geqslant 0$, for $x \in X$. If $p(x) + p(-x) > 0$ for $x \neq 0$, then p is called a *half-norm* on X.

A linear operator $A: \mathfrak{D}(A) \subset X \to X$ is called *p-dissipative* if

$$p(x) \leqslant p(x - hAx),\ h > 0,\ x \in \mathfrak{D}(A). \tag{3.29}$$

Then A is dissipative in $(X, \|\cdot\|)$ if and only if A is p-dissipative with respect to $p(x) = \|x\|$.

Note that $\|x\|_p := \max(p(x), p(-x)), x \in X$ defines a norm on X and if p is continuous, then $\|\cdot\|_p$ is also continuous and there is an $M > 0$ such that $\|x\|_p \leqslant M\|x\|$, for every $x \in X$. Conversely if $\|x\|_p \leqslant M\|x\|$ for every $x \in X$, then $\|\cdot\|_p$ is continuous as well as p. As in Proposition 3.7, we have:

PROPOSITION 3.12. *Let $(X, \|\cdot\|)$ be a real Banach space and let $p: X \to \mathbb{R}$ be a real continuous sublinear functional on X. Let $A: \mathfrak{D}(A) \subset X \to X$ be a p-dissipative operator. Then*

(i) *if A is closable, then its closure is also p-dissipative.*

(ii) *If, moreover, p is a half norm and if $\overline{\mathfrak{D}(A)} \supseteq \mathfrak{R}(A)$, then A is closable.*

Note that, if A is closed, $\mathfrak{R}(I - hA)$ is not necessarily closed for $h > 0$.

However, if $\|\cdot\|_p$ is an equivalent norm on X, then A is also dissipative with respect to the norm $\|\cdot\|_p$ and we can apply Proposition 3.7 with $(X,\|\cdot\|_p)$ instead of $(X,\|\cdot\|)$.

Since a sublinear functional is convex, we can define for $x,y \in X$, $\lambda \neq 0$:

$$p_\lambda'(x,y) := \lambda^{-1}(p(x+\lambda y) - p(x)), \tag{3.30}$$

$$p'(x,y) := \inf_{\lambda>0} p_\lambda'(x,y) = \lim_{\lambda\downarrow 0} p_\lambda'(x,y). \tag{3.31}$$

Next we observe that

$$\sup_{\lambda<0} p_\lambda'(x,y) = \lim_{\lambda\uparrow 0} p_\lambda'(x,y) = -p'(x,-y) \tag{3.32}$$

(see Appendix 1). Then Proposition 2.16 holds with $[x,y]_+$ (resp. $[x,y]_-$) replaced by $p'(x,y)$ (resp. $-p'(x,-y)$).

DEFINITION 3.13. Let $(X,\|\cdot\|)$ be a real Banach space and let $p: X\to\mathbb{R}$ be a continuous sublinear functional. Then an operator $A: \mathfrak{D}(A)\subset X\to X$ is called *strictly p-dissipative* if

$$p'(x,Ax) \leqslant 0 \text{ for every } x\in\mathfrak{D}(A). \tag{3.33}$$

Clearly if A is strictly p-dissipative, then A is p-dissipative.

THEOREM 3.14. *Let p be a continuous sublinear functional on X.*

(i) *Let A be p-dissipative in X, such that* $\overline{\mathfrak{D}(A)}\supseteq\mathfrak{R}(A)$, *then A is strictly p-dissipative.*

(ii) *Let A be strictly p-dissipative in X and let B be p-dissipative in X, then* $A+B$ *is p-dissipative in X.*

(iii) *Let A be densely defined and dissipative in X, and let B be dissipative in X. Then* $A+B$ *is dissipative.*

PROOF. (i) Let $x,y\in\mathfrak{D}(A)$, $\lambda>0$. Then

$$\begin{aligned}
p(x+\lambda Ax) &= p(x+\lambda y+\lambda(Ax-y))\\
&\leqslant p(x+\lambda y)+\lambda p(Ax-y) \leqslant p(x+\lambda y)+\lambda M\|Ax-y\|\\
&\leqslant p((I-\lambda A)(x+\lambda y))+\lambda M\|Ax-y\|\\
&= p(x+\lambda(y-Ax)-\lambda^2 Ay)+\lambda M\|Ax-y\|\\
&\leqslant p(x)+\lambda p(y-Ax)+p(-\lambda^2 Ay)+\lambda M\|Ax-y\|\\
&\leqslant p(x)+2M\lambda\|Ax-y\|+M\lambda^2\|Ay\|.
\end{aligned}$$

Then for $\lambda>0$:

$$\lambda^{-1}[p(x+\lambda Ax)-p(x)] \leqslant 2M\|Ax-y\|+M\lambda\|Ax\|,$$

hence

$$p'(x,Ax) \leqslant 2M\|Ax - y\|.$$

Since $\overline{\mathfrak{D}(A)} \supseteq \mathfrak{R}(A)$, we obtain $p'(x,Ax) \leqslant 0$.
(ii) For $x,y,z \in X$, we have

$$p'(x,y+z) \geqslant p'(x,y) - p'(x,-z).$$

Indeed

$$p'(x,y) = p'(x,-z+y+z) \leqslant p'(x,-z) + p'(x,y+z).$$

Hence for $x \in \mathfrak{D}(A) \cap \mathfrak{D}(B)$, we have

$$p'(x,-(Ax+Bx)) \geqslant p'(x,-Bx) - p'(x,Ax) \geqslant 0.$$

(iii) is a direct consequence of (i) and (ii). □

Let $(X,\|\cdot\|)$ be a real Banach space and let $p: X \to \mathbb{R}$ be a continuous sublinear functional. Then a linear operator $E: X \to X$ is called a *p-contraction* if $p(Ex) \leqslant p(x)$ for every $x \in X$. We have

THEOREM 3.15. *Let $\{T(t)\}_{t \geqslant 0}$ be a C_0-semigroup of bounded linear operators in X and let A_T be its infinitesimal generator. Then $\{T(t)\}_{t \geqslant 0}$ are p-contractions if and only if A_T is p-dissipative.*

PROOF. Let $T(t), t \geqslant 0$ be p-contractions in X. Then for $x \in \mathfrak{D}(A_T)$, $h > 0$, $t > 0$, we have

$$p(x - ht^{-1}(T(t)x - x)) = p((1+ht^{-1})x - ht^{-1}T(t)x) \geqslant$$
$$p((1+ht^{-1})x) - p(ht^{-1}T(t)x) = (1+ht^{-1})p(x) - ht^{-1}p(T(t)x),$$

since p is a half-norm. Moreover $p(T(t)x) \leqslant p(x)$ implies

$$(1+ht^{-1})p(x) - ht^{-1}p(T(t)x) \geqslant (1+ht^{-1})p(x) - ht^{-1}p(x) = p(x).$$

Since p is continuous, we obtain $p(x - hA_Tx) \geqslant p(x)$, and A_T is p-dissipative.
Let A_T be p-dissipative. From Section 3.2 we know that T satisfies (3.12) for some $M \geqslant 1$ and $\omega \in \mathbb{R}$. Moreover, for $\omega h < 1$, $I - hA_T$ is invertible and $\mathfrak{R}(I - hA_T) = X$. Since A_T is dissipative, $p((I - hA_T)^{-1}x) \leqslant p(x)$ for every h such that $h\omega < 1$ and every $x \in X$. By induction, we obtain

$$p((I - hA_T)^{-n}x) \leqslant p(x), \ x \in X \text{ and } h\omega < 1.$$

Let $t > 0$. Then $p((I - t/n \cdot A_T)^{-n}x) \leqslant p(x)$ for $n > t\omega$. By Theorem 3.6 and the continuity of p, we obtain:

$$p(T(t)x) = \lim_{n \to \infty} p((I - \frac{t}{n}A)^{-n}x) \leqslant p(x),$$

for every $x \in X$. This completes the proof of the theorem. □

4. Dual semigroups

Let $\{T(t)\}_{t\geqslant 0}$ be a C_0-semigroup of bounded linear operators on a Banach space $(X,\|\cdot\|)$ satisfying (3.12). Then $\{T^*(t)\}_{t\geqslant 0}$, where $T^*(t)$ is the adjoint of $T(t)$ in the dual space $(X^*,\|\cdot\|)$, is also a semigroup of bounded linear operators satisfying

$$\|T^*(t)\|\leqslant Me^{\omega t}, \quad t\geqslant 0. \tag{3.34}$$

$\{T^*(t)\}_{t\geqslant 0}$ is not necessarily a C_0-semigroup. For example, if $X=L_p(\mathbb{R})$, $1\leqslant p<\infty$, and

$$(T(t)u)(x):=u(t+x), \quad t,x\in\mathbb{R},$$

then $\{T^*(t)\}_{t\in\mathbb{R}}$ is a C_0-group of contractions on $(L_p(\Omega))^*$ if $p>1$, but not for $p=1$. From Propositions 3.8-(ii) and 3.11-(i), it follows that the adjoint of a strongly continuous semigroup on a *reflexive* Banach space is again strongly continuous. In general, this does not hold for a non-reflexive Banach space X. In that case we only know that $T^*(t)$ is w^*-continuous (weak * continuous, that is continuous with respect to the weak * topology), and we define $X^{\odot}$ to be the subspace of X^* on which $T^*(t)$ is strongly continuous. In the theorem below we show that $X^{\odot}$ is closed and invariant under $T^*(t)$, and we define $T^{\odot}(t)=T^*(t)|_{X^{\odot}}$. Then $\{T^{\odot}(t)\}_{t\geqslant 0}$ is again a strongly continuous semigroup on $X^{\odot}$.

Theorem 3.16. *Let $\{T(t)\}_{t\geqslant 0}$ be a C_0-semigroup of bounded linear operators on a Banach space $(X,\|\cdot\|)$, with infinitesimal generator A. Let $\{T^*(t)\}_{t\geqslant 0}$ be the dual semigroup of $\{T(t)\}_{t\geqslant 0}$, and finally, let $X^{\odot}$ be the subspace on which $T^*(t)$ is strongly continuous. Then:*

(i) *$t\to T^*(t)x^*$ is weak * continuous on $[0,\infty)$.*

(ii) *$t\to T^*(t)x^*$ is weak * right differentiable at 0 if and only if $x^*\in\mathcal{D}(A^*)$.*

(iii) *If $x^*\in\mathcal{D}(A^*)$, then, for $t>0$, $T^*(t)x^*\in\mathcal{D}(A^*)$, the function $t\to T^*(t)x^*$ is weak * differentiable and*

$$\frac{d}{dt}<x,T^*(t)x^*> = <x,A^*T^*(t)x^*> = <x,T^*(t)A^*x^*>,$$

for every $x\in X$ and $t\geqslant 0$.

(iv) *$X^{\odot}=\overline{\mathcal{D}(A^*)}$.*

(v) *$X^{\odot}$ is invariant under $T^*(t)$, $t\geqslant 0$, and $\{T^{\odot}(t)\}_{t\geqslant 0}$, the restriction of $\{T^*(t)\}_{t\geqslant 0}$ to $X^{\odot}$, is a C_0-semigroup of bounded operators on $X^{\odot}$.*

(vi) *The infinitesimal generator of $\{T^{\odot}(t)\}_{t\geqslant 0}$, which we shall denote by $A^{\odot}$, is the part of A^* in $X^{\odot}$.*

(vii) *$\mathcal{D}(A^{\odot})$ is weak * dense in X^*.*

(viii) *If X is reflexive, then $X^{\odot}=X^*$ and $A^{\odot}=A^*$*

Proof. We know from Section 3.2 that $\{T(t)\}_{t\geqslant 0}$ satisfies (3.12) for some $M\geqslant 1$ and $\omega\in\mathbb{R}$. Without loss of generality we assume $\omega=0$. Moreover, by using the equivalent norm defined in (3.16) we may assume $M=1$. Then A is m-dissipative and densely defined in X, by Theorem 3.1. We shall denote the

norm (3.16) again by $\|\cdot\|$.
(i) Clearly $t \to <x, T^*(t)x^*> = <T(t)x, x^*>$ is continuous for every $x^* \in X^*$ and $t \geqslant 0$.
(ii) Suppose $x^* \in \mathcal{D}(A^*)$. From (3.5) we have $R_{T,h}x \in \mathcal{D}(A)$ for every $x \in X$, and $AR_{T,h}x = A_{T,h}x$. Then

$$<x, (A_{T,h})^* x^*> = <R_{T,h}x, A^* x^*>,$$

and

$$<x, A^* x^*> = \lim_{h \downarrow 0} <R_{T,h}x, A^* x^*> = \lim_{h \downarrow 0} <x, (A_{T,h})^* x^*>.$$

If

$$\mathcal{D}(\tilde{B}) := \{x \in X^* : w^* - \lim_{h \downarrow 0} h^{-1}(T^*(h) - I)x^* \text{ exists}\},$$

and

$$\tilde{B}x^* := w^* - \lim_{h \downarrow 0} h^{-1}(T^*(h)x^* - x^*),$$

for $x^* \in \mathcal{D}(\tilde{B})$, then $A^* \subseteq \tilde{B}$. Conversely if $x^* \in \mathcal{D}(\tilde{B})$, then for every $x \in \mathcal{D}(A)$, we have

$$\begin{aligned} <Ax, x^*> &= \lim_{h \downarrow 0} <h^{-1}(T(h)x - x), x^*> \\ &= \lim_{h \downarrow 0} <x, h^{-1}(T^*(h)x^* - x^*)> = <x, \tilde{B}x^*>. \end{aligned}$$

Then $\tilde{B} \subseteq A^*$ and thus $A^* = \tilde{B}$, which proves (ii).
(iii) is proven like Proposition 3.3-(i). Note that $T^*(t): X^* \to X^*$ is weak * continuous for every $t > 0$, being the adjoint of a bounded operator in X.
(iv)-(v) If $t \to T^*(t)x^*$ is right continuous at 0, then $t \to T^*(t)x^*$ is continuous on $[0, \infty)$ and for $t > 0$, $x \in \mathcal{D}(A)$ we have:

$$<Ax, \frac{1}{t}\int_0^t T^*(s)x^* ds> = <R_{T,t}Ax, x^*> = <A_{T,t}x, x^*>,$$

by 3.6. Then

$$|<Ax, \frac{1}{t}\int_0^t T^*(s)x^* ds>| \leqslant 2t^{-1}\|x^*\| \, \|x\|,$$

for every $x \in \mathcal{D}(A)$, and

$$t^{-1}\int_0^t T^*(s)x^* ds \in \mathcal{D}(A^*).$$

Moreover

$$x^* = \lim_{t \downarrow 0} \frac{1}{t}\int_0^t T^*(s)x^* ds,$$

and $x^* \in \overline{\mathfrak{D}(A^*)}$. A^* is m-dissipative and $E_h^{A^*} = (E_h^A)^*$ by Proposition 3.11-(i), (ii). For $x \in X$ and $x^* \in \mathfrak{D}(A^*)$, we have

$$<x, T_{A^*}(t)x> = \lim_{n \to \infty} <x, (E_{t/n}^{A^*})^n x^*> = \lim_{n \to \infty} <(E_{t/n}^{A})^n x, x^*>$$
$$= <T(t)x, x^*> = <x, T^*(t)x^*>,$$

for $t \geqslant 0$, by using Theorem 2.3. It follows that $T^*(t)x^* = T_{A^*}(t)x^*$ for $x^* \in \mathfrak{D}(A^*)$, and thus that $t \to T^*(t)x^*$ is strongly continuous on $[0, \infty)$. This proves (iv) and (v).
(vi) We have $T^{\odot}(t) = T_{A^*}(t)$, $t \geqslant 0$ and (vi) follows from Corollary 3.9.
(vii) $\mathfrak{D}(A^{\odot})$ is dense in $X^{\odot}$ by Proposition 3.8 and hence weak * dense in $X^{\odot} = \overline{\mathfrak{D}(A^*)}$. But $\mathfrak{D}(A^*)$ is weak * dense in X^* by Proposition 3.11-(iii).
(viii) is a direct consequence of Proposition 3.8-(ii). □

As an application of the preceding theorem we shall give a characterization of the *Favard class* of a C_0-semigroup of contractions on X, and by the standard procedure described in Section 3.2, of C_0-semigroups of bounded operators in X.

DEFINITION 3.17. Let $T \in C(X)$. Then the Favard class $\mathrm{Fav}(T)$ is defined as

$$\mathrm{Fav}(T) := \{x \in X : \sup_{h>0} \|h^{-1}(T(h)x - x)\| < \infty\}.$$

We have

PROPOSITION 3.18. *Let $T \in C(X)$ and let A_T be its infinitesimal generator. Then, for $x \in X$,*

$$\sup_{h>0} \|A_{T,h}x\| = \sup_{h>0} \|(A_T)_h x\|, \tag{3.35}$$

where $A_{T,h}$ is defined in (3.1) *and $(A_T)_h$ is the Yosida approximation of A_T.*

PROOF. For $\mu > 0$ and $x \in X$, we have

$$T_\mu(t)x - x = \int_0^t \frac{d}{ds} T_\mu(s)x ds,$$

where $T_\mu(t)$ is the semigroup generated by $(A_T)_\mu$. We have

$$\|\frac{d}{ds} T_\mu(s)x\| \leqslant \|\frac{d}{ds} T_\mu(0)x\| = \|(A_T)_\mu x\|,$$

for $s \geqslant 0$. Then

$$t^{-1}\|T_\mu(t)x - x\| \leqslant \|(A_T)_\mu x\|,$$

for $t > 0$, $\mu > 0$ and $x \in X$. Thus

$$\frac{1}{t}\|T(t)x - x\| \leqslant \sup_{\mu>0} \|(A_T)_\mu x\|,$$

and

$$\sup_{t>0} \|A_{T,t}x\| \leq \sup_{\mu>0} \|(A_T)_\mu x\|.$$

Conversely we have

$$\lim_{\mu\downarrow 0} (A_{T,h})_\mu x = \lim_{\mu\downarrow 0} A_{T,h} E_\mu^{A_{T,h}} x = A_{T,h}x,$$

and

$$\|(A_{T,h})_\mu x\| = \|E_\mu^{A_{T,h}} A_{T,h}x\| \leq \|A_{T,h}x\|.$$

Thus

$$\sup_{\mu>0} \|(A_{T,h})_\mu x\| = \|A_{T,h}x\|.$$

Then

$$\sup_{\mu>0} \|(A_T)_\mu x\| = \sup_{\mu>0} \lim_{h\downarrow 0} \|(A_{T,h})_\mu x\| \leq \sup_{\mu>0} \sup_{h>0} \|(A_{T,h})_\mu x\|$$
$$= \sup_{h>0} \sup_{\mu>0} \|(A_{T,h})_\mu x\| = \sup_{h>0} \|A_{T,h}x\|. \quad \square$$

Note that

$$\sup_{h>0} \|(A_T)_h x\| = |A_T x|$$

by (2.13). Thus

$$\mathrm{Fav}(T) = \{x \in X : |A_T x| < \infty\}.$$

Clearly

$$\mathfrak{D}(A_T) \subseteq \mathrm{Fav}(T).$$

It may happen that Fav(T) is larger then $\mathfrak{D}(A_T)$. For example, if $X = \mathrm{BUC}(\mathbb{R})$ equipped with the supremum norm and

$$(T(t)u)(x) := u(t+x), \ t,x \in \mathbb{R},$$

then

$$\mathfrak{D}(A_T) = \{u \in \mathrm{BUC}(\mathbb{R}) : u'(x) \text{ exists for all } x \in \mathbb{R} \text{ and } u' \in \mathrm{BUC}(\mathbb{R})\}.$$

But $u \in \mathrm{Fav}(T)$ iff $u \in \mathrm{Lip}(\mathbb{R}) \cap \mathrm{BUC}(\mathbb{R})$. For $T \in C(X)$, let $T^*, X^\odot, T^\odot$ be as in Theorem 3.16. Then $(X^\odot, \|\cdot\|)$ is a Banach space since $X^\odot$ is closed in X^*. Moreover $T^\odot \in C(X^\odot)$. Let $X^{\odot *}$ be the dual of $X^\odot$ and $T^{\odot *}$ be the dual semigroup of $T^\odot$. Since $X^\odot$ is weak * dense in X^* (Theorem 3.16) we can imbed X into $X^{\odot *}$ by defining

$$i(x)(x^\odot) := \langle x, x^\odot \rangle,$$

for every $x \in X$ and $x^\odot \in X^\odot$. Then, if $i(x) = 0$, we have $x = 0$, so i is injective. Moreover

$$\|i(x)\| = \sup_{\substack{x^{\odot} \in X^{\odot} \\ \|x^{\odot}\| \leqslant 1}} |\langle x, x^{\odot} \rangle| \leqslant \|x\|,$$

for every $x \in X$. For $x^* \in X^*$, with $\|x^*\| \leqslant 1$, and for $t > 0$, we have

$$(R_{T,t})^* x^* \in \mathfrak{D}((A_T)^*) \subset X^{\odot},$$

(see the proof of Theorem 3.16-(iv), (v)), and

$$\|(R_{T,t})^* x^*\| \leqslant 1.$$

Hence

$$\|i(x)\| \geqslant \sup_{\substack{x^* \in X^* \\ \|x^*\| \leqslant 1}} |\langle x, (R_{T,t})^* x^* \rangle|$$

$$= \sup_{\substack{x^* \in X^* \\ \|x^*\| \leqslant 1}} |\langle R_{T,t} x, x^* \rangle| = \|R_{T,t} x\|,$$

for every $x \in X$ and $t > 0$. We obtain

$$\|i(x)\| \geqslant \|x\|$$

by taking the limit as $t \downarrow 0$. Consequently $i: X \to X^{\odot *}$ is an isometry not necessarily surjective, and $i(X)$ is closed in $X^{\odot *}$. Next we identify X with $i(X)$. For $x \in X$ and $x^{\odot} \in X^{\odot}$, we have:

$$\langle T(t)x, x^{\odot} \rangle = \langle x, T^*(t) x^{\odot} \rangle = \langle x, T^{\odot}(t) x \rangle$$
$$= \langle T^{\odot *}(t) x, x^{\odot} \rangle, \; t \geqslant 0.$$

Thus $T(t) \subseteq T(t)^{\odot *}$ in $X^{\odot *}$. By Theorem 3.16, $A_T^{\odot *}$ is the weak * infinitesimal generator of $T^{\odot *}$, hence $A_T \subseteq A_T^{\odot *}$ in $X^{\odot *}$.If $X^{\odot\odot}$ denotes the closure of $\mathfrak{D}(A_T^{\odot *})$ in $X^{\odot *}$, then we know (see Theorem 3.16) that $X^{\odot\odot}$ is the maximal invariant subspace of $X^{\odot *}$ on which $T^{\odot *}$ is strongly continuous. We denote by $T^{\odot\odot}$ the restriction of $T^{\odot *}(t)$ to $X^{\odot\odot}$. Then $T^{\odot\odot} \in C(X^{\odot\odot})$. Clearly $X \subseteq X^{\odot\odot}$ and the infinitesimal generator of $T^{\odot\odot}$ is the part of $A_T^{\odot *}$ in $X^{\odot\odot}$, which we denote by $A_T^{\odot\odot}$. Clearly $A_T \subseteq A_T^{\odot\odot}$. Then

$$\{x \in X^{\odot *} : \sup_{h > 0} h^{-1} \|T^{\odot *}(h)x - x\| < \infty\} =$$

$$\{x \in X^{\odot\odot} : \sup_{h > 0} h^{-1} \|T^{\odot\odot}(h)x - x\| < \infty\} = \mathrm{Fav}(T^{\odot\odot}).$$

By Proposition 3.18,

$$\mathrm{Fav}(T^{\odot\odot}) = \{x \in X^{\odot\odot} : \sup_{h > 0} \|(A_T^{\odot\odot})_h x\| < \infty\}.$$

Next note that $(A_T^{\odot\odot})_h x = (A_T^{\odot *})_h x$, for $x \in X^{\odot\odot}$. Then by Theorem 3.11-(iv),

$$\mathrm{Fav}(T^{\odot\odot}) = \{x \in X^{\odot\odot} : \sup_{h > 0} \|(A_T^{\odot *})_h x\| < \infty\} \subseteq \mathfrak{D}(A_T^{\odot *}).$$

Conversely, if $x \in \mathfrak{D}(A_T^{\odot *})$, then

$$\sup_{t > 0} t^{-1} \|T^{\odot *}(t)x - x\| < \infty$$

by (2.12), (2.13), (2.15), (2.17). Then, since

$$\mathcal{D}(A_T^{\odot *}) \subseteq \overline{\mathcal{D}(A_T^{\odot *})} = X^{\odot\odot},$$

we have

$$\mathrm{Fav}(T^{\odot\odot}) = \mathcal{D}(A_T^{\odot *}).$$

It follows that

$$\mathrm{Fav}(T) = X \cap \mathcal{D}(A_T^{\odot *}). \tag{3.36}$$

If $X = X^{\odot\odot}$ in which case we say that X is $\odot$-reflexive (pronounce: sunreflexive) with respect to A_T, then

$$\mathrm{Fav}(T) = \mathcal{D}(A_T^{\odot *}).$$

In particular, if X is reflexive then $\mathrm{Fav}(T) = \mathcal{D}(A_T^{**}) = \mathcal{D}(A_T)$. Finally we give another characterization of $\mathrm{Fav}(T)$. We claim that

$$\mathrm{Fav}(T) = \{x \in X : \text{there is a constant } M > 0 \text{ and a sequence } \{x_n\} \text{ in } \mathcal{D}(A_T) \text{such that } \|A_T x_n\| \leqslant M \text{ and } \lim_{n\to\infty} x_n = x\}.$$

Indeed if $\{x_n\} \subseteq \mathcal{D}(A_T)$, with $\|A_T x_n\| \leqslant M$ and $\lim_{n\to\infty} x_n = x$, we have

$$\|T(t)x_n - x_n\| \leqslant t\|A_T x_n\| \leqslant Mt$$

by (2.12). Hence $\|T(t)x - x\| \leqslant Mt$, for $t > 0$ and $x \in \mathrm{Fav}(T)$. Conversely if $x \in \mathrm{Fav}(T)$, then there is an $M > 0$ such that $\|(A_T)_h x\| \leqslant M$, for $h > 0$. Then

$$E_h^{A_T} x \in \mathcal{D}(A_T),$$

$$\|AE_h^{A_T} x\| = \|(A_T)_h x\| \leqslant M,$$

and $\lim_{h\downarrow 0} E_h^{A_T} x = x$.

We conclude this section with some additional remarks about the more general case when $\{T(t)\}_{t\geqslant 0}$ is not a contraction semigroup, but satisfies

$$\|T(t)\| \leqslant Me^{\omega t}, \quad t \geqslant 0.$$

We introduce a new norm on X as follows:

$$\|x\|' = \sup_{\substack{x^{\odot} \in X^{\odot} \\ \|x^{\odot}\| \leqslant 1}} |<x, x^{\odot}>|, \quad x \in X,$$

and in the sequel we shall refer to this norm $\|\cdot\|'$ as the prime norm. Using similar argument as above (to be precise: in the proof that i is an isometry if $T \in C(X)$), one shows that

$$\|x\|' \leqslant \|x\| \leqslant M\|x\|', \quad x \in X.$$

So the embedding $i: X \to X^{\odot *}$ is a continuous injection with norm $M^{-1} \leqslant \|i\| \leqslant 1$. If X is equipped with the prime norm $\|\cdot\|'$, then X can be embedded in $X^{\odot\odot}$ by means of the natural mapping, and in this sense the semigroup $T^{\odot\odot}$ is an extension of T. Finally, it can easily be shown that the prime norm on $X^{\odot}$,

$$\|x^{\odot}\|' = \sup_{\substack{x^{\odot\odot}\in X^{\odot\odot}\\ \|x^{\odot\odot}\|\leqslant 1}} |<x^{\odot},x^{\odot\odot}>|, \quad x^{\odot}\in X^{\odot},$$

is the same as the original norm. We leave this as an exercise to the reader.

5. Bounded perturbations of dual semigroups

Let X be an arbitrary Banach space, and let A_0 be the infinitesimal generator of a strongly continuous semigroup $\{T_0(t)\}_{t\geqslant 0}$, which we shall call the *unperturbed semigroup.* Let $B:X\to X$ be a bounded linear operator. A standard perturbation result in semigroup theory says that the closed operator $A=A_0+B$ with domain $\mathfrak{D}(A)=\mathfrak{D}(A_0)$ also generates a strongly continuous semigroup. This so-called *perturbed semigroup* $\{T(t)\}_{t\geqslant 0}$ can be computed from the integral equation

$$T(t)x = T_0(t)x + \int_0^t T_0(t-s)BT(s)x\,ds\,,\ t\geqslant 0\,,\ x\in X, \tag{3.37}$$

which is sometimes called the *Duhamel formula,* or also the *variation-of-constants formula.* By the method of successive approximation one gets the following norm convergent series expansion for $T(t)$:

$$T(t) = \sum_{k=0}^{\infty} T_k(t), \quad t\geqslant 0,$$

where for every $k\geqslant 1$ and $x\in X$,

$$T_k(t)x = \int_0^t T_0(t-s)BT_{k-1}(s)x\,ds. \tag{3.38}$$

In this expression as well as in (3.37) the integral has to be considered as a Riemann integral.

In this section we will show, when X is nonreflexive, that we can extend this perturbation result by using the theory of dual semigroups, described in the previous section.

Assume that $(X,\|\cdot\|)$ is a Banach space, and let $\{T_0(t)\}_{t\geqslant 0}$ be a C_0-semigroup on X with infinitesimal generator A_0. Throughout the rest of this section we assume that *X is $\odot$-reflexive with respect to A_0.* We identify X and $X^{\odot\odot}$ by means of the embedding i. Now let B be a bounded linear operator mapping X into $X^{\odot *}$. Then B^* maps $X^{\odot **}$ into X^*, but here we are only interested in its restriction to $X^{\odot}$. The situation is summarized in the diagram below.

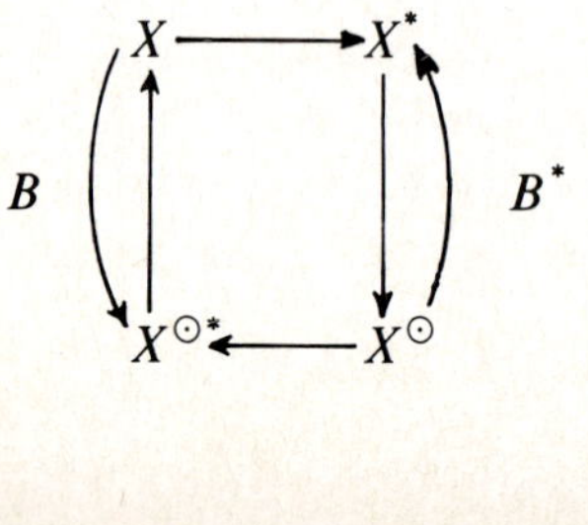

In this form, the variation-of-constants formula (3.37) makes no sense if B maps into $X^{\odot *}$. To give it a meaning we introduce the notion of a *weak * Riemann integral.*

Let Z be a Banach space, and let Z^* be its dual. If $u^*:[a,b]\to Z^*$ is a function which is continuous with respect to the weak * topology of Z^*, then we can define the Riemann integral

$$\int_a^b \langle z,u^*(t)\rangle dt$$

for every $z\in Z$. Obviously, the following estimate holds:

$$\left|\int_a^b \langle z,u^*(t)\rangle dt\right| \leqslant (b-a)\|z\|\cdot \sup_{a\leqslant t\leqslant b}\|u^*(t)\|.$$

Note that, by the uniform boundedness principle, every weak * continuous function is norm bounded. Therefore the mapping

$$z\to\int_a^b \langle z,u^*(t)\rangle dt$$

defines a continuous linear functional on Z, i.e. an element of Z^*, and we denote this element by

$$\int_a^b u^*(t)dt.$$

For obvious reasons we refer to this integral as the weak * Riemann integral.

Now, let us return to our framework. If the function $f:[0,t]\to X^{\odot *}$ is norm continuous, then $s\to T_0^{\odot *}(t-s)f(s)$ is a weak * continuous function on $[0,t]$ with values in $X^{\odot *}$, thus we can define its weak * Riemann integral. The following lemma holds.

LEMMA 3.19. *Let $f:[0,\tau]\to X^{\odot *}$ be a norm continuous function, then*

$$t\to\int_0^t T_0^{\odot *}(t-s)f(s)ds$$

is a norm continuous X-valued function on $[0,\tau]$. If, in addition, $T_0(t)$ obeys (3.12), *then*

$$\left\|\int_0^t T_0^{\odot *}(t-s)f(s)ds\right\| \leqslant \frac{M}{\omega}(e^{\omega t}-1)\sup_{0\leqslant s\leqslant t}\|f(s)\|,$$

for every $t\in[0,\tau]$ ($\leqslant M.\ \sup_{0\leqslant s\leqslant t}\|f(s)\|$ if $\omega=0$).

Proof. Define, for $t \in [0,\tau]$,

$$F(t) = \int_0^t T_0^{\odot *}(t-s) f(s) ds.$$

First we show that $F(t) \in X$, $t \in [0,\tau]$. In order to do so, we must prove that

$$T_0^{\odot *}(h) F(t) \to F(t), \quad h \downarrow 0,$$

with respect to the norm topology. A straightforward calculation shows that

$$\begin{aligned} \|T_0^{\odot *}(h)F(t) - F(t)\| \leqslant \| \int_0^h T_0^{\odot *}(s) f(t-s) ds \| \\ + \| \int_t^{t+h} T_0^{\odot *}(s) f(t+h-s) ds \| \\ + \| \int_h^t T_0^{\odot *}(s) \{ f(t+h-s) - f(t-s) \} ds \|. \end{aligned}$$

Obviously, both the first and the second term are $\mathcal{O}(h)$ as $h \downarrow 0$. The last term is less than

$$K \sup_{0 \leqslant s \leqslant t} \|f(t+h-s) - f(t-s)\|,$$

if K is taken large enough, and by the continuity of f, this also converges to 0 as $h \downarrow 0$. This proves that $F(t) \in X$.
To prove norm continuity, we note that

$$\|F(t+h) - F(t)\| \leqslant \|T_0^{\odot *}(h)F(t) - F(t)\| + \| \int_0^h T_0^{\odot *}(s) f(t+h-s) ds \|.$$

Obviously, both terms converge to 0 as $h \downarrow 0$. □

Now suppose that B maps X continuously into $X^{\odot *}$. We can construct a semigroup $\{T(t)\}_{t \geqslant 0}$ on X by means of the variation-of-constants formula

$$T(t)x = T_0(t)x + \int_0^t T_0^{\odot *}(t-s) B T(s) x ds, \; x \in X, \; t \geqslant 0, \tag{3.39}$$

with the integral interpreted as a weak * Riemann integral. With Lemma 3.19 the following perturbation result is easily proved.

Theorem 3.20. *Let $\{T_0(t)\}_{t \geqslant 0}$ be a C_0-semigroup on X with generator A_0, and assume that X is $\odot$-reflexive with respect to A_0. Let $B: X \to X^{\odot *}$ be a bounded linear operator. Then there exists a unique C_0-semigroup $\{T(t)\}_{t \geqslant 0}$ on X such that (3.39) is satisfied for every $t \geqslant 0$ and every $x \in X$. Furthermore, if $\{T_0(t)\}_{t \geqslant 0}$ obeys (3.12) then $\{T(t)\}_{t \geqslant 0}$ obeys*

$$\|T(t)\| \leqslant M e^{\bar{\omega} t}, \quad t \geqslant 0,$$

where $\bar{\omega} = \omega + M \|B\|$.

As in the original case, where B maps into X, we can get $\{T(t)\}_{t\geqslant 0}$ explicitly by the method of successive approximations:

$$T(t) = \sum_{k=0}^{\infty} T_k(t), \quad t\geqslant 0, \tag{3.40}$$

where

$$T_k(t) = \int_0^t T_0^{\odot *}(t-s)BT_{k-1}(s)ds\,,\, t\geqslant 0\,,\, k\geqslant 1. \tag{3.41}$$

Let $U(t)=T(t)-T_0(t)=\sum_{k=1}^{\infty} T_k(t)$, or alternatively,

$$U(t)x = \int_0^t T_0^{\odot *}(t-s)BTxds\,,\, t\geqslant 0\,,\, x\in X. \tag{3.42}$$

Then, by Lemma 3.19:

$$\|U(t)\|\leqslant M(e^{\bar{\omega}t} - e^{\omega t}), \quad t\geqslant 0. \tag{3.43}$$

An important consequence of this estimate is the following result.

THEOREM 3.21. *The spaces $X^{\odot}$ and $X^{\odot\odot}$ for the perturbed semigroup $\{T(t)\}_{t\geqslant 0}$ are the same as for the unperturbed semigroup $\{T_0(t)\}_{t\geqslant 0}$. In particular, X is $\odot$-reflexive with respect to A, the generator of T.*

PROOF. From $\|U^*(t)\|=\|U(t)\|\to 0$, $t\downarrow 0$, we get that $X^{\odot}$ is "not affected" by the perturbation B. Now, since $T_0^*(t)$ and $T^*(t)$ leave $X^{\odot}$ invariant, the same hold for $U^*(t)$. Let $U^{\odot}(t)$ be the restriction of $U^*(t)$ to $X^{\odot}$, then $T^{\odot}(t)=T_0^{\odot}(t)+U^{\odot}(t)$, and $T^{\odot *}(t)=T_0^{\odot *}(t)+U^{\odot *}(t)$. Clearly $\|U^{\odot}(t)\| =\|U^{\odot *}(t)\|$, $t\downarrow 0$, and therefore $t\to T^{\odot *}(t)x^{\odot *}$ is norm continuous if and only if $x^{\odot *}\in X^{\odot\odot}$. Since $X^{\odot\odot}=X$ by assumption, the result follows. □

It remains to determine the infinitesimal generator of the perturbed semigroup. This is taken care of in the following theorem.

THEOREM 3.22. *Let A be the infinitesimal generator of $\{T(t)\}_{t\geqslant 0}$. Then*

(a) $\mathfrak{D}(A) = \{x\in\mathfrak{D}(A_0^{\odot *}) : A_0^{\odot *}x + Bx\in X\}$,
$Ax = A_0^{\odot *}x + Bx, \quad x\in\mathfrak{D}(A)$.

(b) $\mathfrak{D}(A^*) = \mathfrak{D}(A_0^*)$,
$A^*x^{\odot} = A_0^*x^{\odot} + B^*x^{\odot}\,,\, x^{\odot}\in\mathfrak{D}(A^*)$.

(c) $\mathfrak{D}(A^{\odot}) = \{x\in\mathfrak{D}(A^*) : A_0^*x^{\odot} + B^*x^{\odot}\in X^{\odot}\}$
$A^{\odot}x^{\odot} = A_0^*x^{\odot} + B^*x^{\odot}\,,\, x^{\odot}\in\mathfrak{D}(A^{\odot})$.

(d) $\mathfrak{D}(A^{\odot *}) = \mathfrak{D}(A_0^{\odot *})$,
$A^{\odot *}x = A_0^{\odot *}x + Bx\,,\, x\in\mathfrak{D}(A^{\odot *})$.

Proof. We subsequently prove (b), (c), (d) and (a).
(b) From Theorem 3.16-(iv) we know that $\mathcal{D}(A^*)\subseteq X^{\odot}$. It is easy to see that for every $x^{\odot}\in X^{\odot}$,

$$\frac{1}{h}U^*(h)x^{\odot}\rightarrow B^*x^{\odot}, \text{ as } h\downarrow 0,$$

relative to the weak * topology of X^*. Thus we get from Theorem 3.16-(iii) that $x^{\odot}\in\mathcal{D}(A^*)$ if and only if $h^{-1}(T_0^*(h)x^{\odot}-x^{\odot})$ converges with respect to the weak * topology as $h\downarrow 0$. This is the case if and only if $x^{\odot}\in\mathcal{D}(A_0^*)$, and for such $x^{\odot}$, we get

$$\frac{1}{h}(T^*(h)x^{\odot}-x^{\odot})=\frac{1}{h}(T_0^*(h)x^{\odot}-x^{\odot})+\frac{1}{h}U^*(h)x^{\odot}$$
$$\rightarrow A_0^*x^{\odot}+B^*x^{\odot}, \text{ as } h\downarrow 0.$$

This proves (b).
(c) is now a direct consequence of Theorem 3.16-(v).
(d) follows by similar arguments as (b).
(a) follows from (d) and the fact that $A=A^{\odot\odot}$. □

From this proof it follows that the infinitesimal generator A of $\{T(t)\}_{t\geqslant 0}$ is easily obtained by making a detour via X^*, $X^{\odot}$ and $X^{\odot *}$. An important implication of this last theorem is that the domain $\mathcal{D}(A^*)$ is independent of the perturbation B. The same remark applies to $\mathcal{D}(A^{\odot *})$. However, the domain of A (and also $A^{\odot}$) in general does depend on B. In Section 10.2 we discuss an example where all information about B is contained in the domain of A, a situation which is not so exceptional as it might seem at first sight.

A posteriori, it is clear that we may as well start with $\{T_0^{\odot}(t)\}_{t\geqslant 0}$ as the unperturbed semigroup (on $X^{\odot}$), and consider $B^*:X^{\odot}\rightarrow X^*$ as the perturbation. If we compute $S(t)$ from the variation-of-constants formula

$$S(t)x^{\odot}=T_0^{\odot}(t)x^{\odot}+\int_0^t T_0^*(t-s)B^*S(s)x^{\odot}ds, \tag{3.44}$$

then $\{S(t)\}_{t\geqslant 0}$ defines a C_0-semigroup on $X^{\odot}$ (also if X is not $\odot$-reflexive with respect to A_0), and $S(t)=T^{\odot}(t)$, if X is $\odot$-reflexive.

From Theorem 3.22-(d) it follows that the Favard class is not affected by the perturbation:

$$\mathrm{Fav}(T_0)=\mathrm{Fav}(T)=\mathcal{D}(A_0^{\odot *}).$$

We conclude this section by noting that compactness and irreducibility properties of the perturbed semigroup are studied in Section 9.3.

6. COMMUTATIVE SUM OF m-DISSIPATIVE OPERATORS

In this section we shall study the inhomogeneous initial value problem

$$\begin{aligned} &\frac{du}{dt}(t) = A_0 u(t) + f(t), \quad t>0, \\ &u(0) = x, \end{aligned} \tag{3.45}$$

where A_0 is an m-dissipative operator in a real Banach space X, and $x \in X$, $f:\mathbb{R}^+ \to X$ are given. This will be done by using a result of DA PRATO and GRISVARD [1975] on the sum of two "commuting" m-dissipative operators.

Let A and B be two m-dissipative operators in a real Banach space X. We shall assume that "A and B commute", i.e. that A and B satisfy:

$$E_\lambda^A E_\mu^B = E_\mu^B E_\lambda^A \text{ for every } \lambda,\mu>0. \tag{3.46}$$

We have

PROPOSITION 3.23. *Let A and B be m-dissipative operators satisfying (3.46). Then $A+B$ is closable.*

PROOF. Let $x_n \in \mathfrak{D}(A) \cap \mathfrak{D}(B)$, $n \in \mathbb{Z}^+$, be such that $\lim_{n\to\infty} x_n = 0$ and $\lim_{n\to\infty}(A+B)x_n = y \in X$.. We have to prove that $y=0$. Clearly

$$E_1^A E_1^B (A+B)x_n = E_1^B E_1^A A x_n + E_1^A E_1^B B x_n = E_1^B A_1 x_n + E_1^A B_1 x_n,$$

and hence

$$E_1^A E_1^B y = \lim_{n\to\infty} E_1^A E_1^B (A+B)x_n = \lim_{n\to\infty} E_1^B A_1 x_n + \lim_{n\to\infty} E_1^A B_1 x_n = 0.$$

Then $E_1^B y = (I-A)E_1^A E_1^B y = 0$, and similarly $y = (I-B)E_1^B y = 0$. □

If A and B are m-dissipative in X and B is densely defined, then $A+B$ is dissipative according to Theorem 3.14-(iii). If, in addition A and B "commute" we have

THEOREM 3.24. *Let A and B be m-dissipative operators in X, satisfying (3.46). If B is densely defined then*

(i) $\mathfrak{R}(I-h(A+B)) \supseteq \mathfrak{D}(A)+\mathfrak{D}(B)$, *for every* $h>0$,

(ii) $\overline{A+B}$ *is m-dissipative in X.*

(iii) *For every* $\mu>0$, $A+B_\mu$ *is m-dissipative and*

$$\lim_{\mu\downarrow 0} E_h^{A+B_\mu} x = E_h^{\overline{A+B}} x,$$

for every $h>0$ *and every* $x \in X$.

(iv) *For every* $\lambda>0$, $A_\lambda + B$ *is m-dissipative and*

$$\lim_{\lambda\downarrow 0} E_h^{A_\lambda+B} x = E_h^{\overline{A+B}} x,$$

for every $h>0$ *and every* $x \in X$.

PROOF. The proof will be done in different steps.
(a) For every $x \in \mathfrak{D}(A)$ and every $\mu > 0$, $E_\mu^B x \in \mathfrak{D}(A)$ and $AE_\mu^B x = E_\mu^B Ax$. Indeed if $x \in \mathfrak{D}(A)$, set $y = x - Ax$, then $x = E_1^A y$, and

$$E_\mu^B x = E_\mu^B E_1^A y = E_1^A E_\mu^B y$$

belongs to $\mathfrak{D}(A)$. Then

$$AE_\mu^B x = AE_1^A E_\mu^B y = A_1 E_\mu^B y = (E_1^A - I)E_\mu^B y = E_\mu^B A_1 y = E_\mu^B Ax.$$

Similarly, one proves:
(b) For every $x \in \mathfrak{D}(B)$ and every $\lambda > 0$, $E_\lambda^A x \in \mathfrak{D}(B)$ and $BE_\lambda^A x = E_\lambda^A Bx$.
(c) For every $\mu > 0$, $A + B_\mu$ is m-dissipative. Now $A + B_\mu$ is dissipative by Theorem 3.14-(iii). According to Proposition 3.7-(iv), it suffices to prove that $\mathfrak{R}(I - (A + B_\mu)) = X$. For $y \in X$, the equality $x = Ax + B_\mu x + y$ is equivalent to $(1+\mu)x = \mu Ax + E_\mu^B x + \mu y$, or

$$x = R_\mu x, \tag{3.47}$$

where

$$R_\mu x = E_{\mu/(1+\mu)}^A \left(\frac{1}{1+\mu} E_\mu^B x + \frac{\mu}{1+\mu} y\right). \tag{3.48}$$

Clearly $R_\mu \in \mathrm{Lip}(X)$ and $\|R_\mu\|_{\mathrm{Lip}} \leqslant (1+\mu)^{-1} < 1$. Thus by the Banach fixed-point theorem, there is an $x \in X$ such that $x = R_\mu x$. Hence $A + B_\mu$ is m-dissipative.
(d) Similarly $A_\lambda + B$ is m-dissipative for every $\lambda > 0$.
(e) For $y \in \mathfrak{D}(B)$ and $\mu > 0$, let x_μ satisfy

$$x_\mu = Ax_\mu + B_\mu x_\mu + y. \tag{3.49}$$

We claim that $\lim_{\mu \downarrow 0} x_\mu$ exists. Set

$$L_\mu = A + B_\mu, \quad L = A + B, \tag{3.50}$$

and

$$S_\mu = I - \frac{1}{1+\mu} E_{\mu/(1+\mu)}^A E_\mu^B. \tag{3.51}$$

From (3.48) and (3.51), it follows that

$$x_\mu = S_\mu^{-1}\left(\frac{\mu}{1+\mu} y\right). \tag{3.52}$$

Moreover, since

$$E_\lambda^A E_\nu^B S_\mu = S_\mu E_\lambda^A E_\nu^B$$

for every $\lambda, \mu, \nu > 0$, we have

$$E_\lambda^A E_\nu^B S_\mu^{-1} = S_\mu^{-1} E_\lambda^A E_\nu^B, \quad \lambda, \mu, \nu > 0. \tag{3.53}$$

From (3.53), (3.46) and $y \in \mathfrak{D}(B)$ it follows, like in (a), that $x_\mu \in \mathfrak{D}(B)$ and

$$Bx_\mu = (I - L_\mu)^{-1} By. \tag{3.54}$$

Since $x_\mu \in \mathfrak{D}(A)$ we have $x_\mu \in \mathfrak{D}(L) = \mathfrak{D}(A) \cap \mathfrak{D}(B)$ and

$$\begin{aligned}(I-L)(I-L_\mu)^{-1}y - y \\ &= (I-L_\mu)(I-L_\mu)^{-1}y + (L_\mu - L)(I-L_\mu)^{-1}y - y \\ &= (B_\mu - B)(I-L_\mu)^{-1}y = (I-L_\mu)^{-1}(B_\mu - B)y.\end{aligned}$$

Since $\|(I-L_\mu)^{-1}\| \leqslant 1$, we obtain

$$\|(I-L)(I-L_\mu)^{-1}y - y\| \leqslant \|By - B_\mu y\|.$$

Since $\overline{\mathfrak{D}(B)} = X$, we have $\lim_{\mu\downarrow 0}\|By - B_\mu y\| = 0$ and $\lim_{\mu\downarrow 0}\|(I-L)x_\mu - y\| = 0$. Moreover

$$\begin{aligned}&\lim_{\mu\downarrow 0}\|B_\mu x_\mu - Bx_\mu\| = \\ &\lim_{\mu\downarrow 0}\|(x_\mu - Ax_\mu - Bx_\mu) - (x_\mu - Ax_\mu - B_\mu x_\mu)\| = \\ &\lim_{\mu\downarrow 0}\|(I-L)x_\mu - y\| = 0.\end{aligned}$$

Set $z_\mu = B_\mu x_\mu - Bx_\mu$. We have $x_\mu = Ax_\mu + Bx_\mu + y + z_\mu$. Since $A+B$ is dissipative, we obtain

$$\|x_\mu - x_\nu\| \leqslant \|(y+z_\mu) - (y+z_\nu)\| = \|z_\mu - z_\nu\|.$$

Hence $\{x_\mu\}$ is a Cauchy net and $\lim_{\mu\downarrow 0} x_\mu$ exists.
(f) For $\tilde{y} \in X$ and $\mu > 0$, let $\tilde{x}_\mu$ satisfy

$$\tilde{x}_\mu = Ax_\mu + B_\mu \tilde{x}_\mu + \tilde{y}. \tag{3.55}$$

We claim that $\lim_{\mu\downarrow 0}\tilde{x}_\mu$ exists. For $y \in \mathfrak{D}(B)$ and $\mu > 0$, let x_μ satisfy (3.49). Then, since $A + B_\mu$ is dissipative, we obtain

$$\begin{aligned}\|\tilde{x}_\mu - \tilde{x}_\nu\| &\leqslant \|\tilde{x}_\mu - x_\mu\| + \|x_\mu - x_\nu\| + \|x_\nu - \tilde{x}_\nu\| \\ &\leqslant 2\|\tilde{y} - y\| + \|x_\mu - x_\nu\|.\end{aligned}$$

Since $\overline{\mathfrak{D}(B)} = X$ and $\{x_\mu\}$ is a Cauchy net, $\{\tilde{x}_\mu\}$ is also a Cauchy net and $\lim_{\mu\downarrow 0}\tilde{x}_\mu$ exists.
(g) If $y \in \mathfrak{D}(B)$, then $x := \lim_{\mu\downarrow 0}(I-L_\mu)^{-1}y$ belongs to $\mathfrak{D}(A) \cap \mathfrak{D}(B)$ and

$$x = Ax + Bx + y. \tag{3.56}$$

Set $x_\mu = (I-L_\mu)^{-1}y$. We know that $\lim_{\mu\downarrow 0}x_\mu$ exists. Define $x = \lim_{\mu\downarrow 0}x_\mu$. We also know that $x_\mu \in \mathfrak{D}(B)$ and satisfies equation (3.54). It follows that $\lim_{\mu\downarrow 0}(I-L_\mu)^{-1}By$ exists, thus $x \in \mathfrak{D}(B)$ and $Bx = \lim_{\mu\downarrow 0}Bx_\mu$, since B is closed. Since $\lim_{\mu\downarrow 0}\|B_\mu x_\mu - Bx_\mu\| = 0$, we obtain

$$\lim_{\mu\downarrow 0} Ax_\mu = \lim_{\mu\downarrow 0}(x_\mu - B_\mu x_\mu - y) = x - Bx - y.$$

Since A is closed, $x \in \mathfrak{D}(A)$ and satisfies (3.56).
(h) $\mathfrak{R}(I - h(A+B)) \supseteq \mathfrak{D}(A) + \mathfrak{D}(B)$, for every $h > 0$. Since hA and hB satisfy the same assumptions as A and B, and since $\mathfrak{D}(hA) = \mathfrak{D}(A)$ and $\mathfrak{D}(hB) = \mathfrak{D}(B)$,

it suffices to prove the assertion only for $h=1$. Because of (g), we have $\mathfrak{R}(I-(A+B))\supseteq\mathfrak{D}(B)$. Let $y\in\mathfrak{D}(A)$ and let x_μ satisfy (3.49). Set $x=\lim_{\mu\downarrow 0}x_\mu$. From (3.53), (3.46) and $y\in\mathfrak{D}(A)$, it follows

$$Ax_\mu=(I-L_\mu)^{-1}Ay.$$

Therefore $\lim_{\mu\downarrow 0}Ax_\mu$ exists and since A is closed $x\in\mathfrak{D}(A)$ and $Ax=\lim_{\mu\downarrow 0}Ax_\mu$. Then

$$\lim_{\mu\downarrow 0}B_\mu x_\mu=\lim_{\mu\downarrow 0}(x_\mu-Ax_\mu-y)=x-Ax-y,$$

and

$$\lim_{\mu\downarrow 0}\|x_\mu-E_\mu^B x_\mu\|=\lim_{\mu\downarrow 0}\mu\|B_\mu x_\mu\|=0.$$

It follows that $\lim_{\mu\downarrow 0}E_\mu^B x_\mu=x$. Since

$$\lim_{\mu\downarrow 0}BE_\mu^B x_\mu=x-Ax-y$$

and B is closed, we have $x\in\mathfrak{D}(B)$ and x satisfies (3.56).

(i) $\overline{A+B}$ is m-dissipative. Indeed $A+B$ is closable, dissipative and for $h>0$ $\mathfrak{R}(I-h(A+B))\supseteq\mathfrak{D}(B)$. Then $\overline{A+B}$ is dissipative, $\mathfrak{R}(I-h(\overline{A+B}))$ is closed by Proposition 3.7-(i), (iii). Moreover $R(I-h(A+B))\supseteq\mathfrak{D}(B)$ by (h), and $\overline{\mathfrak{D}(B)}=X$ by assumption. Thus

$$\mathfrak{R}(I-h(\overline{A+B}))\supseteq\overline{\mathfrak{R}(I-h(A+B))}\supseteq\overline{\mathfrak{D}(B)}=X.$$

Hence $\overline{A+B}$ is m-dissipative.

(j)

$$\lim_{\mu\downarrow 0}E_h^{A+B_\mu}x=E_h^{\overline{A+B}}x,$$

for every $x\in X$ and every $h>0$. It is sufficient to prove the assertion for $h=1$ by Theorem 2.7-(i). From (g) we have

$$\lim_{\mu\downarrow 0}E_1^{A+B_\mu}x=E_1^{\overline{A+B}}x,$$

for $x\in\mathfrak{D}(B)$ by noting that $A+B\subseteq\overline{A+B}$. Since $E_1^{A+B_\mu}, E_1^{\overline{A+B}}$ are contractions and $\mathfrak{D}(B)$ is dense in X, the assertion holds.

(k) $\lim_{\lambda\downarrow 0}E_h^{A_\lambda+B}x$ exists for every $h>0$ and $x\in X$. Again it is sufficient to prove the assertion for $h=1$. Since $A_\lambda+B$ is m-dissipative we have

$$E_1^{A_\lambda+B}x=\int_0^\infty e^{-t}T_{A_\lambda+B}(t)x\,dt,$$

by Theorem 3.1. Moreover it is easily verified that

$$\lim_{\mu\downarrow 0}E_h^{A_\lambda+B_\mu}x=E_h^{A_\lambda+B}x,$$

for every $h>0$ and $x\in X$. (In fact it is a special case of Theorem 3.24-(iii) but a simpler proof can be given since A_λ is bounded.) It follows that

$$\lim_{\mu\downarrow 0}\max_{t\in[0,s]}\|T_{A_\lambda+B}(t)x-T_{A_\lambda+B_\mu}(t)x\|=0,$$

for every $s>0$ and every $x\in X$ by Theorem 3.4. But $T_{A_\lambda+B_\mu}(t)x=T_{A_\lambda}(t)T_{B_\lambda}(t)$, for every $x\in X$ and $t\geq 0$, $\lambda,\mu>0$, since $A_\lambda B_\mu=B_\mu A_\lambda$. Therefore

$$x_\lambda:=E_1^{A_\lambda+B}x=\int_0^\infty e^{-t}T_{A_\lambda}(t)T_B(t)xdt.$$

We claim that $\{x_\lambda\}$ is a Cauchy net. Let $\lambda,\nu,\epsilon>0$. We have:

$$\begin{aligned}\|x_\lambda-x_\nu\|&\leq\Big\|\int_0^M e^{-t}(T_{A_\lambda}(t)-T_{A_\nu}(t))T_B(t)xdt\Big\|+2\int_M^\infty e^{-t}dt\|x\|\\&=\Big\|\int_0^M e^{-t}(T_{A_\lambda}(t)-T_{A_\nu}(t))T_B(t)xdt\Big\|+2e^{-M}\|x\|,\end{aligned}$$

for every $M>0$. Choose M such that $2e^{-M}\|x\|<\epsilon/2$. It remains to prove that

$$x_{\lambda,M}:=\int_0^M T_{A_\lambda}(t)e^{-t}T_B(t)xdt$$

defines a Cauchy net. Since $t\to e^{-t}T_B(t)x\in C([0,M];X)$, there exist $n\in\mathbb{Z}^+$ and $y_1,\ldots,y_n\in X$ such that

$$\|e^{-t}T_B(t)x-y_i\|\leq(2M)^{-1}\epsilon/4,\text{ for } t\in[(i-1)M/n,iM/n], i=1,\ldots,n.$$

Then

$$\|x_{\lambda,M}-x_{\nu,M}\|\leq\epsilon/4+\sum_{i=1}^n\Big\|\int_{(i-1)M/n}^{iM/n}(T_{A_\lambda}(t)y_i-T_{A_\nu}(t)y_i)dt\Big\|.$$

It suffices to prove that

$$\int_{(i-1)M/n}^{iM/n}T_{A_\lambda}(t)y_idt$$

is a Cauchy net, for $i=1,\ldots,n$. Note that

$$\int_{(i-1)M/n}^{iM/n}T_{A_\lambda}(t)y_idt=\int_0^{iM/n}T_{A_\lambda}(t)y_idt-\int_0^{(i-1)M/n}T_A(t)y_idt.$$

We are done if we prove that $\int_0^t T_{A_\lambda}(s)yds$ is a Cauchy net for every $t>0, y\in X$. We know (see the proof of Theorem 3.1) that $v_\lambda(t):=\int_0^t T_{A_\lambda}(s)yds$ satisfies

$$\begin{aligned}&\frac{dv_\lambda}{dt}(t)=A_\lambda v_\lambda(t)+y,\quad t\geq 0,\\&v_\lambda(0)=0.\end{aligned}$$

Set $C_\lambda x=A_\lambda x+y$ for $x\in X$. Then C_λ is clearly a nonlinear m-dissipative operator in X, and Lipschitz continuous. By Proposition 2.11, $v_\lambda(t)=T_{C_\lambda}(t)0$, $t\geq 0,\lambda>0$. We claim that

$$\lim_{\lambda\downarrow 0} E_h^{C_\lambda} f = E_h^C f,$$

for every $f \in X$, $h > 0$ with $Cz := Az + y$, for every $z \in \mathfrak{D}(A)$. (Note that C is also nonlinear and m-dissipative in X.) Indeed for $f \in X$, $h > 0$, let x_λ satisfy:

$$x_\lambda = h(A_\lambda x_\lambda + y) + f,$$

or equivalently,

$$x_\lambda = hA_\lambda x_\lambda + hy + f.$$

Then

$$x_\lambda = E_h^{A_\lambda}(hy + f),$$

and

$$x := \lim_{\lambda\downarrow 0} x_\lambda = E_h^A(hy + f),$$

by Proposition 2.8. Then $x \in \mathfrak{D}(A)$ and $x = hAx + hy + f$ or $x = hCx + f$, that is $x = E_h^C f$. This proves the claim. Then by Theorem 2.7, we obtain

$$\lim_{\lambda\downarrow 0} v_\lambda(t) = T_C(t)0, \quad t \geqslant 0.$$

We have proven that $\lim_{\lambda\downarrow 0} E_1^{A_\lambda + B} x$ exists for every $x \in X$.
(l) If $y \in \mathfrak{D}(B)$, then $x = \lim_{\lambda\downarrow 0}(I - (A_\lambda + B))^{-1} y$ belongs to $\mathfrak{D}(A) \cap \mathfrak{D}(B)$ and (3.56) holds. Let x_λ satisfy

$$x_\lambda = A_\lambda x_\lambda + Bx_\lambda + y.$$

Then, as in (c), we prove that

$$x_\lambda = E^B_{\lambda/(1+\lambda)}\left(\frac{1}{1+\lambda} E_\lambda^A x_\lambda + \frac{\lambda}{1+\lambda} y\right),$$

and $x_\lambda \in \mathfrak{D}(B)$. Moreover

$$Bx_\lambda = (I - (A_\lambda + B))^{-1} By,$$

and thus $\lim_{\lambda\downarrow 0} Bx_\lambda$ exists by (k). Since B is closed, we find that $x \in \mathfrak{D}(B)$ and $Bx = \lim_{\lambda\downarrow 0} Bx_\lambda$. Then

$$\lim_{\lambda\downarrow 0} A_\lambda x_\lambda = \lim_{\lambda\downarrow 0} (x_\lambda - Bx_\lambda - y) = x - Bx - y.$$

Then

$$\|E_\lambda^A x_\lambda - x_\lambda\| = \lambda \|A_\lambda x_\lambda\|$$

tends to 0 as $\lambda \downarrow 0$. Since $A_\lambda x_\lambda = AE_\lambda^A x_\lambda$ and A is closed we obtain $x \in \mathfrak{D}(A)$, $Ax = \lim_{\lambda\downarrow 0} A_\lambda x_\lambda$ and $x = Ax + Bx + y$.
(m) For every $x \in X$ and $h > 0$,

$$\lim_{\lambda\downarrow 0} E_h^{A_\lambda + B} x = E_h^{\overline{A + B}} x,$$

The proof is similar to the proof of (j). This completes the proof of the theorem. □

REMARK. Actually, part (iv) of Theorem 3.24 is not contained in DA PRATO & GRISVARD [1975].

REMARK. If A and B are m-dissipative in X and satisfy (3.46), then B_0, the part of B in $X_0 := \overline{\mathfrak{D}(B)}$, is m-accretive in X, and densely defined according to Proposition 3.8. Moreover A_0, the part of A in X, is m-accretive in X_0. Moreover A_0 and B_0 satisfy (3.46) in X_0. Therefore we can apply Theorem 3.24 to A_0 and B_0 in X. In particular $\overline{A_0 + B_0}$ is m-accretive in X_0.

In this last part of this section, we indicate how Theorem 3.24 can be used to solve problem (3.45).

We consider the inhomogeneous initial value problem (3.45) where A_0 is a closed operator in a real Banach space $(X, \|\cdot\|)$, $x \in X$ and $f \in L_1(0,T;X)$ for some $T>0$. Here $L_1(0,T;X)$ is the space of Bochner integrable functions on $[0,T]$ equipped with the usual norm $\|\cdot\|_{L_1(0,T;X)}$ defined by

$$\|f\|_{L_1(0,T;X)} = \int_0^T \|f(t)\| dt.$$

We shall say that u is a *strict solution* of (3.45) in $L_1(0,T;X)$ if
(i) $u \in W_{1,1}(0,T;X)$ (hence $u \in C([0,T];X)$),
(ii) $u(t) \in \mathfrak{D}(A_0)$, $t \in [0,T]$,
(iii) $u'(t) = A_0 u(t) + f(t)$ a.e. on $[0,T]$,
(iv) $u(0) = x$.

REMARKS.
(i) As in Proposition 3.3-(ii) one proves that if A_0 is m-dissipative, $x=0$ and $f=0$, then $u(t)=0$ on $[0,T]$. Hence (3.45) possesses at most one strict solution in $L_1(0,T;X)$, if A_0 is m-dissipative.
(ii) If u is a strict solution of (3.45), then $v(t)=e^{\omega t}u(t)$, $\omega \in \mathbb{R}, t \in [0,T]$ is a strict solution of (3.45) with A_0 replaced by $A_0+\omega I$ and f replaced by $t \to e^{\omega t} f(t)$.
(iii) If u is a strict solution of (3.45) in $L_1(0,T;X)$ then u is also a strict solution of (3.45) in $L_1(0,T;X)$ where X is endowed with an equivalent norm.

It will be convenient to rewrite (3.45) as

$$\begin{aligned} u(t) &= A_0 u(t) - \frac{du}{dt}(t) + f(t), \\ u(0) &= x, \end{aligned} \tag{3.57}$$

and from now on, we shall assume that A_0 is m-dissipative in X.

In order to prove the existence of a strict solution of (3.57) in $L_1(0,T;X)$, we introduce the following operators A and B in $L_1(0,T;X)$. Set

$$\mathfrak{D}(A):=\{u\in L_1(0,T;X): u(t)\in\mathfrak{D}(A_0) \text{ a.e. in } [0,T] \text{ and there exists } v\in L_1(0,T;X) \text{ such that } v(t)=A_0u(t) \text{ a.e. in } [0,T]\}.$$

For $u\in\mathfrak{D}(A)$, $(Au)(t)=A_0u(t)$ a.e. in $[0,T]$. One verifies that A is m-dissipative in $L_1(0,T;X)$. Set

$$\mathfrak{D}(B):=\{u\in W_{1,1}(0,T;X): u(0)=0\},$$

and

$$Bu=-\frac{du}{dt}, \text{ for } u\in\mathfrak{D}(B).$$

One verifies that B is m-dissipative and densely defined in $L^1(0,T;X)$. Assuming $x\in\mathfrak{D}(A_0)$ and setting $v(t)=u(t)-x$, we can write (3.57) as

$$\begin{aligned} &v(t)=A_0v(t)-\frac{dv}{dt}(t)+f(t)-x+A_0,\ t\in[0,T],\\ &v(0)=0. \end{aligned} \tag{3.58}$$

It is convenient to define $g(t)=f(t)-x+A_0x$. Then u is a strict solution of (3.57) in $L_1(0,T;x)$ iff v satisfies

$$v=Av+Bv+g. \tag{3.59}$$

It is easy to verify that (3.57) holds, and then by Theorem 3.24, (3.59) possesses a solution $v\in\mathfrak{D}(A)\cap\mathfrak{D}(B)$ whenever $g\in\mathfrak{D}(A)+\mathfrak{D}(B)$. In particular, if $f\in W_{1,1}(0,T;X)$ and $f(0)+A_0x\in\mathfrak{D}(A_0)$ then (3.57) possesses a strict solution in $L_1(0,T;X)$. Note that since A_0 is closed, $\mathfrak{D}(A_0)$ endowed with the graph norm is a Banach space which we denote by X_{A_0} and $\mathfrak{D}(A)=L_1(0,T;X_{A_0})$. If $f\in L_1(0,T;X_{A_0})$ and $A_0x\in\mathfrak{D}(A_0)$, then (3.57) possesses a strict solution in $L_1(0,T;X)$. If $x\notin\mathfrak{D}(A_0)$ and $f=0$, one may consider the function $t\to e^{-t}T_{A_0}(t)x$ as a generalized solution to (3.57). Note that

$$e^{-t}T_{A_0}(t)x=\lim_{\lambda\downarrow 0}e^{-t}T_{A_{0,\lambda}}(t)x,$$

where $A_{0,\lambda}$ is the Yosida approximation of A_0. If $x=0$ and $f\in L_1(0,T;X)$, then, for every $\lambda>0$, the problem

$$\begin{aligned} &u(t)=A_{0,\lambda}u(t)-\frac{du}{dt}(t)+f(t),\\ &u(0)=0, \end{aligned}$$

possesses a unique solution u_λ in $W_{1,1}(0,T;X)$, and by Theorem 3.24-(iv), $\lim_{\lambda\downarrow 0}u_\lambda$ exists in $L_1(0,T;X)$. In fact for $\lambda>0$, it is not difficult to verify that

$$u_\lambda(t)=\int_0^t T_{A_{0,\lambda}}(t-s)f(s)ds,\quad t\in[0,T].$$

If $\overline{\mathcal{D}(A_0)} = X$, then

$$u(t) = \lim_{\lambda \downarrow 0} u_\lambda(t) = \int_0^t T_{A_0}(t-s)f(s)ds.$$

Then, for $x \in X$ and $f \in L_1(0,T;X)$,

$$u(t) := T_{A_0}(t)x + \int_0^t T_{A_0}(t-s)f(s)ds$$

is usually called a *mild solution* of (3.45). If $\overline{\mathcal{D}(A_0)} \neq X$, then for $x \in \overline{\mathcal{D}(A_0)}$ and $f \in L_1(0,T;X)$,

$$u(t) = \lim_{\lambda \downarrow 0} [T_{A_{0,\lambda}}(t)x + \int_0^t T_{A_{0,\lambda}}(t-s)f(s)ds]$$

is well-defined,

$$\lim_{\lambda \downarrow 0} T_{A_{0,\lambda}}(t)x = T_{A_0}(t)x,$$

but

$$\lim_{\lambda \downarrow 0} \int_0^t T_{A_{0,\lambda}}(t-s)f(s)ds \neq \int_0^t T_{A_0}(t-s)f(s)ds,$$

since $T_{A_0}(t): \overline{\mathcal{D}(A_0)} \to \overline{\mathcal{D}(A_0)}$. This "solution" is a limit of strict solutions in $L_1(0,T;X)$ by Theorem 3.24, that is, there are $x_n \in \mathcal{D}(A_0)$, $f_n \in L_1(0,T;X)$ such that $\lim_{n\to\infty} x_n = x$ in X, $\lim_{n\to\infty} f_n = f$ in $L_1(0,T;X)$ and strict solutions u_n in $L_1(0,T;X)$ of

$$\frac{du}{dt} = A_0 u + f_n,$$

$$u(0) = x_n,$$

converging to u in $L_1(0,T;X)$. It can be shown that the function u belongs to $C([0,T];X)$. Obviously, one may apply Theorem 3.24 in other spaces than $L_1(0,T;X)$ and obtain further regularity results, but we shall not do that here.

We conclude this section by noting a connection between Theorem 3.24 and the notion of n-parameter semigroups.

DEFINITION 3.25. A family of bounded linear operators $\{T(t): t \in \mathbb{R}^n_+\}$ on a Banach space X is called a *C_0-n-parameter semigroup* of bounded linear operators in X if

(i) $T(t+s) = T(t)T(s), \quad t,s \in \mathbb{R}^n_+$,

(ii) $T(0) = I$,

(iii) $\lim_{\substack{t\to 0 \\ t\in\mathbb{R}^n_+}} T(t)x = x$ for every $x \in X$.

We mention the following result.

PROPOSITION 3.26. *Let* $\{T(t): t \in \mathbb{R}^n_+\}$ *be a* C_0*-n-parameter semigroup of bounded linear operators in X. Then*

(i) $T(t) = \prod_{k=1}^{n} T_k(t_k)$ *where* $t=(t_1, \ldots, t_k, \ldots, t_n)$, *and* $T_k(t_k): X \to X$ *is a one-parameter* C_0*-semigroup in X for* $k=1, \ldots, n$. *Let* A_k *denote the infinitesimal generator of* T_k, *with domain* $\mathfrak{D}(A_k)$, $k=1, \ldots, n$.

(ii) *If* $f \in \mathfrak{D}(A_k)$, *then* $T(t)f \in \mathfrak{D}(A_k)$ *for* $f \in \mathbb{R}^n_+$, *and* $A_k T(t) f = T(t) A_k f$ *for* $t \in \mathbb{R}^n_+$, $k=1, \ldots, n$.

(iii) $\bigcap_{k=1}^{n} \mathfrak{D}(A_k)$ *is dense in X and is a Banach space with the norm*

$$|||f||| = \|f\| + \sum_{k=1}^{n} \|A_k f\|.$$

(iv) *If* $f \in \mathfrak{D}(A_j) \cap \mathfrak{D}(A_j A_k)$, *then* $f \in \mathfrak{D}(A_k A_j)$ *and* $A_k A_j f = A_j A_k f$, $j \neq k$.

In particular, if A and B are m-dissipative and densely defined, then A and B are the "partial" generators (derivatives) of the C_0-2-parameter semigroup of contractions:

$$T(t_1, t_2) := T_A(t_1) T_B(t_2) = T_B(t_2) T_A(t_1).$$

Moreover by Theorem 3.24, $\overline{A+B}$ is m-dissipative and densely defined, and $\overline{A+B}$ is the infinitesimal generator $t \to T(t,t) =: T(t)$. Then

$$T(t)x = \lim_{n \to \infty} [T_A(\frac{t}{n}) T_B(\frac{t}{n})]^n x = \lim_{n \to \infty} [E^A_{t/n} E^B_{t/n}]^n x.$$

More generally one has

THEOREM 3.27. *Let A and B be densely defined m-dissipative operators in a Banach space X. If* $\overline{A+B}$ *is also densely defined and m-dissipative then*

$$T_{\overline{A+B}}(t)x = \lim_{n \to \infty} [T_A(\frac{t}{n}) T_B(\frac{t}{n})]^n x, \tag{3.60}$$

$$T_{\overline{A+B}}(t)x = \lim_{n \to \infty} [E^A_{t/n} E^B_{t/n}]^n x, \tag{3.61}$$

for every $x \in X$, *uniformly on bounded t-intervals.*

PROOF. By Theorem 2.14, it is sufficient to prove that

$$\lim_{\lambda \downarrow 0} \|E_h^{C_\lambda} x - E_h^{\overline{A+B}} x\| = 0, \tag{3.62}$$

for every $x \in X$ and some $h > 0$. Here $C_\lambda = \lambda^{-1}(T_A(\lambda) T_B(\lambda) - I)$, $\lambda > 0$ by (3.60), and $C_\lambda := \lambda^{-1}(E^A_\lambda E^B_\lambda - I)$, $\lambda > 0$, by (3.61). Note that $T_A(\lambda) T_B(\lambda)$ and $E^A_\lambda E^B_\lambda$ are contractions. The proof of (3.62) will be give in Section 3.7, to be precise in Propositions 3.33 and 3.34. □

A simple example where the assumptions of Theorems 3.27 are satisfied is the case of relative bounded perturbation of A. We have:

THEOREM 3.28. *Let A be m-dissipative and densely defined in X and let B be dissipative in X. If $\mathfrak{D}(B) \subseteq \mathfrak{D}(A)$ and $\|Bx\| \leqslant a\|Ax\| + b\|x\|$, for every $x \in \mathfrak{D}(A)$ for some $a \in [0,1)$ and $b \geqslant 0$, then $A+B$ is m-dissipative and densely defined.*

PROOF. $A+B$ is clearly densely defined and dissipative by Theorem 3.14. By Proposition 3.7, it suffices to prove that $\mathfrak{R}(I-(A+B)) = X$. Assume that $A+\lambda B$ is m-dissipative for some $\lambda \in [0,1)$. We consider the equation

$$x = (A + \lambda B)x + hBx + f, \tag{3.63}$$

for $h > 0$ and $f \in X$. Setting $C = A + \lambda B$ and $y = x - Cx$, (3.63) is equivalent to

$$y = hBE_1^C y + f. \tag{3.64}$$

If BE_1^C is bounded in X and $\|BE_1^C\|$ is bounded independently of $\lambda \in [0,1)$, then (3.64) is uniquely solvable for every $f \in X$, whenever

$$h\|BE_1^C\| < 1. \tag{3.65}$$

Then by choosing first $\lambda = 0$ and $h_0 > 0$ satisfying (3.65), we can solve (3.64), and then (3.63) with $\lambda = h_0$, $h = h_0$, and by induction $A+\lambda B$ is m-dissipative for all $\lambda \in [0, 1+h_0)$. It remains to estimate the norm of BE_1^C. For $x \in X$, we have

$$\|BE_1^C x\| \leqslant a\|AE_1^C x\| + b\|E_1^C x\|,$$

by assumption. But

$$\begin{aligned} \|AE_1^C x\| &= \|-AE_1^C x\| = \|(I - A - \lambda B)E_1^C x - (I - \lambda B)E_1^C x\| \\ &= \|x + \lambda BE_1^C x - E_1^C x\| \\ &\leqslant \lambda\|BE_1^C x\| + 2\|x\| \leqslant \|BE_1^C x\| + 2\|x\|. \end{aligned}$$

Hence

$$\|BE_1^C x\| \leqslant a\|BE_1^C x\| + (2a + b)\|x\|,$$

and

$$\|BE_1^C\| \leqslant (1-a)^{-1}(2a + b),$$

which is independent of λ. □

COROLLARY 3.29. *Let A be m-dissipative and densely defined in X. Let B be dissipative with $\mathfrak{D}(B) \supseteq \mathfrak{D}(A)$. If there are $a, b \geqslant 0$ and $\alpha \in (0,1)$ such that*

$$\|Bx\| \leqslant a\|Ax\|^{\alpha}\|x\|^{1-\alpha} + b\|x\|,$$

for every $x \in \mathfrak{D}(A)$, then $A+B$ is m-dissipative and densely defined.

PROOF. From Young's inequality, we obtain

$$\|Ax\|^{\alpha}\|x\|^{1-\alpha}\leq \alpha\epsilon^{\alpha^{-1}}\|Ax\|+(1-\alpha)\epsilon^{-(1-\alpha)^{-1}}\|x\|,$$

for every $\epsilon>0$ and $x\in\mathfrak{D}(A)$. By choosing $\epsilon>0$ such that $a\alpha\epsilon^{\alpha^{-1}}<1$, we can apply Theorem 3.28. □

7. C_0-SEMIGROUPS ON COMPLEX BANACH SPACES

Let $\{T(t)\}_{t\geq 0}$ be a semigroup of bounded linear operators in a real Banach space $(X,\|\cdot\|)$, satisfying (3.11), hence also (3.12). We have already observed in Section 3.2, that if $\{T(t)\}_{t\geq 0}$ is strongly continuous, then (3.11) is a consequence of the uniform boundedness theorem. We define

$$\omega_0(T):=\inf\{\omega\in\mathbb{R}:\exists_{M\geq 1}\forall_{t\geq 0}\|T(t)\|\leq Me^{\omega t}\}. \tag{3.66}$$

$\omega_0(T)$ is called the *type of the semigroup* $T(t)$. Clearly $\omega_0(T)<\infty$, but $\omega_0(T)$ may be $-\infty$. Note that $\omega_0(T^*)=\omega_0(T)$, where T^* is the dual semigroup of T. Note also that, if T is a contraction semigroup, then $\omega_0(T)\leq 0$. Observe that if X is equipped with a different but equivalent norm, then the type of the semigroup T does not change. The type can also be defined in another way.

PROPOSITION 3.30. *Let $\{T(t)\}_{t\geq 0}$ be a semigroup satisfying (3.12). Then*

$$\omega_0(T)=\inf_{t>0}\frac{1}{t}\log\|T(t)\|=\lim_{t\to\infty}\frac{1}{t}\log\|T(t)\|.$$

PROOF. If $\omega_0(T)=-\infty$, then for every $n\in\mathbb{N}$ there is an $M_n\geq 1$ such that $\|T_0(t)\|\leq M_ne^{-nt}$. Hence

$$\limsup_{t\to\infty}\frac{1}{t}\log\|T(t)\|\leq\limsup_{t\to\infty}\frac{1}{t}\log M_n-n=-n.$$

Hence

$$\inf_{t>0}\frac{1}{t}\log\|T(t)\|=\lim_{t\to\infty}\frac{1}{t}\log\|T(t)\|=-\infty.$$

If $\omega_0(T)>-\infty$, then for every $\epsilon>0$ and for every $n\in\mathbb{Z}^+$, there is $t_n>0$ such that $\|T(t_n)\|>ne^{(\omega_0(T)-\epsilon)t_n}$. Clearly $\sup_{n\in\mathbb{Z}^+}t_n=\infty$. Thus

$$\limsup_{t\to\infty}\frac{1}{t}\log\|T(t)\|\geq\omega_0(T)-\epsilon.$$

Hence

$$\omega_0(T)\leq\limsup_{t\to\infty}\frac{1}{t}\log\|T(t)\|.$$

Set $p(t)=\log\|T(t)\|$. Observe that

$$\begin{aligned}p(t+s)&=\log\|T(t+s)\|=\log\|T(t)T(s)\|\leq\log(\|T(t)\|\|T(s)\|)\\&=\log\|T(t)\|+\log\|T(s)\|=p(t)+p(s).\end{aligned}$$

A function with such a property is called subadditive. Now let $t>t_0$ and $n\in\mathbb{Z}^+$ be such that $nt_0\leq t<(n+1)t_0$, then we have

$$\frac{p(t)}{t} \leqslant \frac{p(nt_0)}{t} + \frac{p(t-nt_0)}{t} \leqslant \frac{nt_0}{t}\frac{p(t_0)}{t_0} + \frac{p(t-nt_0)}{t}.$$

Note that $t-nt_0 \leqslant t_0$, hence $\|T(t-nt_0)\| \leqslant \sup_{s\in[0,t_0]} \|T(s)\| < \infty$. We obtain

$$\limsup_{t\to\infty} \frac{p(t)}{t} \leqslant \limsup_{t\to\infty} \left(\frac{nt_0}{t}\right) \cdot \frac{p(t_0)}{t_0} = \frac{p(t_0)}{t_0}.$$

Hence

$$\limsup_{t\to\infty} \frac{1}{t} \log \|T(t)\| = \inf_{t>0} \frac{1}{t} \log \|T(t)\|.$$

Finally for $\epsilon>0$, there is $M_\epsilon>0$, such that $t^{-1}p(t) \leqslant t^{-1} \log M_\epsilon + (\omega_0(T)+\epsilon)$, $t>0$. Hence $\inf_{t>0} t^{-1} \log \|T(t)\| \leqslant \omega_0(T)+\epsilon$, for every $\epsilon>0$. □

If A is an m-dissipative operator in $(X, \|\cdot\|)$ and T_A is the semigroup on $\overline{\mathcal{D}(A)}$ generated by A (Theorem 2.3), then $\omega_0(T_A) \leqslant 0$. If $\omega_0(T_A)<0$, then for every $\omega \in (\omega_0(T_A), 0)$, there exists an equivalent norm (see Section 3.2) for which $A-\omega I$ is also m-dissipative, and the type of $T_{A-\omega I}$ is $\omega_0(T_A)-\omega$. If $T_{A-\omega_0(T_A)I}$ is bounded, that is if

$$\sup_{t\geqslant 0} \|T_{A-\omega_0(T_A)I}(t)\| < \infty,$$

then there is an equivalent norm for which $A-\omega_0(T_A)I$ is m-dissipative, and the type of the semigroup generated by $A-\omega_0(T_A)I$ is 0. This motivates the following definition. An m-dissipative operator A in a real Banach space $(X, \|\cdot\|)$ is called *completely m-dissipative* if $A-\omega_0(T_A)I$ is also m-dissipative. Then for every $\omega \in [\omega_0(T_A), 0)$, $A-\omega I$ is also m-dissipative in X with respect to the original norm $\|\cdot\|$.

REMARK. If $\{T(t)\}_{t\geqslant 0}$ is a strongly continuous semigroup with generator A, then we sometimes write $\omega_0(A)$ instead of $\omega_0(T)$.

We now consider the complexification of the real Banach space $(X, \|\cdot\|)$, that is $X_{\mathbf{C}} = X + iX$ equipped with the norm $\|z\|_{\mathbf{C}} := \max_{0\leqslant\theta\leqslant 2\pi} \|\cos\theta \cdot x - \sin\theta \cdot y\|$ with $z = x+iy$, $x,y \in X$ (see Appendix 2). Then, if $\{T(t)\}_{t\geqslant 0}$ is a semigroup of bounded operators in X, we define

$$T_{\mathbf{C}}(t)z := T(t)x + iT(t)y,$$

for $z=x+iy$, $x,y\in X$ and $t \geqslant 0$. $T_{\mathbf{C}}$ is a semigroup of bounded operators and we have $\|T(t)\| \leqslant \|T_{\mathbf{C}}(t)\| \leqslant M\|T(t)\|$, $t\geqslant 0$, for some $M \geqslant 1$. In particular $\omega_0(T_{\mathbf{C}}) = \omega_0(T)$. Note that if we choose an equivalent norm on $X_{\mathbf{C}}$ then the type of $T_{\mathbf{C}}$ is also equal to $\omega_0(T)$. If $\{T(t)\}_{t\geqslant 0}$ is a C_0-semigroup of operators in X, then $t \to T_{\mathbf{C}}(t)z$ is strongly continuous in $(X_{\mathbf{C}}, \|\cdot\|_{\mathbf{C}})$, and $\{T_{\mathbf{C}}(t)\}_{t\geqslant 0}$ is also called a C_0-semigroup on $X_{\mathbf{C}}$.

From now on, we assume that $(X, \|\cdot\|)$ is a complex Banach space and that $\{T(t)\}_{t\geqslant 0}$ is a C_0-semigroup of bounded linear operators on X, that is $T(t): X \to X$ are bounded linear operators on X for $t \geqslant 0$ and $\{T(t)\}_{t\geqslant 0}$ is a C_0-

semigroup of operators on $(X,\|\cdot\|)$ viewed as a real Banach space. Clearly the norm of $T(t)$ in $(X,\mathbb{R},\|\cdot\|)$ and the norm of $T(t)$ in $(X,\mathbb{C},\|\cdot\|)$ are the same, and $\omega_0(T)$, the type of T, is defined by (3.66). Therefore, for every $\omega>\omega_0(T)$, there is an $M_\omega\geqslant 1$ such that (3.12) holds and for every $\lambda\in\mathbb{C}$ with $\mathrm{Re}\lambda>\omega_0(T)$, and every $f\in X$, $\int_0^\infty e^{-\lambda t}T(t)fdt$ is well-defined. Moreover it follows from a basic lemma on Laplace transforms (see e.g. BUTZER and BERENS [1967]) that $\lambda\to\int_0^\infty e^{-\lambda t}T(t)fdt$ is holomorphic in the half-plane $\mathrm{Re}\lambda>\omega_0(T)$. We have

$$\left(\frac{d}{d\lambda}\right)^n\int_0^\infty e^{-\lambda t}T(t)dt=\int_0^\infty(-t)^n e^{-\lambda t}T(t)fdt, \tag{3.67}$$

for $\mathrm{Re}\lambda>\omega_0(T)$, $n\geqslant 1$. Next, if A_T is the infinitesimal generator of T defined by (2.0), then clearly A_T is a linear operator (with respect to the field $\mathbb{C}$). For $\omega>\omega_0(T)$, we have $\omega I-A_T:\mathfrak{D}(A_T)\to X$ is a bijection and

$$(\omega I-A_T)^{-1}f=\int_0^\infty e^{-\omega t}T(t)fdt,\quad f\in X, \tag{3.68}$$

by Theorem 3.5, where X is viewed as a real Banach space. Clearly $(\omega I-A_T)^{-1}$ is linear (with respect to the field $\mathbb{C}$) and bounded. Therefore $\omega\in\rho(A_T)$, where $\rho(A_T)$ denotes the resolvent set of A_T.
For $\gamma\in\mathbb{R}$, $\{e^{-i\gamma t}T(t)\}_{t\geqslant 0}$ is a C_0-semigroup of bounded operators in X, (with respect to the field $\mathbb{R}$), with type equal to $\omega_0(T)$ and infinitesimal generator $A=A_T-i\gamma I$. Then from (3.68) we obtain $\lambda\in\rho(A_T)$, and

$$(\lambda I-A_T)^{-1}f=\int_0^\infty e^{-\lambda t}T(t)fdt,\quad f\in X,$$

for $\mathrm{Re}\lambda>\omega_0(T)$. In particular, if $\sigma(A_T)$, the spectrum of A_T is not empty, we have $\mathrm{Re}\lambda\leqslant\omega_0(T)$, for every $\lambda\in\sigma(A_T)$. We summarize these results in the following

THEOREM 3.31. *Let $\{T(t)\}_{t\geqslant 0}$ be a C_0-semigroup of bounded linear operators in a complex Banach space $(X,\|\cdot\|)$. Let A_T be its infinitesimal generator and let $\omega_0(T)$ be the type of $\{T(t)\}_{t\geqslant 0}$. Then*
(i) *For every $\lambda\in\mathbb{C}$ with $\mathrm{Re}\lambda>\omega_0(T)$, $\lambda\in\rho(A_T)$, the resolvent set of A_T.*
(ii) $(\lambda I-A_T)^{-1}f=\int_0^\infty e^{-\lambda t}T(t)fdt$, *$f\in X$ and $\mathrm{Re}\lambda>\omega_0(T)$.*
Moreover, if $\|T(t)\|\leqslant Me^{\omega t}$, $t\geqslant 0$, then for $\mathrm{Re}\lambda>\omega$

$$\|(\lambda I-A_T)^{-n}\|\leqslant M(Re\,\lambda-\omega)^{-n},\quad n\in\mathbb{Z}^+. \tag{3.69}$$

PROOF. It remains to prove the last assertion. If (3.12) holds, then $\omega\geqslant\omega_0(T)$ and for $\mathrm{Re}\lambda>\omega$, (3.69) follows from (3.14), by considering again the semigroup $\{e^{it\,\mathrm{Im}\lambda}T(t)\}_{t\geqslant 0}$. (3.69) can also be proven by using the identity

$$\left(\frac{d}{d\lambda}\right)^n(\lambda I-A_T)^{-1}f=(-1)^n n!(\lambda I-A_T)^{-(n+1)}f,$$

for $n\in\mathbb{N}$ and $f\in X$, and (3.67). Then

$$\begin{aligned}\|(\lambda I - A_T)^{-(n+1)}f\| &= \frac{1}{n!}\|(\frac{d}{d\lambda})^n(\lambda I - A_T)^{-1}f\| \\ &= \frac{1}{n!}\|\int_0^\infty (-t)^n e^{-\lambda t} T(t) f dt\| \\ &= \frac{1}{n!}\|\int_0^\infty (-t)^n e^{-(\lambda-\omega)t} e^{-\omega t} T(t) f dt\| \\ &\leqslant M\frac{1}{n!}\|f\| \int_0^\infty t^n e^{-(\mathrm{Re}\lambda-\omega)t} dt \\ &= M(\mathrm{Re}\lambda - \omega)^{-n-1}\|f\|. \quad \square\end{aligned}$$

Now we are in a position to state and prove a more general version of Theorem 3.6-(iv).

THEOREM 3.32. *Let $M \geqslant 1$ and $\omega \in \mathbb{R}$: Let $\{T_n\}$ be a sequence of C_0-semigroups of bounded linear operators in a complex Banach space $(X, \|\cdot\|)$ satisfying $\|T_n(t)\| \leqslant Me^{\omega t}$, $t \geqslant 0$, with infinitesimal generators A_n. Suppose that for some $\lambda_0 \in \mathbb{C}$ with $\mathrm{Re}\lambda_0 > \omega$*

(i) *$\lim_{n\to\infty} (\lambda_0 I - A_n)^{-1} f$ exists for every $f \in X$.*
Set $Rf := \lim_{n\to\infty} (\lambda_0 I - A_n)^{-1} f$, $f \in X$.

(ii) *Suppose that R has a dense range in X. Then there exists a unique C_0-semigroup of bounded linear operators $\{T(t)\}$ satisfying*

$$\|T(t)\| \leqslant Me^{\omega t}, \quad t \geqslant 0,$$

$$\lim_{n\to\infty} \max_{t\in[0,s]} \|T_n(t)f - T(t)f\| = 0, s > 0, f \in X.$$

Moreover the infinitesimal generator A_T of T satisfies $(\lambda_0 I - A_T)^{-1} = R$.

PROOF. By replacing A_n by $A_n - (i \cdot \mathrm{Im}\lambda_0 + \omega)I$, we may assume $\omega = 0$, $\lambda_0 \in \mathbb{R}$ and $\lambda_0 > 0$. By Theorem 3.31, we have

$$\|(\lambda I - A_n)^{-m}\| \leqslant M\lambda^{-m}, m, n \in \mathbb{Z}^+, \lambda > 0.$$

Equivalently $\|(I - hA_n)^{-m}\| \leqslant M$, $m, n \in \mathbb{Z}^+$, $h > 0$. For $n \in \mathbb{Z}^+$, we define the Banach spaces $(X_n, \|\cdot\|_n)$ by setting $X_n = X$ and

$$\|x\|_n = \sup_{\mu>0} \sup_{m\geqslant 0} \|(I - \mu A_n)^{-m} x\|, \quad x \in X_n. \tag{3.70}$$

It follows from the proof of Theorem 3.6 that A_n is m-dissipative and densely defined in $(X_n, \|\cdot\|)$ and that

$$\|x\| \leqslant \|x\|_n \leqslant M\|x\|, \quad x \in X, n \in \mathbb{Z}^+. \tag{3.71}$$

Hence X_n is a Banach space. Next we define E the vector space of all sequences $\{x_n\}$ with $x_n \in X_n$, and such that $\lim_{n\to\infty} x_n$ exists in X. Clearly E is a Banach space with the norm

$$\|\hat{x}\|_E := \sup_{n\geq 0} \|x_n\|_n,$$

where $\hat{x}=\{x_0,x_1,x_2 \cdots\}$. Then we define

$$\mathfrak{D}(\hat{A}) := \{\hat{x}\in E : x_n\in\mathfrak{D}(A_n), n\in\mathbb{Z}^+ \text{ and } \{A_n x_n\}\in E\}.$$

For $\hat{x}\in\mathfrak{D}(\hat{A})$, we set

$$\hat{A}\hat{x} := \{A_n x_n\}.$$

Since $\|x_n\|_n \leq \|x_n - hA_n x_n\|_n$ for every $n\in\mathbb{Z}^+$, $x_n\in\mathfrak{D}(A_n)$, and $h>0$, we obtain

$$\|\hat{x}\|_E \leq \|\hat{x} - h\hat{A}\hat{x}\|_E,$$

for every $\hat{x}\in\mathfrak{D}(\hat{A})$ and $h>0$. Hence $\hat{A}$ is dissipative. We claim that $\mathfrak{R}(I-h_0\hat{A})=E$ for $h_0=\lambda_0^{-1}$. Indeed if $\hat{f}\in E$, then there exist $x_n\in\mathfrak{D}(A_n)$ such that $x_n - h_0 A_n x_n = f_n$ for every $n\in\mathbb{Z}^+$, since A_n are m-dissipative. Next we prove that $\lim_{n\to\infty} x_n$ exists. Let $f=\lim_{n\to\infty} f_n$, and $y_n\in\mathfrak{D}(A_n)$ be such that $y_n - h_0 A_n y_n = f$. By assumption $\lim_{n\to\infty} y_n = \lim_{n\to\infty}(I-h_0 A_n)^{-1} f$ exists. Moreover $x_n - y_n$ satisfies

$$(x_n - y_n) = h_0 A_n (x_n - y_n) + f_n - f.$$

Hence

$$\|x_n - y_n\|_n \leq \|f_n - f\|_n \leq M\|f_n - f\| \to 0, \text{ as } n\to\infty.$$

It follows that $\lim_{n\to\infty} x_n$ exists. Moreover $\lim_{n\to\infty} A_n x_n = h_0^{-1}\lim_{n\to\infty}(x_n - f_n)$ exists, thus $\hat{x} := \{x_n\}\in\mathfrak{D}(\hat{A})$, and satisfies $\hat{x} - h_0\hat{A}\hat{x} = \hat{f}$. By Proposition 3.7-(iv), $\hat{A}$ is m-dissipative in E. This implies in particular that $\lim_{n\to\infty}(I-hA_n)^{-1}y$ exists for all $y\in X$ and all $h>0$ (choose $\hat{f}=\{y,y,y, \cdots\}$). Finally we prove that $\mathfrak{D}(\hat{A})$ is dense in E. Let $\hat{f}=\{f_n\}$ with $\lim_{n\to\infty} f_n = f$, and let $\epsilon>0$. By assumption there exists $\tilde{y}\in X$ such that

$$\|\lim_{n\to\infty}(\lambda_0 I - A_n)^{-1}\tilde{y} - f\| < \frac{\epsilon}{3M}.$$

Then $y := h_0\tilde{y}$ satisfies

$$\|\lim_{n\to\infty}(I - h_0 A_n)^{-1} y - f\| < \frac{\epsilon}{3M}.$$

Choose $N\in\mathbb{Z}^+$ such that $\|f_n - f\| < \epsilon(3M)^{-1}$ and

$$\|(I - h_0 A_n)^{-1} y - \lim_{m\to\infty}(I - h_0 A_m)^{-1} y\| < \frac{\epsilon}{3M},$$

for $n>N$. Set $x_n := (I-h_0 A_n)^{-1} y$ for $n\geq N$. Then clearly

$$\|x_n - f_n\|_n \leq M\|x_n - f_n\| < \epsilon \text{ for } n>N.$$

We have $x_n\in\mathfrak{D}(A_n)$, $n>N$ and $\lim_{n\to\infty} A_n x_n$ exists. For $n=0,\ldots,N$, choose $x_n\in\mathfrak{D}(A_n)$ such that $\|x_n - f_n\| < \epsilon M^{-1}$. This is possible since $\overline{\mathfrak{D}(A_n)}=X$. Then $\|x_n - f_n\|_n < \epsilon$ for $n\in\mathbb{Z}^+$ and $\|\hat{x}-\hat{f}\| \leq \epsilon$. Since $\hat{x}\in\mathfrak{D}(\hat{A})$, $\mathfrak{D}(\hat{A})$ is dense in E. Therefore $\hat{A}$ is the infinitesimal generator of a C_0-semigroup of contractions in

E, which we denote by $\hat{T}=\{T_n\}$. Let us define for $\hat{x}=\{x_n\}$, $P_n\hat{x}=x_n$, $n\in\mathbb{Z}^+$. Clearly $P_n^2=P_n$ and $\|P_n\|=1$. Moreover,

$$P_n(I-h\hat{A})^{-1}\hat{x}=(I-h\hat{A})^{-1}P_n\hat{x}=(I-hA_n)^{-1}x_n,$$

for every $\hat{x}\in E$ and $h>0$. Therefore $P_n\hat{T}(t)P_n\hat{x}=T_n(t)x_n$, $t\geqslant 0$, $n\in\mathbb{Z}^+$ and $\hat{x}\in E$, by Theorem 2.3. We also have that $\lim_{n\to\infty}T_n(t)x$ exists for every $x\in X$ and every $t\geqslant 0$. Set $T(t)x:=\lim_{n\to\infty}T_n(t)x$. Clearly $T(t):X\to X$ is a semigroup of linear operators in X satisfying $\|T(t)\|\leqslant M$, $t\geqslant 0$. Moreover from Theorem 2.3 we have

$$\|\hat{T}(t)\hat{x}-\hat{T}(s)\hat{x}\|_E\leqslant\|\hat{A}\hat{x}\|_E\cdot|t-s|,$$

for every $\hat{x}\in\mathfrak{D}(\hat{A})$ and $t,s\geqslant 0$. Choose $\hat{x}=(I-h_0\hat{A})^{-1}\hat{f}$, $\hat{f}\in E$, with $\hat{f}=\{f,f,f,\cdots\}$, and set $x=\lim_{n\to\infty}(I-h_0A_n^{-1}f)$. We have

$$\|T_n(t)x_n-T_n(s)x_n\|\leqslant\sup_{n\geqslant 1}\|T_n(t)x_n-T_n(s)x_n\|,$$

where $x_n:=(I-h_0A_n)^{-1}f$. Then

$$\sup_{n\geqslant 1}\|(T_n(t)-T_n(s))x_n\|\leqslant\sup_{n\geqslant 1}\|(T_n(t)-T_n(s))x_n\|_n$$
$$=\|(\hat{T}(t)-\hat{T}(s))\hat{x}\|_E\leqslant\|A\hat{x}\|_E\cdot|t-s|.$$

It follows that

$$\|T_n(t)x-T_n(s)x\|\leqslant 2M\|x-x_n\|+\|A\hat{x}\|_E\cdot|t-s|,$$

thus $T_n(t)x$ converges uniformly to $T(t)x$ on bounded t-intervals, on a dense subset of X. Since $\|T(t)\|\leqslant M$, $T_n(t)x$ converges uniformly to $T(t)x$ on bounded t-intervals for every $x\in X$, and $t\mapsto T(t)x$ is strongly continuous. Let A_T be the infinitesimal generator of T. Then $A_T\in G(M,0)$, and

$$(\lambda I-A_T)^{-1}y=\int_0^\infty e^{-\lambda t}T(t)y\,dt\,,\ \text{for } \lambda>0,$$

by (3.15). It follows that

$$Ry=\lim_{n\to\infty}(\lambda I-A_n)^{-1}y=\lim_{n\to\infty}\int_0^\infty e^{-\lambda t}T_n(t)y\,dt=\int_0^\infty e^{-\lambda t}T(t)y\,dt,$$

for $\lambda>0$ and $y\in X$. Here we used Theorem 3.4. □

We now consider a special situation where the assumptions of Theorem 3.32 are satisfied.

PROPOSITION 3.33. *Let $A:\mathfrak{D}(A)\subset X\to X$ be a densely defined dissipative operator such that its closure A is m-dissipative. Let $A_n:X\to X$ be a sequence of bounded dissipative operators satisfying*

$$\lim_{n\to\infty}A_nx=Ax \text{ for every } x\in\mathfrak{D}(A). \tag{3.72}$$

Then

$$\lim_{n\to\infty} \|E_h^{A_n}x - E_h^{\bar{A}}x\| = 0 \tag{3.73}$$

for every $h>0$ and every $x\in X$.

PROOF. First note that every A_n is m-dissipative. Indeed for $n\in\mathbb{Z}^+$, $I-hA_n$ is bijective provided $h\|A_n\|<1$. By Proposition 3.7-(iv), A_n is m-dissipative. By Theorem 2.7-(i), it suffices to prove (3.73) for $h=1$. Since $E_1^{A_n}$ and $E_1^{\bar{A}}$ are contractions , it is sufficient to prove (3.73) for a dense subset of X. Since $\mathfrak{D}(A)$ is dense and $\bar{A}$ is m-dissipative, $\mathfrak{R}(I-A)$ is dense in X, by Proposition 3.10 (or by direct verification). Therefore it is sufficient to verify

$$\lim_{n\to\infty} \|E_1^{A_n}(x - Ax) - E_1^{\bar{A}}(x - Ax)\| = 0,$$

for every $x\in\mathfrak{D}(A)$. Since $Ax=\bar{A}x$, we have

$$\|E_1^{A_n}(x - Ax) - E_1^{\bar{A}}(x - \bar{A}x)\| =$$

$$\|(E_1^{A_n}x - x) - E_1^{A_n}Ax\| = \|(A_n)_1 x - E_1^{A_n}Ax\|$$

$$= \|E_1^{A_n}A_n x - E_1^{A_n}Ax\| \leqslant \|A_n x - Ax\| \to 0, \text{ as } n\to\infty.$$

Here $(A_n)_1$ denotes the Yosida approximation of A_n. □

We now complete the proof of Theorem 3.27.

PROPOSITION 3.34. *Let A and B be densely defined m-dissipative operators in X. Then for every $x\in\mathfrak{D}(A)\cap\mathfrak{D}(B)$,*

$$\lim_{t\downarrow 0} t^{-1}(T_A(t)T_B(t)x - x) = (A + B)x, \tag{3.74}$$

and

$$\lim_{h\downarrow 0} h^{-1}(E_h^A E_h^B x - x) = (A + B)x.$$

PROOF.

$$t^{-1}(T_A(t)T_B(t)x - x) =$$

$$T_A(t)[t^{-1}(T_B(t)x - x) + t^{-1}(T_A(t)x - x)] \to Bx + Ax, \text{ as } t\downarrow 0,$$

since $\|T_A(t)\|\leqslant 1$ and $\lim_{t\downarrow 0} T_A(t)y = y$, $y\in X$. Similarly

$$h^{-1}(E_h^A E_h^B x - x) = E_h^A[h^{-1}(E_h^B x - x) + h^{-1}(E_h^A x - x)]$$

$$= E_h^A(B_h x + A_h x) \to Bx + Ax \text{ as } h\downarrow 0. \quad \square$$

Chapter 4

L_1-semigroup Theory of Nonlinear Diffusion

1. Introduction

Consider the initial value problem

$$
\text{(I)} \quad \begin{cases} u_t = (\phi(u))_{xx}, & (t,x) \in (0,\infty) \times \mathbb{R}, \\ u(0,x) = u_0(x), & x \in \mathbb{R}, \end{cases}
$$

where ϕ is a maximal monotone graph in $\mathbb{R}^2$ with domain $\mathcal{D}(\phi)$: see Section 2.1 and Appendix 1. Equations of this type arise in many problems of physical interest.

EXAMPLE 4.1.
(a) The Porous Media Equation, where $\phi(u) = |u|^m \cdot \operatorname{sign}(u)$, $m > 0$.
(b) The Stefan Problem, where ϕ is constant on an interval $I \subset \mathcal{D}(\phi)$ and strictly increasing on $\mathcal{D}(\phi) \setminus I$.

It is well-known from the field of partial differential equations that Problem (I) has a unique solution, denoted by $u(t;u_0)$, when u_0 and ϕ are appropriately chosen. This solution satisfies $u(t+s;u_0) = u(t;u(s;u_0))$ and $u(0;u_0) = u_0$. In addition, solutions of Problem (I) often satisfy a comparison principle

$$
u_0 \geqslant v_0 \text{ implies } u(t;u_0) \geqslant u(t;v_0), \text{ for all } t \geqslant 0, \tag{4.1}
$$

and mass-conservation (throughout this chapter we denote by $\int_{\mathbb{R}} u$ the integral $\int_{\mathbb{R}} u(x)dx$)

$$
\int_{\mathbb{R}} u(t;u_0) = \int_{\mathbb{R}} u_0, \text{ for all } t \geqslant 0. \tag{4.2}
$$

Below we will make precise that if $u(t;u_0)\in L_1(\mathbb{R})$ for all $t\geqslant 0$, such that (4.1) and (4.2) hold, then $u(t;\cdot)$ defines a semigroup of contractions on $L_1(\mathbb{R})$.

Let L be a mapping in $L_1(\mathbb{R})$ which conserves the integral, i.e.

$$\int_{\mathbb{R}} L(f) = \int_{\mathbb{R}} f, \tag{4.3}$$

and let $f\vee g=\max(f,g)$ and $r^+=r\vee 0$. The following result is due to CRANDALL and TARTAR [1980].

PROPOSITION 4.2. *Let $K\subset L_1(\mathbb{R})$ have the property that $f,g\in K$ implies $f\vee g\in K$. Let $L:K\to L_1(\mathbb{R})$ satisfy* (4.3) *for $f\in K$. Then the following three properties of L are equivalent.*

(a) *$f,g\in K, f\leqslant g$ a.e. implies $Lf\leqslant Lg$ a.e.*

(b) $\int_{\mathbb{R}}(L(f)-L(g))^+\leqslant\int_{\mathbb{R}}(f-g)^+$, *for $f,g\in K$*

(c) $\|L(f)-L(g)\|_1\leqslant\|f-g\|_1$.

PROOF. We show (a)$\Rightarrow$(b)$\Rightarrow$(c)$\Rightarrow$(a). Let $f,g\in K$. Then $f\vee g\in K$ and from (a) we obtain that $L(f\vee g)\geqslant L(g)$, which implies that $L(f\vee g)-L(g)\geqslant 0$, and $L(f\vee g)\geqslant L(f)$, which implies that $L(f\vee g)-L(g)\geqslant L(f)-L(g)$. Thus we get that $(L(f)-L(g))^+\leqslant L(f\vee g)-L(g)$, and

$$\int_{\mathbb{R}}(L(f)-L(g))^+\leqslant\int_{\mathbb{R}}(L(f\vee g)-L(g))=\int_{\mathbb{R}}(f\vee g-g)=\int_{\mathbb{R}}(f-g)^+,$$

because $f\vee g=g+(f-g)^+$. Using the identity $|s|=s^++(-s)^+$ gives

$$\begin{aligned}\int_{\mathbb{R}}|L(f)-L(g)| &= \int_{\mathbb{R}}(L(f)-L(g))^+ + \int_{\mathbb{R}}(L(g)-L(f))^+\\ &\leqslant \int_{\mathbb{R}}(f-g)^+ + \int_{\mathbb{R}}(g-f)^+ = \int_{\mathbb{R}}|f-g|.\end{aligned}$$

Finally, if $f,g\in K, f\leqslant g$, and (c) holds, the identity $2s^+=|s|+s$ gives

$$\begin{aligned}2\int_{\mathbb{R}}(L(f)-L(g))^+ &= \int_{\mathbb{R}}|L(f)-L(g)|+\int_{\mathbb{R}}(L(f)-L(g))\\ &\leqslant \int_{\mathbb{R}}|f-g|+\int_{\mathbb{R}}(f-g)=2\int_{\mathbb{R}}(f-g)^+=0.\end{aligned}$$

Thus $L(f)-L(g)\leqslant 0$ a.e. □

We give an interesting application of this proposition. Consider the scalar conservation law:

$$\frac{\partial u}{\partial t}+\frac{\partial}{\partial x}f(u)=0,\ t>0, x\in\mathbb{R},$$

where the function $f:\mathbb{R}\to\mathbb{R}$ is continuously differentiable. The LAX finite difference approximation gives

$$U_i^{n+1} = U_i^n + \frac{\Delta t}{2\Delta x}\{f(U_{i+1}^n) - f(U_{i-1}^n)\}$$
$$+ \frac{1}{2}\{U_{i+1}^n - 2U_i^n + U_{i-1}^n\},$$

where $i \in \mathbb{Z}$ and $n \in \mathbb{Z}^+$. Write this difference equation as

$$U_i^{n+1} = L(U^n)_i, \ i \in \mathbb{Z}, n \in \mathbb{Z}^+,$$

and set $K = \{U \in l^1 : a \leqslant U_i \leqslant b, \ i \in \mathbb{Z} \text{ and } a,b \in \mathbb{R}\}$. Then $L: K \to l^1$ satisfies the conservation equation

$$\sum_{i=\infty}^{\infty} L(U)_i = \sum_{i=\infty}^{\infty} U_i.$$

Further, if the condition

$$(\Delta t / \Delta x) \cdot \max_{a \leqslant r \in b} |f'(r)| \leqslant 1 \qquad (4.4)$$

is satisfied, then L is also *ordering preserving* on K, i.e.

$$L(U)_i \geqslant L(V)_i \text{ when } U_i \geqslant V_i, \text{ for all } i \in \mathbb{Z}.$$

Proposition 4.2 now gives that L is a contraction. Therefore (4.4) is a stability condition.

It is known from Chapter 2 that, if $A: \mathfrak{D}(A) \subseteq X \to X$ is m-dissipative in a Banach space X, then it generates a semigroup of contractions $\{T_A(t)\}_{t \geqslant 0}$ on $\overline{\mathfrak{D}(A)}$. The abstract function $u: [0, \infty) \to X$, defined by

$$u(t) = T_A(t)u_0, \ u_0 \in \overline{\mathfrak{D}(A)} \text{ and } t \geqslant 0,$$

is the unique integral solution of the abstract problem

$$\text{(II)} \quad \begin{aligned} &\frac{du}{dt} \in Au \\ &u(0) = u_0. \end{aligned}$$

See Chapter 2 or CRANDALL [1976] and EVANS [1977] for survey papers on this subject.

Here, we first put the nonlinear diffusion Problem (I) into the abstract framework of Problem (II) in the case $X = L_1(\mathbb{R})$. Then we show, under appropriate conditions on u_0 and ϕ, that the semigroup solution $u(t)$ is a weak solution of of Problem (I). More precisely, if

$$u(t,x) := u(t)(x) = T_A(t)u_0(x), \quad t > 0, x \in \mathbb{R},$$

then the function u satisfies

$$u_t - (\phi(u))_{xx} \text{ in } \mathfrak{D}'((0, \infty) \times \mathbb{R}),$$

(where $\mathcal{D}'((0,\infty)\times\mathbb{R})$ denotes the space of distributions on $(0,\infty)\times\mathbb{R}$) and

$$\lim_{t\downarrow 0} u(t) = u_0 \text{ in } L_1(\mathbb{R}).$$

We shall consider the question of continuous dependence of u on ϕ and finally we discuss some smoothness properties of the semigroup solution of the Porous Media Equation.

REMARK 4.3. It follows from Proposition 4.2 that there is no hope that the theory related to m-dissipative operators is applicable to evolution problems whose solutions satisfy a mass conservation property but no comparison principle.

To give the abstract formulation of the equation $u_t-(\phi(u))_{xx}=0$ in the Banach space $L_1(\mathbb{R})$, we define the operator $A_\phi:\mathcal{D}(A_\phi)\subseteq L_1(\mathbb{R})\to L_1(\mathbb{R})$, which is formally given by $A_\phi(u)=(\phi(u))''$, as follows. Set

$$\mathcal{D}(A_\phi)=\{u\in L_1(\mathbb{R}): \exists w\in L_1^{loc}(\mathbb{R}) \text{ such that } w(x)\in\phi(u(x)) \text{ a.e.}$$
$$\text{and } w''\in L_1(\mathbb{R}) \text{ in the distributional sense}\},$$

and define

$$A_\phi u = w'', \text{ for all } u\in\mathcal{D}(A_\phi).$$

We first show that the operator A_ϕ is m-dissipative in $L_1(\mathbb{R})$. This means that we have to prove that

(*) $E_\lambda^{A_\phi}$ is a contraction in $L_1(\mathbb{R})$ for all $\lambda>0$, i.e. when $u_i\in\mathcal{D}(A_\phi)$ and $f_i\in L_1(\mathbb{R})$ satisfy $u_i-\lambda A_\phi u_i=f_i, i=1,2$, then we must prove that

$$\|u_1-u_2\|_1\leqslant\|f_1-f_2\|_1.$$

(**) $\mathcal{R}(I-\lambda A_\phi)=L_1(\mathbb{R})$. Thus we must prove that for any $f\in L_1(\mathbb{R})$, there exists a $u\in\mathcal{D}(A_\phi)$ such that

$$u-A_\phi u=f.$$

In solving this equation in $L_1(\mathbb{R})$, we look for functions u and w such that

$$u-w''=f \text{ in } \mathcal{D}'(\mathbb{R}) \text{ and } w(x)\in\phi(u(x)) \text{ a.e. on } \mathbb{R}.$$

It is convenient to work with the inverse of ϕ, that is $\beta=\phi^{-1}$. This is also a maximal monotone graph in $\mathbb{R}^2$. In terms of β we have

$$u-w''=f \text{ in } \mathcal{D}'(\Omega) \text{ and } u(x)\in\beta(w(x)) \text{ a.e. on } \mathbb{R},$$

or simply

$$-w''+\beta(w)\ni f \text{ on } \mathbb{R}. \tag{P}$$

This equation is studied in the following two sections. We consider the case where $0\in\beta(0)$ with $0\in\operatorname{int}\mathcal{R}(\beta)$ and the case where $\mathcal{R}(\beta)\subseteq(0,\infty)$.

2. THE CASE $0\in\beta(0)$ AND $0\in\text{int}\,\mathcal{R}(\beta)$
Instead of Problem (P) consider the perturbed problem

$$\epsilon w - w'' + \beta(w)\ni f, \quad \text{for } \epsilon>0. \tag{P_ϵ}$$

Let $Aw:=w''-\epsilon w$ for $w\in\mathcal{D}(A)=\{u\in L_1(\mathbb{R})\colon u''\in L_1(\mathbb{R})\}$. Then the linear operator A satisfies the properties:

(I) A is a densely defined and m-dissipative in $L_1(\mathbb{R})$
(II) for any $\lambda>0$ and $f\in L_1(\mathbb{R})\cap L_\infty(\mathbb{R})$

$$\sup_{\mathbb{R}}(I-\lambda A)^{-1}f\leqslant \max\{0,\sup_{\mathbb{R}} f\}$$

(III) $\epsilon\|u\|_1\leqslant\|Au\|_1$, for all $u\in\mathcal{D}(A)$.

THEOREM 4.4. *Let A be a linear operator which satisfies I, II and III. For every $f\in L_1(\mathbb{R})$, there exists a unique $w\in\mathcal{D}(A)$ such that*

$$-Aw(x)+\beta(w(x))\ni f(x) \text{ a.e. on } \mathbb{R}. \tag{4.5}$$

Moreover if $f,\hat f\in L_1(\mathbb{R})$ and $w,\hat w$ are corresponding solutions of (4.5) then

$$\|(f+Aw)-(\hat f+A\hat w)\|_1\leqslant\|f-\hat f\|_1, \tag{4.6}$$

and

$$\|A(w-\hat w)\|_1\leqslant 2\|f-\hat f\|_1. \tag{4.7}$$

For the proof we need the following lemma.

LEMMA 4.5. *Let γ be a maximal monotone graph in $\mathbb{R}^2$ with $0\in\gamma(0)$. Suppose A satisfies I and II. Let $1\leqslant p\leqslant\infty$ and let $w\in\mathcal{D}(A)\cap L_p(\mathbb{R})$ with $Aw\in L_{p'}(\mathbb{R})$. Let $g\in L_p(\mathbb{R})$ such that $g(x)\in\gamma(w(x))$. Then*

$$\int_{\mathbb{R}} Aw(x)g(x)dx \leqslant 0.$$

PROOF OF THEOREM 4.4. We first prove the inequalities (4.6) and (4.7) from which the uniqueness follows when using property III. We write $f\in Bu$ if the conditions $f,u\in L_1(\mathbb{R})$ and $f(x)\in\beta(u(x))$ a.e. on $\mathbb{R}$ are satisfied. Let $g=f+Aw\in B(w)$ and $\hat g=\hat f+A\tilde w\in B(\hat w)$. Then $g-\hat g=f-\hat f+A(w-\hat w)$. Define the function

$$h(x)=\begin{cases}+1, & \text{if } g(x)>\hat g(x) \text{ or } w(x)>\hat w(x)\\ 0, & \text{if } g(x)=\hat g(x) \text{ or } w(x)=\hat w(x)\\ -1, & \text{if } g(x)<\hat g(x) \text{ or } w(x)<\hat w(x).\end{cases}$$

Then h is well-defined on $\mathbb{R}$ and $h(x)\in\text{sign}[w(x)-\hat w(x)]$, where $\text{sign}[r]=1$ for $r>0$, $\text{sign}[r]=-1$ for $r<0$, and $\text{sign}[0]=[-1,1]$. Apply Lemma 4.5 for $p=1$, $p'=\infty$, $w=w-\hat w$ and $\gamma=$ sign. Then

$$(h,A(w-\hat{w}))\leqslant 0, \tag{4.8}$$

where $(u,v)=\int_{\mathbb{R}} u(x)v(x)dx$. From (4.8) we get, since $h(x)\in \operatorname{sign}[g(x)-\hat{g}(x)]$,

$$\|g-\hat{g}\|_1 = (h,g-\hat{g})\leqslant(f-\hat{f},h)\leqslant\|f-\hat{f}\|_1.$$

Clearly, (4.7) follows immediately. □

REMARK 4.6. In proving the estimate (4.6) we only used properties I and II of A, which also hold when $\epsilon=0$. This gives precisely the dissipativity of the operator A_ϕ.

Next we consider the existence of solutions of (P_ϵ). First we show that

$\mathcal{R}(-A+B)$ *is closed.*

Let $\{f_n\}\subset\mathcal{R}(-A+B)$ such that $f_n\to f$ in $L_1(\mathbb{R})$. Then, there exists a sequence $\{w_n\}\subset\mathcal{D}(A)$ such that $-Aw_n+Bw_n\ni f_n$. Using inequality (4.7) and III, we find that both $\{w_n\}$ and $\{Aw_n\}$are Cauchy sequences in $L_1(\mathbb{R})$. By the closedness of $A, w_n\to w$ and $Aw_n\to Aw$. Since $f_n+Aw_n\in Bw_n$ and β is maximal, one finds $f+Aw\in Bw$. Thus $f\in\mathcal{R}(-A+B)$.

$\mathcal{R}(-A+B)$ *is dense in* $L_1(\mathbb{R})$.

Consider the Yosida-approximation of β:

$$\beta_\lambda = \frac{1}{\lambda}(I-(I+\lambda\beta)^{-1}),\quad \lambda>0.$$

For $\mu>0$, consider the equation

$$\mu w-Aw+\beta_\lambda(w)=f,\ f\in L_1(\mathbb{R}). \tag{4.9}$$

Write (4.9) as

$$\mu\lambda w-\lambda Aw+w=\lambda f+(I+\lambda\beta)^{-1}w,$$

or

$$w=\frac{1}{1+\lambda\mu}(I-\frac{\lambda}{1+\lambda\mu}A)^{-1}(\lambda f+(I+\lambda\beta)^{-1}w). \tag{4.10}$$

The operator on the right is the product of two contractions in $L_1(\mathbb{R})$ and a number less than 1. Thus it is a strict contraction and (4.10) has a fixed point $w\in\mathcal{D}(A)$. If, in addition, $f\in L_1(\mathbb{R})\cap L_\infty(\mathbb{R})$ then we use II and a fixed point argument in $L_1(\mathbb{R})\cap L_\infty(\mathbb{R})$ to find that $w, Aw\in L_\infty(\mathbb{R})$.

Let $f\in L_1(\mathbb{R})\cap L_\infty(\mathbb{R})$ and denote the solution of (4.9) by w_λ. Then from (4.10)

$$\|w_\lambda\|\leqslant\frac{1}{(1+\lambda\mu)}\{\lambda\|f\|+\|w_\lambda\|\},$$

or

$$\|w_\lambda\| \leqslant \frac{1}{\mu}\|f\|, \tag{4.11}$$

where $\|\cdot\|$ denotes the norm in $L_1(\mathbb{R}) \cap L_\infty(\mathbb{R})$. Multiply equation (4.9) by sign (w_λ). Then

$$\mu|w_\lambda| - Aw_\lambda \operatorname{sign}(w_\lambda) + |\beta_\lambda(w_\lambda)| \leqslant |f|.$$

Using Lemma 4.5 gives $(Aw_\lambda, \operatorname{sign}(w_\lambda)) \leqslant 0$, and thus

$$\|\beta_\lambda(w_\lambda)\|_1 \leqslant \|f\|_1, \ \forall \lambda > 0.$$

Next multiply equation (4.9) by $|\beta_\lambda(w_\lambda)|^{p-2}\beta_\lambda(w_\lambda)$. Then

$$\mu \cdot |\beta_\lambda(w_\lambda)|^{p-2} \cdot \beta_\lambda(w_\lambda) w_\lambda - Aw_\lambda |\beta_\lambda(w_\lambda)|^{p-2} \beta_\lambda(w_\lambda) + |\beta_\lambda(w_\lambda)|^p$$
$$= f \cdot |\beta_\lambda(w_\lambda)|^{p-2} \cdot \beta(w_\lambda).$$

Taking the integral in this equation gives nonnegative numbers for the first two terms. Therefore

$$\int_{\mathbb{R}} |\beta_\lambda(w_\lambda)|^p \leqslant \int_{\mathbb{R}} |f| \cdot |\beta_\lambda(w_\lambda)|^{p-1} \leqslant \left(\int_{\mathbb{R}} |f|^p\right)^{1/p} \left(\int_{\mathbb{R}} |\beta_\lambda(w_\lambda)|^p\right)^{1-1/p}.$$

Thus

$$\|\beta_\lambda(w_\lambda)\|_p \leqslant \|f\|_p, \ \forall p \geqslant 1. \tag{4.12}$$

In particular $\{w_\lambda\}$ and $\{\beta_\lambda(w_\lambda)\} \in L_2(\mathbb{R})$. We show that they are both Cauchy sequences. Subtract the equations for w_α and w_δ and multiply by $(w_\alpha - w_\delta)$ This gives:

$$\mu(w_\alpha - w_\delta)^2 - A(w_\alpha - w_\delta)(w_\alpha - w_\delta)$$
$$+ (\beta_\alpha(w_\alpha) - \beta_\delta(w_\delta))(w_\alpha - w_\delta) = 0.$$

Using Lemma 4.5 once more we get:

$$\mu\|w_\alpha - w_\delta\|_2^2 + (\beta_\alpha(w_\alpha) - \beta_\delta(w_\delta), w_\alpha - w_\delta) \leqslant 0.$$

We want to rewrite the last term. Since

$$w_\alpha - w_\delta = \{w_\alpha - (I + \alpha\beta)^{-1} w_\alpha\}$$
$$+ \{(I + \alpha\beta)^{-1} w_\alpha - (I + \delta\beta)^{-1} w_\delta\} - \{w_\delta - (I + \delta\beta)^{-1} w_\delta\},$$

and

$$((I + \alpha\beta)^{-1} w_\alpha - (I + \delta\beta)^{-1} w_\delta)(\beta_\alpha(w_\alpha) - \beta_\delta(w_\delta)) \geqslant 0,$$

it follows that

$$\mu\|w_\alpha - w_\delta\|_2^2 + (\beta_\alpha(w_\alpha) - \beta_\delta(w_\delta), \alpha\beta_\alpha(w_\alpha) - \delta\beta_\delta(w_\delta)) \leqslant 0.$$

Hence $\{w_\lambda\}$ is a Cauchy sequence in $L_2(\mathbb{R})$. One can also show that $\{\beta_\lambda(w_\lambda)\}$ is a Cauchy sequence in $L_2(\mathbb{R})$. Therefore $w_\lambda \to w$ and $\beta_\lambda(w_\lambda) \to g$ in $L_2(\mathbb{R})$. Since the estimates (4.11) and (4.12) are uniform in $\lambda > 0$, it follows that

$$w, g \in L_1(\mathbb{R}) \cap L_\infty(\mathbb{R}) \text{ and } g \in \beta(w),$$

because β is maximal. Thus we have $(\mu I-A)w_\lambda\to f-g$ in $L_2(\mathbb{R})$ as $\lambda\downarrow 0$. Let $v=(\mu I-A)^{-1}(f-g)$. Then $v\in\mathfrak{D}(A)\cap L_\infty(\mathbb{R})$ and

$$(\mu I - A)(v - w_\lambda)\to 0 \text{ in } L_2(\mathbb{R}).$$

Multiplying this expression by $(v-w_\lambda)$ and using Lemma 4.5 again, gives

$$\mu\|v - w_\lambda\|_2^2 + \text{ pos. contribution}\to 0, \quad \text{as } \lambda\downarrow 0.$$

Thus $w_\lambda\to v$ in $L_2(\mathbb{R})$ and $v=w$. The definition of v gives

$$\mu w - Aw + g = f \text{ or } \mu w - Aw + \beta(w)\ni f.$$

Next we let $\mu\to 0$. Let $f\in L_1(\mathbb{R})$ and let $f^\mu\to f$ in $L_1(\mathbb{R})$ as $\mu\downarrow 0$, where $f^\mu\in L_1(\mathbb{R})\cap L_\infty(\mathbb{R})$. Consider the problem

$$\mu w^\mu + Aw^\mu + Bw^\mu\ni f^\mu.$$

Use estimate (4.7) with $\hat{w}=0$ and $\beta\to\beta+\mu I$. This gives

$$\epsilon\|w^\mu\|_1\leqslant\|Aw^\mu\|_1\leqslant 2\|f^\mu\|_1.$$

Hence $\mu w^\mu\to 0$ in $L_1(\mathbb{R})$. Therefore $f=\lim(f^\mu - w^\mu)\in\lim(-Aw^\mu+Bw^\mu)$. This shows that f belongs to the closure of $-A+B$, which completes the proof. □

Below we prove Lemma 4.5 and give some interesting estimates. We first introduce the set

$$J_0 = \{j:\mathbb{R}\to[0,\infty]: j \text{ is convex , lower semicontinuous and } j(0)=0\}.$$

Consider the following lemma.

LEMMA 4.7. *Let $L:L_1(\mathbb{R})\to L_1(\mathbb{R})$ satisfy*

$$\|Lu - Lv\|_1\leqslant\|u - v\|_1 \text{ for } u,v\in L_1(\mathbb{R}),$$

and

$$\min\{0,\inf u\}\leqslant Lu(x)\leqslant\max\{0,\sup u\}, \text{ for } u\in L_1(\mathbb{R}),$$

and let $j\in J_0$. Then

$$\int_{\mathbb{R}} j(Lu(x))dx\leqslant\int_{\mathbb{R}} j(u(x))dx, \tag{4.13}$$

for all $u\in L_1(\mathbb{R})$ such that $j\circ y\in L_1(\mathbb{R})$.

PROOF. Let $t\geqslant 0$ and consider the functions $j_1,j_2\in J_0$ given by

$$j_1(r)=(r-t)^+ \text{ and } j_2(r)=(-r-t)^+.$$

Further let $y(x)=\min\{u(x),t)\}$ for $u\in L_1(\mathbb{R})$. By the maximum principle for L we have $Ly(x)\leqslant t$. Thus

$$(Lu(x)-t)^+\leqslant(Lu(x) - Ly(x))^+\leqslant|Lu(x) - Ly(x)|.$$

Integration over $\mathbb{R}$ and using the contraction property gives:

$$\int_{\mathbb{R}}(Lu(x)-t)^{+}\leqslant\int_{\mathbb{R}}|u(x)-y(x)| = \int_{\mathbb{R}}(u(x)-t)^{+},$$

which is (4.13) with $j=j_1$. Similarly one deduces inequality (4.13) with $j=j_2$. Combined, they give

$$\int_{\mathbb{R}}[t(Lu(x)-t)]^{+}\leqslant\int_{\mathbb{R}}[t(u(x)-t)]^{+}. \tag{4.14}$$

Next consider convex combinations of j_1 and j_2. Observe that any smooth $j\in J_0$ can be written as

$$j(r)=\int_{\mathbb{R}}\frac{j''(t)}{|t|}[t(r-t)]^{+}\,dt.$$

Then multiplying (4.14) by $j''(t)/|t|$, integrating, and using Fubini's Theorem gives (4.13) for any smooth $j\in J_0$. When $j\in J_0$ is arbitrary, we construct a smooth approximation $j_\lambda\in J_0$ such that $j_\lambda\uparrow j$ when $\lambda\to 0$. For instance, let

$$j_\lambda(r)=\inf_t\{\frac{1}{2\lambda}|r-t|^2+j(t)\}.$$

EXAMPLE 4.8. Let $j\in J_0$ be given by:

$$j(r)=\begin{cases}0, & r\leqslant 0\\ \infty, & r>0.\end{cases}$$

Then

$$j_\lambda(r)=\begin{cases}0, & r\leqslant 0\\ \dfrac{r^2}{2\lambda}, & r>0.\end{cases}$$

Now

$$\int_{\mathbb{R}}j_\lambda(Lu(x))dx\leqslant\int_{\mathbb{R}}j_\lambda(u(x))\leqslant\int_{\mathbb{R}}j(u(x)).$$

This gives $j(Lu)\in L_1(\mathbb{R})$ and inequality (4.13). □

PROOF OF LEMMA 4.5. Let $j\in J_0$ such that $\partial_j=\gamma$ and let $L_\lambda=(I-\lambda A)^{-1}$. Then L_λ satisfies the contraction property and the maximum principle from Lemma 4.7. Let $u\in\mathfrak{D}(A)\cap L_p(\mathbb{R})$ and $g\in L_{p'}(\mathbb{R})$ such that $g(x)\in\partial j(u(x))$ a.e. We have:

$$\begin{aligned}j(L_\lambda u(x))-j(u(x))&\geqslant g(x)[L_\lambda u(x)-u(x)]\\ &=\lambda g(x)L_\lambda(Au(x)) \text{ a.e.,}\end{aligned} \tag{4.15}$$

and also

$$j(0)-j(u(x))\geqslant g(x)(0-u(x)) \text{ a.e.,}$$

or

$$0\leqslant j(u(x))\leqslant g(x)u(x) \text{ a.e..}$$

This shows that $j\circ u\in L_1(\mathbb{R})$. Applying Lemma 4.7 gives

$$\int_{\mathbb{R}} j(L_\lambda u(x))\leqslant \int_{\mathbb{R}} j(u(x))dx,$$

and using this in (4.15) results in

$$\int_{\mathbb{R}} g(x)L_\lambda(Au(x))dx\leqslant 0.$$

To complete the proof we distinguish:

$p=1$. Then $g\in L_\infty(\mathbb{R})$ and $Au\in L_1(\mathbb{R})$. Since A is m-dissipative in $L_1(\mathbb{R})$ and $\mathfrak{D}(A)=L_1(\mathbb{R})$ we have $L_\lambda(Au)=E_\lambda^A(Au)\rightarrow Au$ in $L_1(\mathbb{R})$ when $\lambda\downarrow 0$. This establishes the result for this case.

$1<p<\infty$. Then $Au\in L_p(\mathbb{R})$ and $g\in L_{p'}(\mathbb{R})$. Apply Lemma 4.7 with $j(r)=|r|^p$. This gives $\|L_\lambda Au\|_p\leqslant\|Au\|_p, 1<p<\infty$. Thus $L_\lambda Au\rightarrow Au$ weakly in $L_p(\mathbb{R})$, which completes the proof.

$p=\infty$. Using property II gives $\|L_\lambda Au\|_\infty\leqslant\|Au\|_\infty$. Then there is a subsequence $\{\lambda_k\}$ such that $L_{\lambda_k}Au\rightarrow Au$ a.e.. Applying Lebesgue's dominated convergence theorem gives the desired result. □

We proof two more estimates before returning to Problem (P).

Proposition 4.9. *Let $j\in J_0$ and let $f\in L_1(\mathbb{R})$ such that $j\circ f\in L_1(\mathbb{R})$. Let w be a solution of (P_ϵ). Then*

$$\int_{\mathbb{R}} j(f+Aw)\leqslant \int_{\mathbb{R}} j(f).$$

Proof. Define the map $L:L_1(\mathbb{R})\rightarrow L_1(\mathbb{R})$ by $Lf=f+Aw$, where w is the solution of (P_ϵ). L is a contraction in $L_1(\mathbb{R})$. Thus the proposition follows from Lemma 4.7 as soon as we prove that L satisfies the maximum principle. Let $k=\max\{0,\sup f\}$ and let $g=f+Aw$. We must show that $g(x)\leqslant k$ a.e. on $\mathbb{R}$. We assume that $k\in\mathfrak{R}(\beta)$, otherwise this inequality is obvious. Thus suppose there exists an $l\geqslant 0$ such that $k\in\beta(l)$. Let

$$h(x)=\begin{cases}1, & \text{if } u(x)>l \text{ or } g(x)>k\\ 0, & \text{elsewhere.}\end{cases}$$

Clearly $h(x)\in\gamma(u(x))$ where $\gamma(r)=[\text{sign}(r-l)]^+$. Using Lemma 4.5 with $p=1$, gives

$$0\leqslant(-Aw,h)=(f-g,h)\leqslant(k-g,h).$$

But $g(x)\geqslant k$ if $u(x)>l$. Hence $g(x)\leqslant k$ a.e.. Similarly $g(x)\geqslant\min\{0,\inf f\}$. □

Observe that it follows from Proposition 4.9 and its proof that $\|f-Aw\|_p\leq\|f\|_p$ when $f\in L_1(\mathbb{R})\cap L_p(\mathbb{R})$ and $1\leq p\leq\infty$.

PROPOSITION 4.10. *Let $f,\hat{f}\in L_1(\mathbb{R})$ and let $w,\hat{w}$ be the corresponding solutions of (P_ϵ). Then*

$$\|[(f+Aw)-(\hat{f}+A\hat{w})]^+\|_1\leq\|[f-\hat{f}]^+\|_1.$$

Further, if $f\leq\hat{f}$, then $f+Aw\leq\hat{f}+A\hat{w}$ and $w\leq\hat{w}$ a.e. on $\mathbb{R}$.

PROOF. Let $g=f+Aw$ and $\hat{g}=\hat{f}+A\hat{w}$. Then $g\in Bw$ and $\hat{g}\in B\hat{w}$. Let h be the characteristic function of the set E, defined by

$$E=\{\hat{w}(x)>w(x)\}\cup\{\hat{g}(x)>g(x)\}.$$

Then multiply the equation $\hat{g}-g-A(\hat{w}-w)=\hat{f}-f$ by h and integrate. Again we use Lemma 4.5 to find

$$\int_E(\hat{g}-g)\leq\int_E(\hat{f}-f)=\int_{\mathbb{R}}[\hat{f}-f]^+.$$

By the monotonicity, we have $\hat{g}(x)\leq g(x)$ on $\mathbb{R}\setminus E$. Thus

$$\int_E\hat{g}-g=\int_{\mathbb{R}}[\hat{g}-g]^+\leq\int_{\mathbb{R}}[\hat{f}-f]^+.$$

Next, let $f\leq\hat{f}$. Then clearly $f+Aw\leq\hat{f}+A\hat{w}$. Let $w_\mu,\hat{w}_\mu$ be the solutions of

$$\mu w_\mu+Aw_\mu+\beta(w_\mu)\ni f \text{ and } \mu\hat{w}_\mu+A\hat{w}_\mu+\beta(\hat{w}_\mu)\ni\hat{f}.$$

We just proved that $g_\mu=f+Aw_\mu\leq\hat{f}+A\hat{w}_\mu=\hat{g}_\mu$, where $g_\mu\in\beta(w_\mu)+\mu w_\mu$ and $\hat{g}_\mu\in\beta(\hat{w}_\mu)+\mu\hat{w}_\mu$. The monotonicity of β gives $w_\mu\leq\hat{w}_\mu$. Letting $\mu\to0$ gives $w\leq\hat{w}$ a.e. □

Next we let ϵ tend to zero, and return to Problem (P):

$$-w''+\beta(w)\ni f, \text{ on } \mathbb{R}. \tag{P}$$

We first give some definitions.

DEFINITION 4.11. A function $w:\mathbb{R}\to\mathbb{R}$ is called a *solution* of the problem (P) if: (i) $w\in L_1^{loc}(\mathbb{R})$, (ii) $w''\in L_1(\mathbb{R})$ in sense of distributions, (iii) $w''(x)+f(x)\in\beta(w(x))$ a.e. on $\mathbb{R}$.

DEFINITION 4.12. The problem (P) is said to be *well-posed* in a subset $\mathfrak{L}\subseteq L_1^{loc}(\mathbb{R})$ if:

(i) for every $f\in L_1(\mathbb{R})$, (P) has at least one solution $w\in\mathfrak{L}$. We set

$$G_\beta(f)=\{w\in\mathfrak{L}: w \text{ is a solution of } (P)\}.$$

(ii) $L_\beta f:=\{f+w'':w\in\mathcal{G}_\beta(f)\}$ has exactly one element for $f\in L_1(\mathbb{R})$.
(iii) $\int_{\mathbb{R}}j(L_\beta f)\leq\int_{\mathbb{R}}j(f)$, for every $f\in L_1(\mathbb{R})$ and $j\in J_0$.

(iv) $\int_{\mathbb{R}}[L_\beta f - L_\beta g]^+ \leqslant \int_{\mathbb{R}}[f-g]^+$, for $f,g \in L_1(\mathbb{R})$.

REMARKS 4.13.
a) (iii) implies, taking $j(r)=|r|$, $\|L_\beta f\|_1 \leqslant \|f\|_1$.
b) (iv) implies, taking $g \geqslant f$, $L_\beta g \geqslant L_\beta f$.
c) (iv) implies, interchanging g and f, $\|L_\beta f - L_\beta g\|_1 \leqslant \|f-g\|_1$.
Thus if (P) is well-posed in $\mathfrak{L}$, then L_β is an *order preserving contraction* on $L_1(\mathbb{R})$.

In the first part of this section we proved that for any $f \in L_1(\mathbb{R})$, there exists a unique $w_\epsilon \in L_1(\mathbb{R})$ such that $w_\epsilon'' \in L_1(\mathbb{R})$ and

$$-w_\epsilon'' + \epsilon w_\epsilon + \beta(w_\epsilon) \ni f \text{ a.e. on } \mathbb{R}.$$

The operator $Aw = w''$ satisfies conditions I and II of Theorem 4.4. This is enough to prove Propositions 4.9 and 4.10. Therefore, in the perturbed case, (iii) and (iv) hold with β replaced by $\beta+\epsilon I$. Thus (P_ϵ), which is (P) with β replaced by $\beta+\epsilon I$, is well-posed in $\mathfrak{L}=L_1(\mathbb{R})$.

We consider the convergence as $\epsilon \downarrow 0$. We have the following lemma.

LEMMA 4.14. *Let $f \in L_1(\mathbb{R})$ and w_ϵ the corresponding solution of (P_ϵ). Let $u_\epsilon = w_\epsilon'' + f = L_{\beta+\epsilon I} f$. Suppose that $\{w_\epsilon\}_{\epsilon>0}$ is uniformly bounded in $L_\infty(\mathbb{R})$. Then $\{(w_\epsilon, u_\epsilon)\}_{\epsilon>0}$ is precompact in $(L_1^{loc}(\mathbb{R}))^2$. In addition, if $\epsilon_k \downarrow 0$ and $(w_{\epsilon_k}, u_{\epsilon_k}) \to (w,u)$, then*

$$u = w'' + f \in L_1(\mathbb{R}) \text{ and } u \in \beta(w).$$

Also $\int_{\mathbb{R}} j(u) \leqslant \int_{\mathbb{R}} j(f)$, for every $j \in J_0$.

PROOF. Observe that
(i) $L_{\beta+\epsilon I}$ is a contraction on $L_1(\mathbb{R})$,
(ii) $L_{\beta+\epsilon I}$ is translation invariant,
(iii) $L_{\beta+\epsilon I}(0)=0$.
From these observations it follows that $u_\epsilon = L_{\beta+\epsilon I} f$ satisfies

$$\|u_\epsilon\|_1 \leqslant \|f\|_1$$

and

$$\int_{\mathbb{R}} |u_\epsilon(x+h) - u_\epsilon(x)| \leqslant \int_{\mathbb{R}} |f(x+h) - f(x)|, \quad \forall h \in \mathbb{R}.$$

Since translation is a continuous operation on $L_1(\mathbb{R})$, we get that $\{u_\epsilon\}_{\epsilon>0}$ is precompact in $L_1^{loc}(\mathbb{R})$. Also $\int_{\mathbb{R}} j(u_\epsilon) \leqslant \int_{\mathbb{R}} j(f)$. If $\epsilon_k \to 0$ and $u_{\epsilon_k} \to u$ in $L_1^{loc}(\mathbb{R})$ then it follows from Fatou's lemma that $\int_{\mathbb{R}} j(u) \leqslant \int_{\mathbb{R}} j(f)$. The functions w_ϵ satisfy for any $\epsilon>0$:

$$\epsilon\|w_\epsilon\|_1\leqslant\|f\|_1 \text{ and } \|w_\epsilon''\|_1\leqslant 2\|f\|_1.$$

Then from

$$w_\epsilon'(x)-w_\epsilon'(y)=\int_y^x w_\epsilon''(s)ds \text{ and } w_\epsilon''\in L_1(\mathbb{R}),$$

one deduces that

$$\|w_\epsilon'\|_\infty\leqslant\|w_\epsilon''\|_1\leqslant 2\|f\|_1.$$

Thus the set $\{w_\epsilon\}_{\epsilon>0}$ is equicontinuous on $\mathbb{R}$. Moreover, by assumption, $\{w_\epsilon\}_{\epsilon>0}$ is bounded in $L_\infty(\mathbb{R})$. Thus there exist $\epsilon_k\to 0$ and $\{w_{\epsilon_k}\}$ such that $w_{\epsilon_k}\to w$ uniformly on compact subsets of $\mathbb{R}$. Now since $w_{\epsilon_k}''=u_{\epsilon_k}-f$, $(u_{\epsilon_k}-f,\phi)\to(u-f,\phi)$ and $(w_{\epsilon_k}'',\phi)=(w_{\epsilon_k},\phi'')\to(w,\phi'')$ as $\epsilon_k\downarrow 0$ for all $\phi\in C_0^\infty(\mathbb{R})$, it follows that

$$w''=u-f, \text{ in } \mathcal{D}'(\mathbb{R}).$$

Moreover, since $u_\epsilon\in\beta(w_\epsilon)$ and β is maximal, one has $u\in\beta(w)$ a.e. □

Remark 4.15. If we know that $L_\beta f=w''+f$, w a solution of (P), is unique, then $u_\epsilon=L_{\beta+\epsilon I}f\to L_\beta f$. Because

$$\int_{\mathbb{R}}[L_{\beta+\epsilon I}f-L_{\beta+\epsilon I}g]^+\leqslant\int_{\mathbb{R}}[f-g]^+, \text{ for all } \epsilon>0,$$

we use Fatou's lemma again to obtain (iv).

We will show below that two solutions of (P) in $L_1^{loc}(\mathbb{R})$ differ by a constant. Then $L_\beta f$ is still unique, so that the above holds.

First consider some a priori estimates on the solution w of (P). Since $w''\in L_1(\mathbb{R})$, it follows that $w'\in L_\infty(\mathbb{R})$ and $w'(\pm\infty)$ exist. Now suppose, e.g. $w'(+\infty)\neq 0$. Then $|w(x)|\to\infty$ for $x\to\infty$. However, $0\in$int $\mathcal{R}(\beta)$, which contradicts the fact that $w''+f\in\beta(w)$ and $w''+f\in L_1(\mathbb{R})$. Thus $w'(\pm\infty)=0$ and $\|w'\|_\infty\leqslant\|w''\|_1$.

Let $j\in J_0$ be such that $\partial j=\beta$. Then $j(w)'=uw'$ a.e. and $\|j(w)'\|_1\leqslant\|u\|_1\|w'\|_\infty$. Thus $j(w)\in L_\infty(\mathbb{R})$ and $j(w)(\pm\infty)$ exist. In particular $j(w)(\pm\infty)=0$. Then $\|j(w)\|_\infty\leqslant\|u\|_1\|w'\|_\infty$. But on the other hand $j(r)\to\infty$ as $|r|\to\infty$ since $0\in$int $\mathcal{R}(\beta)$. Thus $w\in L_\infty(\mathbb{R})$.
If $w''\in L_1(\mathbb{R})$ and $w\in L_\infty(\mathbb{R})$ we have for any $p\in L_\infty(\mathbb{R})\cap C^1(\mathbb{R})$, nondecreasing,

$$\int_{\mathbb{R}}p'(w(x))\{w'(x)\}^2+\int_{\mathbb{R}}p(w(x))w''(x)=0. \tag{4.16}$$

Let w_1,w_2 be two solutions of Problem (P). Then

$$u_1=w_1''+f,\ u_1\in\beta(w_1), \tag{4.17}$$

$$u_2 = w_2'' + f,\ u_2 \in \beta(w_1). \tag{4.18}$$

By the monotonicity of β, $(u_1 - u_2)(w_1 - w_2) \geq 0$ on $\mathbb{R}$. Using (4.17) and (4.18) gives $(w_1'' - w_2'')(w_1 - w_2) \geq 0$ and also $(w_1'' - w_2'')p(w_1 - w_2) \geq 0$ if we choose $p(0)=0$. The first term of (4.16) now gives $w_1' - w_2' = 0$, which implies that $w_1 - w_2 = c$ on $\mathbb{R}$.

Using a similar argument to the solution w_ϵ of (P_ϵ) gives:

$$\|\frac{\epsilon}{2}w_\epsilon^2 + j(w_\epsilon)\|_\infty \leq \|u_\epsilon\|_1 \|w_\epsilon'\|_\infty \leq 2\|f\|_1^2.$$

This shows that $\{w_\epsilon\}$ is uniformly bounded on $L_\infty(\mathbb{R})$.

Conclusion: (P) is well-posed in $L_1^{loc}(\mathbb{R})$.

We summarize the estimates for the solution w of (P).

$$\|w'\|_\infty \leq \|w''\|_1 \leq 2\|f\|_1.$$

$$\|j(w)\|_\infty \leq \|u\|_1 \|w'\|_\infty \leq 2\|f\|_1^2, \quad \text{if } \partial j = \beta.$$

Let $f, \hat{f} \in L_1(\mathbb{R})$ and $w \in G_\beta(f)$, $\hat{w} \in G_\beta(\hat{f})$. Because $(w - \hat{w})'(\pm\infty) = 0$, we have

$$\|w' - \hat{w}'\|_\infty \leq \|w'' - \hat{w}''\|_1 \leq 2\|f - \hat{f}\|_1.$$

Finally, observe that

$$\int_{\mathbb{R}} L_\beta f = \int_{\mathbb{R}} f, \text{ for } f \in L_1(\mathbb{R}).$$

3. The case $\mathcal{R}(\beta) \subseteq (0, \infty)$

This case arises when studying the nonlinear diffusion equation

$$u_t = (\log u)_{xx}, \quad u > 0.$$

Then $\phi(u) = \log u$ and $\beta(w) = e^w$. Consider again Problem (P) which we denote here by

$$-w'' + \beta(w) \ni f, \text{ on } \mathbb{R}. \tag{P_f}$$

Based on the knowledge obtained in the previous section we define

Definition 4.16. A function $w: \mathbb{R} \to \mathbb{R}$ is a *solution* of (P_f) for $f \in L_1(\mathbb{R})$ if
(i) w, w' are locally absolutely continuous on $\mathbb{R}$,
(ii) $f(x) + w''(x) \in \beta(w(x))$ a.e. on $\mathbb{R}$,
(iii) $w'(\pm\infty) = 0$.

We also introduce

DEFINITION 4.17. *Minimal Section:* $\beta^0(r)=\inf\{s : s\in\beta(r)\}$ with $r\in\mathfrak{D}(\beta)$.

REMARK 4.18. $\beta^0(r)$ is a nondecreasing function on $\mathfrak{D}(\beta)$.

THEOREM 4.19. *The following statements are equivalent.*
(i) *For $f\in L_1(\mathbb{R})$ the problem (P_f) has a solution exactly when $\int f>0$.*
(ii) *There exists some $f\in L_1(\mathbb{R})$ for which (P_f) has a solution.*
(iii) *There exists some $a\in\mathbb{R}$ such that*

$$(-\infty,a]\subseteq\mathfrak{D}(\beta) \text{ and } \int_{-\infty}^{a}\beta^0(r)dr<\infty.$$

PROOF. We subsequently show that (i)⇒(iii)⇒(ii)⇒(i). First define $L_1(\mathbb{R})_+=\{f\in L_1(\mathbb{R}):\int f>0\}$. Observe that $f\in L_1(\mathbb{R})_+$ is a necessary condition for the existence of solutions of (P_f). This follows from $w''+f>0$ a.e., and

$$0<\|f+w''\|_1=\int_{\mathbb{R}}(f+w'')=\int_{\mathbb{R}}f<\|f\|_1. \tag{4.19}$$

Moreover we have

$$\|w''\|_1\leqslant 2\|f\|_1, \tag{4.20}$$

and

$$\|w'\|_\infty\leqslant\|w''\|_1\leqslant 2\|f\|_1. \tag{4.21}$$

(i)⇒(iii): If $f\in L_1(\mathbb{R})_+$ and if w is a solution of (P_f) then by (4.21), $\|w'\|_\infty\leqslant 2\|f\|_1$. Thus there exists a positive constant c such that

$$w(x)\geqslant cx,\quad x\leqslant -1.$$

Further

$$f(x)+w''(x)\geqslant\beta^0(w(x))\geqslant\beta^0(cx)>0 \text{ for a.e. } x<-1.$$

Hence

$$\|f\|_1\geqslant\|f+w''\|_1\geqslant\int_{-\infty}^{-1}\beta^0((cx))dx=\frac{1}{c}\int_{-\infty}^{-c}\beta^0(s)ds.$$

(iii)⇒(ii): Suppose there exists an $a\in\mathbb{R}$ such that

$$\int_{-\infty}^{a}\beta^0(r)dr<\infty.$$

We claim that there exists a function $g:(-\infty,a]\to\mathbb{R}$ such that $g\geqslant 1$ and nonincreasing on $(-\infty,a]$, $\lim_{x\to-\infty}g(x)=\infty$, and

$$\int_{-\infty}^{a}\beta^0(x)g(x)dx<\infty.$$

Suppose we have such a function g. Then define $v:(-\infty,-1]\to\mathbb{R}$ by

$$v'(x)=\frac{1}{g(v(x))},\quad x<-1$$
$$v(-1)=a-1.$$

We see that $v'>0$ and $v''>0$. Also $\lim_{x\to-\infty}v(x)=-\infty$ and $\lim_{x\to-\infty}v'(x)=0$. Finally $v''\in L_1(-\infty,-1)$ and

$$\int_{-\infty}^{-1}\beta^0(v(x))dx=\int_{-\infty}^{a-1}\beta^0(y)g(y)dy<\infty.$$

Now define the function $\bar{w}:\mathbb{R}\to\mathbb{R}$ such that $\bar{w}$ is even and smooth on $\mathbb{R}$, $\bar{w}(x)=v(x)$ for $-\infty<x\leqslant-1$ and $\bar{w}(x)<a$ everywhere. Finally define

$$f(x):=-\bar{w}''(x)+\beta^0(\bar{w}(x))\in L_1(\mathbb{R}).$$

Hence $\bar{w}$ is a solution of (P_f).

Next, we give a construction of g.
Let $\{a_n\}_{n=1}^{\infty}$ be an increasing sequence $\cdots<a_n<a_{n+1}<\cdots<a$ such that

$$\int_{-\infty}^{a_n}\beta^0(x)dx<\frac{1}{n^2}.$$

Let g be any smooth nonincreasing function such that $g(a_n)=\sqrt{n}$ for $n=1,2,3,\ldots$ and $g=1$ on $[a_1,a]$. Then

$$\int_{-\infty}^{a}\beta^0(x)g(x)dx=\int_{a_1}^{a}\beta^0(x)dx+\sum_{n=1}^{\infty}\int_{a_{n+1}}^{a_n}\beta^0(x)g(x)dx$$

$$\leqslant\int_{a_1}^{a}\beta^0(x)dx+\sum_{n=1}^{\infty}\sqrt{n+1}\int_{a_{n+1}}^{a_n}\beta^0(x)dx<\int_{a_1}^{a}\beta^0(x)dx+\sum_{n=1}^{\infty}\frac{\sqrt{n+1}}{n^2}<\infty.$$

(ii)$\Rightarrow$(i): follows from the Lemma's 4.20, 4.21, and 4.22 below. □

LEMMA 4.20. *Let $f,g,\in L_1(\mathbb{R})_+$ and $\int f>\int g$. If (P_g) has a solution, then so has (P_f).*

LEMMA 4.21. *Suppose (ii) of Theorem 4.19 holds. Let V be the set given by $V:=\{f\in L_1(\mathbb{R})_+:\exists g\in L_1(\mathbb{R})_+,\int f>\int g$ and (P_g) has a solution $\}$. Then V is dense in $L_1(\mathbb{R})_+$.*

LEMMA 4.22. *The set of $f\in L_1(\mathbb{R})_+$ for which (P_f) has a solution is closed in $L_1(\mathbb{R})_+$.*

We prove here only Lemma 4.21. For the proofs of Lemma 4.20 and Lemma 4.22, see CRANDALL & EVANS [1975].

PROOF OF LEMMA 4.21. Choose $f\in L_1(\mathbb{R})_+$ such that (P_f) has a solution $\bar{w}$. By (ii), this can be done. For fixed $\epsilon>0$, we construct a $g\in L_1(\mathbb{R})_+$ such that $\|g\|_1<\epsilon$ and (P_g) has a solution. For $\delta, M>0$, define

$$\overline{w}_{\delta,M}=\overline{w}(\delta x)-M.$$

Then $\overline{w}_{\delta,M}$ solves $(P_{f_{\delta,M}})$, where

$$\begin{aligned} f_{\delta,M}(x) &\equiv -\overline{w}''_{\delta,M}(x)+\beta_0(\overline{w}_{\delta,M}(x)) \\ &= -\delta^2\overline{w}''(\delta x)+\beta^0(\overline{w}(\delta x)-M). \end{aligned}$$

Now

$$\|\overline{w}_{\delta,M}''\|_1=\delta\|\overline{w}''\|_1<\epsilon/2,$$

for δ small enough. Moreover $\beta_0(\overline{w}(\delta x)-M)\to 0$ for $M\to\infty$. This follows from (ii), which holds only if $\int_{-\infty}^{a}\beta_0(r)dr<\infty$. By the monotone convergence theorem, we can select an M sufficiently large such that

$$\|\beta^0(\overline{w}(\delta x)-M)\|_1<M\epsilon/2.$$

Then $g:=f_{\delta,M}$ satisfies $\|g\|_1<\epsilon$. To complete the proof, take $f\in L_1(\mathbb{R})_+$ and set $h=f+g$. Then $h\in L_1(\mathbb{R})_+$ and $\int h>\int g$ and (P_g) has a solution. Therefore $h\in V$. Also $\|h-f\|_1<\epsilon$, implying that V is dense in $L_1(\mathbb{R})_+$. □

REMARK 4.23. Theorem 4.19 says that there exists a solution w of (P_f) exactly when $f\in L_1(\mathbb{R})_+$. Define $Lf:=w''+f$, where w is solution of (P_f). As in Section 4.2 one can prove the following facts:

a) L is a contraction in $L_1(\mathbb{R})$.
b) $\int[Lf-Lg]^+\leqslant\int[f-g]^+$, for $f,g\in L_1(\mathbb{R})_+$ thus L is order preserving.
c) $\int j(Lf(x))\leqslant\int j(f(x))$, for $f\in L_1(\mathbb{R})_+$ and $j\in J_0$.
d) When $m\leqslant f(x)\leqslant M$ a.e., $m\leqslant 0, M\geqslant 0$, then $m\leqslant Lf(x)\leqslant M$ a.e. (see the proof of Proposition 4.10).
e) Solutions of (P_f) are unbounded: $w(x)\to-\infty$ when $|x|\to\infty$. This follows from $f+w''\in\beta(w), f+w''\geqslant\beta^0(w)$ and $f+w''\in L^1(\mathbb{R})$.
f) Solutions of (P_f) are unique. Let w and $\overline{w}$ be two solutions of (P_f). As in Section 4.2, one first proves that $w(x)-\hat{w}(x)=c$. From this fact one obtains that $Lf(x)\in\beta(\hat{w}(x))\cap\beta(\hat{w}(x)+c)$ a.e. Using $\hat{w}(x)\to-\infty$ when $|x|\to\infty$, we can choose x so that $\hat{w}(x)$ is a point of strict increase of β^0 : $\beta^0(\hat{w}(x))<\beta^0(w(x)+r)$ for any $r>0$. For this x it follows that $c=0$.
g) Since $w'(\pm\infty)=0$, we have $\int_{\mathbb{R}}Lf=\int_{\mathbb{R}}f$.

4. THE SEMIGROUP SOLUTION

It follows directly from the results obtained in Sections 4.2 and 4.3 that the operator A_ϕ satisfies:

(i) A_ϕ is m-dissipative in $L_1(\mathbb{R})$
(ii) $E_\lambda^{A_\phi}$ is order preserving in $L_1(\mathbb{R})$
(iii) $E_\lambda^{A_\phi}$ satisfies a maximum principle: i.e. if $f\in L_1(\mathbb{R})\cap L_\infty(\mathbb{R})$ such that

$$m \leqslant f(x) \leqslant M \text{ a.e., then } m \leqslant E_\lambda^{A_\phi} f(x) \leqslant M \text{ a.e.}$$

Let $u_0 \in \overline{\mathfrak{D}(A_\phi)}$ and let $\epsilon>0$. Consider the backward difference scheme

$$\begin{cases} u_\epsilon(t) - u_\epsilon(t-\epsilon) = \epsilon A_\phi u_\epsilon(t), & t>0, \\ u_\epsilon(t) = u_0, & t \leqslant 0. \end{cases}$$

Then $u_\epsilon(t) = E_\epsilon^{A_\phi} u_\epsilon(t-\epsilon)$ for $t>0$. We iterate to find:

$$u_\epsilon(t) = (E_\epsilon^{A_\phi})^{[t/\epsilon]+1} u_0, \text{ for } t>0,$$

where $[t/\epsilon]$ is the largest integer in the interval $(-\infty, t/\epsilon]$. The semigroup generation Theorem 2.3 gives that

$$u(t) = T_{A_\phi}(t) u_0 = \lim_{\epsilon \downarrow 0} u_\epsilon(t)$$

exists uniformly on $[0,T]$, for any $T>0$, i.e.

$$\lim_{\epsilon \downarrow 0} \sup_{[0,T]} \| T_{A_\phi}(t) u_0 - u_\epsilon(t) \|_1 = 0.$$

It follows from the maximum principle for $E_\lambda^{A_\phi}$ that when $u_0 \in \overline{\mathfrak{D}(A_\phi)} \cap L_\infty(\mathbb{R})$ such that $m \leqslant u_0(x) \leqslant M$, then $u(t,x) := T_{A_\phi}(t) u_0(x)$ satisfies

$$m \leqslant u(t,x) \leqslant M, \text{ a.e. in } [0,\infty) \times \mathbb{R}.$$

Moreover, since $E_\lambda^{A_\phi}$ is order preserving we have,

$$u_0(x) \geqslant v_0(x) \text{ a.e. on } \mathbb{R} \text{ implies } u(t,x) \geqslant v(t,x) \text{ a.e., in } [0,\infty) \times \mathbb{R},$$

where $v(t,x) := T_{A_\phi}(t) v_0(x)$ for $(t,x) \in (0,\infty) \times \mathbb{R}$. Finally, we have conservation of mass:

$$\int_{\mathbb{R}} u(t,x) dx = \int_{\mathbb{R}} u_0(x) dx, \text{ for all } t>0.$$

THEOREM 4.24. *Let $u_0 \in \overline{\mathfrak{D}(A_\phi)} \cap L_\infty(\mathbb{R})$ and $\phi : \mathfrak{D}(\phi) \to \mathbb{R}$ both continuous and nondecreasing such that either $\phi(0)=0$ and $0 \in \operatorname{int} \mathfrak{D}(\phi)$ or $\mathfrak{D}(\phi) \subset (0,\infty)$. Then $u(t,x)$ is a weak solution of Problem* (I) *and satisfies* $\lim_{t \downarrow 0} \| u(t) - u_0 \|_1 = 0$.

PROOF. Clearly, $\beta = \phi^{-1}$ satisfies the conditions of Sections 4.2 and 4.3. Then define $u_\epsilon(t) = (E_\lambda^{A_\phi})^{[t/\epsilon]+1} u_0$ for $t>0$ and $u_\epsilon(t) = u_0$ for $t \leqslant 0$. The function $u_\epsilon(t,x)$ satisfies for each $t>0$,

$$\frac{1}{\epsilon} \{ u_\epsilon(t,x) - u_\epsilon(t-\epsilon,x) \} = A_\phi u_\epsilon(t,x), \text{ in } \mathfrak{D}'(0,\infty),$$

or

$$\frac{1}{\epsilon} \{ u_\epsilon(t,x) - u_\epsilon(t-\epsilon,x) \} = w_\epsilon''(t,x), \text{ in } \mathfrak{D}'(0,\infty),$$

or, with $\xi \in C_0^\infty(\mathbb{R})$:

$$\frac{1}{\epsilon}\int_{\mathbb{R}}\{u_\epsilon(t,x)-u_\epsilon(t-\epsilon,x)\}\xi dx - \int_{\mathbb{R}} w_\epsilon(t,x)\xi''(x)dx = 0,$$

where $w_\epsilon = \phi(u_\epsilon)$. Next let $\zeta \in C_0^\infty(0,\infty)$. Multiply the equation by ζ and integrate. This gives

$$\int_0^\infty\int_{\mathbb{R}}\frac{1}{\epsilon}\{u_\epsilon(t,x)-u_\epsilon(t-\epsilon,x)\}\zeta(t)\xi(x)dx = \int_0^\infty\int_{\mathbb{R}}\phi(u_\epsilon(t,x))\zeta(t)\xi''(x)dx.$$

Set $f(t,x)=\zeta(t)\xi(x)$. Then

$$\frac{1}{\epsilon}\int_0^\infty\int_{\mathbb{R}}\{u_\epsilon(t,x)-u_\epsilon(t-\epsilon,x)\}f(t,x) =$$

$$\frac{1}{\epsilon}\int_0^\infty\int_{\mathbb{R}}u_\epsilon(t,x)f(t,x)-\frac{1}{\epsilon}\int_{-\epsilon}^\infty\int_{\mathbb{R}}u_\epsilon(t,x)f(t+\epsilon,x) =$$

$$\int_0^\infty\int_{\mathbb{R}}u_\epsilon(t,x)\frac{f(t,x)-f(t+\epsilon,x)}{\epsilon},$$

since $f \in C_0^\infty((0,\infty)\times\mathbb{R})$. To take the limit for $\epsilon\downarrow 0$, we use that (i) $u_\epsilon \to u$ a.e., (ii) $\phi(u_\epsilon)\to\phi(u)$ a.e. (by the continuity of ϕ), (iii) $\|u_\epsilon\|_\infty \leqslant \|u_0\|_\infty$ for all $\epsilon>0$, and (iv) the dominated convergence theorem. We obtain

$$\int_0^\infty\int_{\mathbb{R}}\{u(t,x)f_t(t,x)-\phi(u(x,t))f_{xx}(t,x)\} = 0,$$

for all $f \in C_0^\infty((0,\infty)\times\mathbb{R})$. □

REMARK 4.25. BREZIS & CRANDALL [1979] showed that if $\phi:\mathbb{R}\to\mathbb{R}$ is continuous and nondecreasing, and if $\phi(0)=0$, then bounded weak solutions of Problem (I) which satisfy $\lim_{t\downarrow 0}\|u(t)-u_0\|_1 = 0$ are unique.

5. CONTINUOUS DEPENDENCE ON ϕ

In this section we study the dependence on ϕ of solutions of Problem (I) where the function $\phi:\mathbb{R}:\to\mathbb{R}$ is continuous and nondecreasing and satisfies $\phi(0)=0$ and $0\in \operatorname{int}\mathfrak{D}(\phi)$, and where $u_0 \in \overline{\mathfrak{D}(A_\phi)}\cap L_\infty(\mathbb{R})$. As we observed in Section 4.4 the semigroup solution $u \in C([0,\infty);L_1(\mathbb{R}))\cap L_\infty((0,\infty)\times\mathbb{R})$, and satisfies the differential equation in the sense of distributions. Let $\{\phi_n\}_{n\geqslant 1}$ be a sequence of continuous and nondecreasing functions $\phi_n:\mathbb{R}\to\mathbb{R}$, satisfying $\phi_n(0)=0$ and $0\in\operatorname{int}\mathfrak{D}(\phi_n)$. Moreover, let $\{u_{0n}\}_{n\geqslant 1}$ be a sequence of functions $u_{0n}:\mathbb{R}\to\mathbb{R}$ satisfying $u_{0n}\in\overline{\mathfrak{D}(A_\phi)}\cap L_\infty(\mathbb{R})$. Then consider for each $n\geqslant 1$ the problem:

$$(I_n)\quad \begin{cases} u_t = (\phi_n(u))_{xx}, & (t,x)\in(0,\infty)\times\mathbb{R}, \\ u(0,x)=u_{0n}(x), & x\in\mathbb{R}. \end{cases}$$

We denote the unique weak solution of Problem (I_n) by u_n. Clearly $u_n \in C([0,T];L_1(\mathbb{R}))\cap L_\infty((0,\infty)\times\mathbb{R})$. We have the following continuous dependence result.

THEOREM 4.26. *Let u_n be the solution of (I_n) for $n\geqslant 1$. Further let*

$$\lim_{n\to\infty}\phi_n(r)=\phi(r),\ \textit{for } r\in\mathbb{R},$$

and

$$\lim_{n\to\infty}u_{0n}=u_0,\ \textit{in } L_1(\mathbb{R}).$$

Then $u_n\to u$ in $C([0,\infty);L_1(\mathbb{R}))$, where u is the weak solution of Problem (I).

REMARK 4.27. When $\phi(r)=|r|^m\cdot\operatorname{sign} r$, $m>1$, and $u_0\geqslant 0$, the classical approach with respect to Problem (I) (e.g. see OLEINIK, KALASHNIKOV and CHZOU [1958]) is to replace u_0 by u_0+1/n and letting $n\to\infty$. In the setting of Theorem 4.26 this approximation corresponds to the choice $\phi_n(r)=\phi(r)+r/n$.

Theorem 4.26 is a direct consequence of Theorem 2.7 as soon as $\phi_n\to\phi$ implies that $A_{\phi_n}\to A_\phi$ in the sense of resolvent: i.e,. if $\lim_{n\to\infty}\phi_n=\phi$ implies that $\lim_{n\to\infty}E_1^{A_{\phi_n}}f=E_1^{A_\phi}f$, for all $f\in L_1(\mathbb{R})$. We shall consider the more general case where $\{\phi_n\}_{n\geqslant 1}$ and ϕ are maximal monotone graphs in $\mathbb{R}^2$ which satisfy $\phi_n\to\phi$ when $n\to\infty$.

REMARK 4.28. When we write $\phi_n\to\phi$, $n\to\infty$, we mean that $\phi_n^0(r)\to\phi^0(r)$ a.e. on $\mathbb{R}$ when $n\to\infty$. Here ϕ^0 denotes the *minimal section:* see Definition 4.17.

First we define a class of maximal monotone graphs in $\mathbb{R}^2$. Let $\delta, r_0>0$ be given. Then define

$$\Phi_{\delta,r_0}=\{\phi \text{ maximal monotone in } \mathbb{R}^2: 0\in\phi(0) \text{ and } |\phi^0(\pm\delta)|<r_0\}.$$

PROPOSITION 4.29. *Let $\delta,r_0>0$ and $f\in L_1(\mathbb{R})$. Then $\{E_1^{A_\phi}f:\phi\in\Phi_{\delta,r_0}\}$ is precompact in $L_1(\mathbb{R})$.*

PROOF. Set $E=E_1^{A_\phi}$. From Section 4.2 we know that $E:L_1(\mathbb{R})\to L_1(\mathbb{R})$ is a contraction which satisfies $E0=0$. Therefore

$$\|Ef\|_1\leqslant\|f\|_1 \text{ for all } \phi\in\Phi_{\delta,r_0}.$$

Moreover, E is invariant under translation: i.e. when $u=Ef$ then $u_h=Ef_h$, where $f_h(x)=f(x+h)$, and $u_h(x)=u(x+h)$ for all $h\in\mathbb{R}$. This implies that

$$\|Ef_h-Ef\|_1\leqslant\|f_h-f\|_1,\ \text{for all } h\in\mathbb{R}.$$

Since translation is continuous on $L_1(\mathbb{R})$, this shows that $\{E_1^{A_\phi}f:\ \phi\in\Phi_{\delta,r_0}\}$ is precompact in $L_1^{loc}(\mathbb{R})$. To prove precompactness in $L_1(\mathbb{R})$, it remains to show that

$$\lim_{R\to\infty}\int_{|x|>R}|Ef(x)|dx=0, \text{ uniformly in } \phi\in\Phi_{\delta,r_0}. \tag{4.22}$$

Let $\epsilon>0$ and choose $M,R_0>0$ such that

$$\|(f-M\chi_{\{|x|\leqslant R_0\}})^+\|_1<\epsilon,$$

where χ_S denotes the characteristic function of the set $S\subseteq\mathbb{R}$. To shorten the notation set $g=M\chi_{\{|x|\leqslant R_0\}}$. Since $f\leqslant f^+\leqslant g+(f-g)^+$, and since E is order preserving we have

$$(Ef)^+\leqslant Ef^+\leqslant E(g+(f-g)^+).$$

Thus

$$\begin{aligned}\int_{|x|>R}(Ef)^+ &\leqslant \int_{|x|>R}Eg+\int_{|x|>R}|(E(g+(f-g)^+)-Eg|\\ &\leqslant \int_{|x|\geqslant R}Eg+\|(f-g)^+\|_1<\int_{|x|>R}Eg+\epsilon.\end{aligned}$$

Hence it is sufficient to show that $\lim_{R\to\infty}\int_{|x|>R}Eg=0$, uniformly in $\phi\in\Phi_{\delta,r_0}$, in order to establish the same for $(Ef)^+$. Treating $(Ef)^-$ similarly, we see that (4.22) holds for general $f\in L_1(\mathbb{R})$ if it holds for functions of the form $f=\pm M\chi_{\{|x|\leqslant R_0\}}$. Consider again the problem:

$$-w''+\beta(w)\ni f \quad (\text{with } \beta=\phi^{-1}). \tag{P}$$

Setting $u=Ef$, we have $u=f+w''$ (w a solution of (P)) and $u\in\beta(w)$. We have to show that

$$\lim_{R\to\infty}\int_{|x|>R}u(x)dx=0, \text{ uniformly in } \phi\in\Phi_{\delta,r_0}.$$

Observe that this holds if $\lim_{R\to\infty}w'(\pm R)=0$, uniformly in $\phi\in\Phi_{\delta,r_0}$. Since the proof of both cases is identical, we only consider $w'(+R)$. From the maximum principle for E it follows that

$$w''(x)+f(x)=Ef(x)\geqslant 0, \text{ a.e. on } \mathbb{R}. \tag{4.23}$$

Hence $w''(x)\geqslant 0$, for a.e. $x>R_0$. Thus we obtain that $w'(r)\geqslant w'(s)$ and $w(r)\geqslant w(s)+(r-s)w'(s)$ for all $r\geqslant s\geqslant R_0$. Then either $w'(s)\leqslant 0$ or $w(\infty)=\infty$. However, because $\phi\in\Phi_{\delta,r_0}$, it follows that $|\beta(\pm r_0)|\geqslant\delta$. This together with $w(+\infty)=\infty$ and $u\in\beta(w)$ contradicts $u\in L_1(\mathbb{R})$. Thus $w'(s)\leqslant 0$ for all $s>R_0$. This implies that $w(s)$ and $w''(s)\in\beta(w(s))$ are decreasing in s for $s>R_0$. Now fix $x>R_0$. Then

$$\begin{aligned}w''(2x)\cdot x &\leqslant \int_x^{2x}w''(s)ds=\int_x^{2x}(w''(s)+f(s))ds\\ &\leqslant \|w''+f\|_1\leqslant\|f\|_1=2MR_0.\end{aligned}$$

Thus

$$w''(x) \leqslant \frac{4MR_0}{x}, \text{ for all } x > 2R_0.$$

Next choose a constant $L > 4MR_0/\delta$. Then, for $x > L$, we have $w''(x) < \delta$ and thus

$$|w(x)| < r_0, \text{ uniformly in } \phi \in \Phi_{\delta, r_0}.$$

Finally we use again (4.23). This gives

$$w(t) - w(r) = \int_r^t w'(s)ds \leqslant (t-r)w'(t), \; t \geqslant r \geqslant R_0,$$

or

$$w(r) \geqslant w(t) + (t-r)(-w'(t)), \quad t \geqslant r \geqslant R_0,$$

and $(t-L)(-w'(t)) \leqslant 2r_0$, for $t > L$. Thus $w'(t) \to 0$ when $t \to \infty$, uniformly in $\phi \in \Phi_{\delta, r_0}$ and the proof of the proposition is complete. □

Next we prove a theorem of which Theorem 4.26 is a direct consequence.

THEOREM 4.30. *Let $\{\phi_n\}_{n \geqslant 1}$ and ϕ be maximal monotone graphs in $\mathbb{R}^2$ all of which contain the origin and satisfy $\phi_n \to \phi$ when $n \to \infty$. Further let $0 \in \operatorname{int} \mathfrak{D}(\phi)$. Then*

$$A_{\phi_n} \to A_{\phi}, \textit{ as } n \to \infty,$$

in the sense of the resolvent.

PROOF. Fix $f \in L_1(\mathbb{R})$. Since $0 \in \operatorname{int} \mathfrak{D}(\phi)$, there exists a $\delta > 0$ so that $(-\delta, \delta) \in \mathfrak{D}(\phi)$. Set $r_0 = \max\{-\phi^0(-\delta/2), \phi^0(+\delta/2)\} + 1$. Then for n sufficiently large, $\phi_n \in \Phi_{\delta, r_0}$ and we may apply Proposition 4.29. We show that every subsequence $\{\phi_{n_k}\}$ for which $u_{n_k} := E_1^{A_{\phi_{nk}}} f \to u$ in $L_1(\mathbb{R})$, implies $u = E_1^{A_\phi} f$. Since $u_{n_k} = f + w_{n_k}''$, we have

$$w_{n_k}'' \to u - f \text{ in } L_1(\mathbb{R}).$$

From Section 4.2 we know that

$$\|j_{n_k}(w_{n_k})\|_\infty \leqslant 2\|f\|_1^2,$$

where $\partial j_{n_k} = \beta_{n_k} = \phi_{n_k}^{-1}$. Then $\phi_{n_k} \in \Phi_{\delta, r_0}$ implies that the sequence $\{w_{n_k}\}$ is uniformly bounded in $L_\infty(\mathbb{R})$. Hence there exists a subsequence $\{w_{n_l}\}$ and a function $w \in L_1^{loc}(\mathbb{R})$ such that $w_{n_l} \to w$ a.e. on $\mathbb{R}$. This implies $w'' = f - u$. Since ϕ is maximal monotone and $u_{n_l} \in \phi_{n_l}(w_{n_l})$ we also have $u \in \phi(w)$. Thus $u = E_1^{A_\phi} f$. □

We have the following important consequence.

THEOREM 4.31. *Let ϕ be a maximal monotone graph in $\mathbb{R}^2$ such that $0 \in \phi(0)$ and $0 \in \operatorname{int} \mathcal{D}(\phi)$. Let A_ϕ and $\mathcal{D}(A_\phi)$ be as in Section 4.1. Then*

$$\overline{\mathcal{D}(A_\phi)} = \{v \in L_1(\mathbb{R}): \inf \mathcal{D}(\phi) \leqslant v(x) \leqslant \sup \mathcal{D}(\phi) \text{ a.e.}\}.$$

REMARK 4.32. In Theorem 4.26 it was assumed that $\mathcal{D}(\phi) = \mathbb{R}$. Therefore $\overline{\mathcal{D}(A_\phi)} = L_1(\mathbb{R})$.

PROOF. Let $f \in L_1(\mathbb{R})$ be such that $\inf \mathcal{D}(\phi) \leqslant f(x) \leqslant \sup \mathcal{D}(\phi)$ a.e. on $\mathbb{R}$. For $\lambda > 0$, we set $u_\lambda = E_\lambda^{A_\phi} f = (I - \lambda A_\phi)^{-1} f = (I - A_{\lambda\phi})^{-1} f = E_1^{A_{\lambda\phi}} f$. Then $u_\lambda \in \mathcal{D}(A_\phi)$. Further $\lambda\phi \to \phi_0$ when $\lambda \downarrow 0$, where

$$\phi_0(r) = \begin{cases} \{0\}, & \inf \mathcal{D}(\phi) < r < \sup \mathcal{D}(\phi) \\ [-\infty, 0], & r = \inf \mathcal{D}(\phi) \\ [0, \infty), & r = \sup \mathcal{D}(\phi). \end{cases}$$

Using Theorem 4.30 we see that $u_\lambda \to E_1^{A_{\phi_0}} f$ if $\lambda \downarrow 0$. Now when $\beta = \phi_0^{-1}$ and w solves $-w'' + \beta(w) \ni f$, it is trivial that $w = 0$ is the unique solution since $f \in \overline{\mathcal{R}(\beta)}$. Thus $E_1^{A_{\phi_0}} f = w'' + f = f$, and the proof is complete. □

6. A REGULARITY RESULT

A formal computation with respect to the problem

$$(I') \quad \begin{cases} u_t = (|u|^m \operatorname{sign} u)_{xx}, & (t,x) \in (0,\infty) \times \mathbb{R} \\ u(0,x) = u_0(x), & x \in \mathbb{R}, \end{cases}$$

when $m \neq 1$ and $m > 0$, shows that, if $u(t,x)$ is the solution corresponding to the initial value u_0, then

$$v(t,x) = \lambda^{1/(m-1)} u(\lambda t, x), \quad \lambda > 0,$$

is the solution corresponding to the initial value $\lambda^{1/(m-1)} u_0$. Thus the associated semigroup $\{T(t)\}_{t \geqslant 0}$ in $L_1(\mathbb{R})$ satisfies

$$T(t)(\lambda^{1/(m-1)} u_0) = \lambda^{1/(m-1)} T(\lambda t) u_0, \quad \text{for } \lambda > 0 \text{ and } t \geqslant 0,$$

for all $u_0 \in L_1(\mathbb{R})$. This motivates the following

THEOREM 4.33. *Let $(X, \|\cdot\|)$ be a normed vector space and $C \subseteq X$ a subset satisfying $rC \subseteq C$, $r \geqslant 0$. Let $T(t): C \to X$, for every $t \geqslant 0$, satisfy*
(i) $T(t)(\lambda^{1/(m-1)} x) = \lambda^{1/(m-1)} T(\lambda t) x$, *for* $\lambda > 0$ *and* $x \in C$ $(m > 0, \neq 1)$,
(ii) $\|T(t)x - T(t)\hat{x}\| \leqslant L \|x - \hat{x}\|$ *for* $x, \hat{x} \in C$,
(iii) $T(t)0 = 0$.
Then for any $x \in C$ and $t, h > 0$,

$$\|T(t+h)x - T(t)x\| \leqslant 2L\|x\| \cdot |1 - (1 + \frac{h}{t})^{1/(1-m)}|, \tag{4.24}$$

which gives

$$\limsup_{h \downarrow 0} \|\frac{T(t+h)x - T(t)x}{h}\| \leqslant \frac{2L\|x\|}{|m-1|} \cdot \frac{1}{t}. \tag{4.25}$$

PROOF. We set $\lambda=(1+h/t)$. For any $x\in C$ we have

$$\begin{aligned}&T(t+h)x-T(t)x=\\&T(\lambda t)x-T(t)x=\lambda^{-1/(m-1)}T(t)(\lambda^{1/(m-1)}x)-T(t)x=\\&\lambda^{-1/(m-1)}\{T(t)(\lambda^{1/(m-1)}x)-T(t)x\}+(\lambda^{-1/(m-1)}-1)T(t)x,\end{aligned}\tag{4.26}$$

where we have used property (i). Next we use (ii) and (iii) to find

$$\begin{aligned}&\|T(t+h)x-T(t)x\|\leqslant\lambda^{-1/(m-1)}\cdot L\cdot|\lambda^{1/(m-1)}-1|\cdot\|x\|\\&+L\cdot|\lambda^{-1/(m-1)}-1|\cdot\|x\|=2L\cdot|1-\lambda^{1/(1-m)}|\cdot\|x\|.\end{aligned}$$

This gives (4.24). Inequality (4.25) follows then immediately. □

Next we assume that X is an ordered vector space.

THEOREM 4.34. *Let $(X,\|\cdot\|)$ be an ordered vector space with order relation $\geqslant$. Let $T(t):C\to X$, $t\geqslant 0$, satisfy*

$$T(t)x\geqslant T(t)y, \text{ if } x,y\in C \text{ and } x\geqslant y,\tag{4.27}$$

where C is as in Theorem 4.33, and $T(t)$ satisfies property (i). For $x\in C, x\geqslant 0$, and $t,h>0$ we have

$$(m-1)\{T(t+h)x-T(t)x\}\geqslant(m-1)\{(1+\frac{h}{t})^{1/(1-m)}-1\}T(t)x.\tag{4.28}$$

PROOF. When $m>1$, we use (4.26). Since $\lambda>1$ and $\lambda^{1/(m-1)}>1$ we have $\lambda^{1/(m-1)}x\geqslant x$ (since $x\geqslant 0$), and

$$\lambda^{-1/(m-1)}\{T(t)(\lambda^{1/(m-1)}x)-T(t)x\}\geqslant 0,$$

by using (4.27). Thus in (4.26) we have:

$$T(t+h)x-T(t)x\geqslant(\lambda^{-1/(m-1)}-1)T(t)x,\tag{4.29}$$

from which (4.28) follows immediately. When $m<1$, we obtain the reverse of (4.29). Multiplying through by $(m-1)$ gives again (4.28). □

When we apply (4.25) and (4.28) to the semigroup solutions of Problem (I') we obtain

$$\limsup_{h\downarrow 0}\int_{\mathbb{R}}\frac{|u(t+h,x)-u(t,x)|}{h}dx\leqslant\frac{1}{t}\frac{2}{|m-1|}\int_{\mathbb{R}}|u_0(x)|dx,$$

and for $u_0\geqslant 0$ (which implies $u(t)\geqslant 0$):

$$-(m-1)\int_0^\infty\int_{\mathbb{R}}u\zeta_t\geqslant-\int_0^\infty\int_{\mathbb{R}}\frac{u}{t}\zeta$$

for $\zeta\in C_0^\infty((0,\infty)\times\mathbb{R}),\zeta\geqslant 0$. This yields that

$$(m-1)u_t\geqslant-\frac{u}{t}, \text{ in } \mathcal{D}'((0,\infty)\times\mathbb{R}).\tag{4.30}$$

REMARK. Similar results can be obtained for solutions of other types of partial differential equations, on which the semigroup theory is applicable: e.g. $u_t + (|u|^m . \operatorname{sign} u)_x = 0$ or $u_t = (|u_x|^m . \operatorname{sign}(u_x))_x$. The rescaling properties of the associated semigroups all satisfy property (i) of Theorem 4.33.

Part II

Analytic Semigroups

Chapter 5

The Calculus of Analytic Semigroups

1. Banach couples: some terminology

Let $\mathcal{C}$ denote the class of all pairs of Banach spaces $E=(E_1,E_0)$ such that E_1 is *densely* embedded in E_0. A pair $E\in\mathcal{C}$ will be called a *Banach couple.* The inclusion mapping from E_1 to E_0 will be denoted by j_E, or just j for short.

Given two Banach couples $E=(E_1,E_0)$ and $F=(F_1,F_0)$, a *map of Banach couples* from E to F is defined to be any bounded linear operator $T{:}E_0\to F_0$ such that $T(E_1)$ is contained in F_1. Thus we have the following diagram:

$$\begin{array}{ccc} E_0 & \xrightarrow{T} & F_0 \\ \cup & & \cup \\ E_1 & \xrightarrow{T} & F_1 \end{array}$$

It follows from the closed graph theorem that for any map of Banach couples $T{:}E\to F$, the linear operator $T|_{E_1}{:}E_1\to F_1$ is bounded. The set of maps of Banach couples from E to F will be denoted by $\mathcal{L}(E,F)$. It is a vector space, and with the norm

$$\|T\|_{\mathcal{L}(E,F)}=\max\,(\|T\|_{\mathcal{L}(E_0,F_0)},\ \|T\|_{\mathcal{L}(E_1,F_1)})$$

it becomes a Banach space. As usual, if $E=F$ then we shall write $\mathcal{L}(E)$ instead of $\mathcal{L}(E,E)$. The space $\mathcal{L}(E)$ is a Banach algebra.

If E is a Banach couple, and X is a Banach space such that $E_1\subset X\subset E_0$ holds (continuous, but not necessarily dense inclusions) then X is called an *intermediate space* of E. The space X is said to be *of exponent θ,* for some $\theta\in(0,1)$ if for all $x\in E_1$

$$\|x\|_X \leqslant C \cdot \|x\|_{E_0}^{1-\theta} \|x\|_{E_1}^{\theta}$$

holds (with C independent of x).

2. THE SETS $Gen(E)$ AND $Hol(E)$

Let $E=(E_1,E_0)$ be a given Banach couple, and let $A:E_1 \to E_0$ be a bounded linear operator. We may regard A as being an unbounded linear operator in E_0, with domain $\mathfrak{D}(A)=E_1$. Define $Gen(E)$ to be the set of $A \in \mathfrak{L}(E_1,E_0)$ such that the corresponding unbounded operator in E_0 is closed, and generates a C_0-semigroup in E_0. It follows from the Hille-Yosida theorem that for all $\lambda \in \mathbb{C}$, with $\mathrm{Re}\lambda$ larger than some constant $\omega_0 \in \mathbb{R}$, the operator $\lambda j - A:E_1 \to E_0$ is an isomorphism.

We shall be concerned with a subset of $Gen(E)$, namely $Hol(E)$. This subset is defined as follows:

DEFINITION. If $A \in \mathfrak{L}(E_1,E_0)$ then $A \in Hol(E)$ if and only if for some $\omega \in \mathbb{R}$

$$(\lambda j - A):E_1 \to E_0$$

is an isomorphism for all λ with $\mathrm{Re}\lambda \geqslant \omega$, and the inverses $(\lambda j - A)^{-1}:E_0 \to E_1$ are uniformly bounded in λ ($\mathrm{Re}\lambda \geqslant \omega$).

By definition $Hol(E) \subset \mathfrak{L}(E_1,E_0)$. It is however not clear that $Hol(E) \subset Gen(E)$. This will be shown to hold in Section 5.4.

The inverse $(\lambda j - A)^{-1}$ will be called the *resolvent of A* at λ, and will be denoted by $R(\lambda,A)$, or $R(\lambda)$ for short. The set of $\lambda \in \mathbb{C}$ such that $R(\lambda)$ exists is called the *resolvent set of A, $\rho(A)$*, and its complement is known as *the spectrum of A, $\sigma(A)$*. It should be noted that the resolvent $R(\lambda)$ is an operator from E_0 to E_1, whereas it is traditionally defined to be a bounded operator on E_0, i.e. one usually considers the operator $j \circ R(\lambda)$. These two operators are in some sense "the same", however it follows from

$$\begin{aligned} j \circ R(\lambda) &= \lambda^{-1}(\lambda j - A + A)R(\lambda) \\ &= \lambda^{-1}(1 + AR(\lambda)) \qquad (\lambda \neq 0) \end{aligned}$$

that for $\lambda \in \mathbb{C}$ with $\mathrm{Re}\lambda \geqslant \omega$ we have

$$\|j \circ R(\lambda)\| \leqslant \frac{C}{|\lambda|},$$

if for those λ's we have $\|R(\lambda)\| \leqslant C'$.

3. THE RESOLVENT

Let $A \in Hol(E)$ be given, and assume that $R(\lambda)$ exists for $\mathrm{Re}\lambda \geqslant \omega$, and that

$$\|R(\lambda)\| \leqslant M$$

holds for these values of λ. Then we have:

LEMMA 5.1. *There is a* θ $(0<\theta<\pi/2)$ *such that* $\Sigma_{\omega,\theta}=\{\lambda\in\mathbb{C}:|\arg(\omega-\lambda)|<\theta\}$ *contains the spectrum of* A, $\sigma(A)$. *Furthermore the resolvent of* A *is uniformly bounded on* $\mathbb{C}\setminus\Sigma_{\omega,\theta}$.

PROOF. Let $\lambda\in\mathbb{C}$ have $\operatorname{Re}\lambda=\omega$. Then for any $\mu\in\mathbb{C}$

$$(\lambda+\mu)j-A=(1+\mu j\circ R(\lambda))(\lambda j-A)$$

will be invertible iff $1+\mu j\circ R(\lambda)$ is invertible. Thus, if $\|\mu j\circ R(\lambda)\|<1$, then $\lambda+\mu\in\rho(A)$, and the series expansion

$$R(\lambda+\mu)=R(\lambda)\sum_{k=0}^{\infty}(-\mu)^k(j\circ R(\lambda))^k$$

implies

$$\|R(\lambda+\mu)\|\leqslant\|R(\lambda)\|(1-|\mu|\cdot\|j\circ R(\lambda)\|)^{-1}.$$

Using the estimate $\|j\circ R(\lambda)\|\leqslant C\cdot|\lambda|^{-1}$ we get

$$\|R(\lambda+\mu)\|\leqslant\|R(\lambda)\|(1-\frac{C|\mu|}{|\operatorname{Im}\lambda|})^{-1},$$

which is uniformly bounded if we require $|\mu|\leqslant\frac{1}{2C}|\operatorname{Im}\lambda|$.

Under this condition $\lambda+\mu$ can achieve any value outside of $\Sigma_{\theta,\omega}$ for some θ, which proves the lemma. ☐

Naturally, if $\lambda\in\mathbb{C}\setminus\Sigma_{\omega,\theta}$, then one has

$$\|j\circ R(\lambda)\|\leqslant C\cdot|\lambda|^{-1},$$

for some constant C.

It is because of the form of the set $\Sigma_{\omega,\theta}$ that contains $\sigma(A)$, that operators $A\in Hol(E)$ are sometimes called *sectorial operators.*

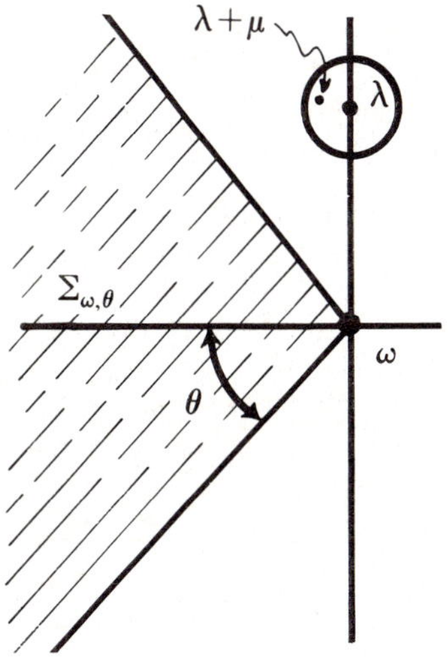

FIGURE 1.

4. PERTURBATION THEOREMS

Let $A\in Hol(E)$ be given, and let $B\in\mathfrak{L}(E_1,E_0)$ be another operator. In this section we give some conditions on A and B which ensure that $A+B\in Hol(E)$. These perturbation results are all based on the following observation:

LEMMA 5.2. *If* $A\in Hol(E)$, $B\in\mathfrak{L}(E_1,E_0)$ *and* $\lambda\in\rho(A)$ *satisfy either*

$$\|R(\lambda,A)B\|\leqslant 1/2$$

or

$$\|BR(\lambda,A)\| \leqslant 1/2$$

then $\lambda \in \rho(A+B)$, *and*

$$\|R(\lambda,A+B)\| \leqslant 2\|R(\lambda,A)\|.$$

In particular, if this holds for all λ *with* $\mathrm{Re}\lambda \geqslant \omega_0$ *for some* ω_0 *then* $A+B \in Hol(E)$.

PROOF. This follows immediately from

$$\begin{aligned}\lambda j - A - B &= (1 - BR(\lambda))(\lambda j - A) \\ &= (\lambda j - A)(1 - R(\lambda)B).\end{aligned}$$

Indeed, under the conditions of the lemma one has

$$R(\lambda,A+B) = \sum_{k=0}^{\infty} (R(\lambda,A)B)^k R(\lambda,A),$$

or

$$R(\lambda,A+B) = R(\lambda,A) \sum_{k=0}^{\infty} (BR(\lambda,A))^k. \quad \square$$

Our first application of this lemma is:

THEOREM 5.3. *Hol(E) is open in* $\mathfrak{L}(E_1,E_0)$ *(in the norm topology).*

PROOF. Since $R(\lambda,A)$ is uniformly bounded for $\mathrm{Re}\lambda \geqslant \omega$, one certainly has $\|BR(\lambda)\| \leqslant 1/2$ if $\|B\|$ is small enough, so that for small B, $A+B$ will be in $Hol(E)$. $\square$

Let X be an intermediate space for the couple E such that

$$\forall \epsilon > 0\ \exists C_\epsilon > 0\ \forall x \in E_1 : \|x\|_X \leqslant \epsilon \|x\|_{E_1} + C_\epsilon \|x\|_{E_0}.$$

Any intermediate space of exponent $\theta \in (0,1)$ satisfies this condition.

THEOREM 5.4. *If* $A \in Hol(E)$, *and* $B \in \mathfrak{L}(X,E_0)$ *then* $A + B|_{E_1} \in Hol(E)$.

PROOF. Let $k: E_1 \hookrightarrow X$ denote the inclusion, and let $R(\lambda,A) = R(\lambda)$ be uniformly bounded for $\mathrm{Re}\lambda \geqslant \omega_0$. Then for arbitrary $\epsilon > 0$ we have:

$$\begin{aligned}\|k \circ R(\lambda)\| &\leqslant \epsilon \|R(\lambda)\| + C_\epsilon \|j \circ R(\lambda)\| \\ &\leqslant K \cdot \left(\epsilon + \frac{C_\epsilon}{|\lambda|}\right)\end{aligned}$$

for some constant K. Thus, given any $\delta > 0$ we can find an $\omega_1 > \omega_0$ such that $\|k \circ R(\lambda)\| < \delta$ for all λ with $\mathrm{Re}\lambda \geqslant \omega_1$. For these λ's we then have

$$\|B \circ k \circ R(\lambda)\| \leqslant \delta \|B\|.$$

Finally we put $\delta=(2\|B\|)^{-1}$ and apply Lemma 5.2 to conclude that $A+B\circ k\in Hol(E)$. □

In order to prove our last perturbation theorem we need a lemma.

LEMMA 5.5. *Let $A\in Hol(E)$ be given. Then for any $x\in E_0$*

$$\lim_{\substack{|\lambda|\to\infty \\ \mathrm{Re}\lambda\geqslant\omega}} \|R(\lambda)x\|_{E_1}=0,$$

if the resolvent $R(\lambda)$ is uniformly bounded on $\mathrm{Re}\lambda\geqslant\omega$.

PROOF. In view of the uniform boundedness of $R(\lambda)$ we only have to prove the lemma for all x in some dense subspace of E_0, such as E_1. One easily verifies that for any $\mu_0\in\mathbb{C}$ the diagram

$$\begin{array}{ccc} & j\circ R(\lambda) & \\ E_0 & \longrightarrow & E_0 \\ \mu_0 j-A \uparrow & & \uparrow \mu_0 j-A \\ E_1 & \longrightarrow & E_1 \\ & R(\lambda)\circ j & \end{array}$$

commutes. Therefore, if we fix a $\mu_0\in\rho(A)$, we get

$$\|R(\lambda)\circ j\|=\|(\mu_0 j-A)^{-1}\circ j\circ R(\lambda)(\mu_0 j-A)\|\leqslant C\cdot|\lambda|^{-1}.$$

This means that $R(\lambda)x\to 0$ as $|\lambda|\to\infty$ for any $x\in E_1$ which completes the proof. □

Using this lemma we now show:

THEOREM 5.6. *Let $A\in Hol(E)$ be given. Then for any compact $B:E_1\to E_0$ we have $A+B\in Hol(E)$*

PROOF. Assume that $R(\lambda,A)=R(\lambda)$ is uniformly bounded on $\mathrm{Re}\lambda\geqslant\omega_0$. For $\lambda\in\mathbb{C}$ with $\mathrm{Re}\lambda\geqslant\omega_0$ we define

$$K=\text{ closure in } E_0 \text{ of } \{Bx:\|x\|_{E_1}\leqslant 1\},$$

$$f_\lambda(y)=\|R(\lambda)y\|_{E_1}\quad (y\in K).$$

By assumption K is a compact metric space, and the functions f_λ on K are uniformly Lipschitz continuous and uniformly bounded. Therefore they form a precompact set in $C(K)$ (by the Ascoli-Arzela theorem). By Lemma 5.5 the $f_\lambda(x)$ converge pointwise to zero as $|\lambda|\to\infty$ so we may use the precompactness in $C(K)$ to conclude that the f_λ converge uniformly to zero. Thus there exists an $\omega_1>\omega_0$ such that $\sup_{y\in K}|f_\lambda(y)|\leqslant 1/2$ for all λ with $\mathrm{Re}\lambda\geqslant\omega_1$. A close examination of the definition of K and f_λ reveals that we have shown that

$$\|R(\lambda)B\| \leqslant 1/2$$

for all λ with $\mathrm{Re}\lambda \geqslant \omega_1$. Hence $A + B \in Hol(E)$. □

5. A FUNCTIONAL CALCULUS

In this section we shall define an operator $f(A)$ for any $A \in Hol(E)$ and any analytic function $f(\lambda)$ which is defined on a neighbourhood of $\sigma(A)$, and vanishes fast enough at $\lambda = \infty$. The device used to define $f(A)$ is a Dunford-integral. The material in this section may be seen as a straightforward generalization of the case where A is a bounded operator in a Banach space.

Let $G \subset \mathbb{C}$ be a region of the form

$$G = \{\lambda \in \mathbb{C} : |\arg(-\lambda)| \leqslant \theta \text{ or } |\lambda| \leqslant \epsilon\}.$$

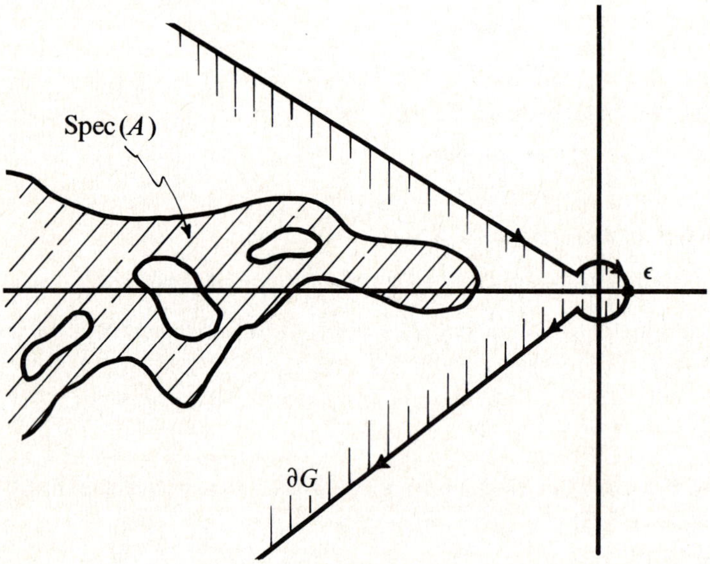

FIGURE 2.

For any function $f: G \to \mathbb{C}$ which is continuous on G and analytic in int G, we define

$$\omega_f(r) = \sup\{|f(\lambda)| : |\lambda| = r, \ \lambda \in G\}.$$

The set of such f's for which

$$\int_\epsilon^\infty \omega_f(r)\frac{dr}{r} + \sup_{r > 0} \omega_f(r) \stackrel{def}{=} \|f\|_{\mathfrak{F}},$$

is finite will be denoted by $\mathfrak{F}$. With the norm $\|\cdot\|_{\mathfrak{F}}$, $\mathfrak{F}$ becomes a commutative Banach algebra without unit.

Define $\mathcal{Q}$ to be the subset of $Hol(E)$ consisting of those $A \in Hol(E)$ for which

$$\{\lambda : |\arg(\lambda)| \leqslant \pi - \theta\} \subset \rho(A),$$

and $\|R(\lambda,A)\| \leqslant C$ for $\lambda \in \mathbb{C}$ with $|\arg(\lambda)| \leqslant \pi - \theta$ holds for some constant C. One easily verifies that $\mathcal{Q}$ is open in $\mathfrak{L}(E_1,E_0)$, e.g. by expanding $(\lambda j - A - B)^{-1}$ in a power series in B, as we did in Section 5.4.

From here on we shall only consider A's in $\mathcal{Q}$. This is no restriction since for any $A \in Hol(E)$ one can find a $k \in \mathbb{R}$ such that $A - k \cdot j \in \mathcal{Q}$ if θ is chosen suitably. For any $f \in \mathfrak{F}$ and $A \in \mathcal{Q}$ we now define

$$f(A) = \frac{1}{2\pi i} \int_{\partial G} f(\lambda) j \circ R(\lambda,A) d\lambda,$$

where ∂G is oriented in the upward direction. The integral should be interpreted as a Bochner integral. The integrand is a continuous function with values in $\mathfrak{L}(E_0)$ and therefore certainly measurable. It follows from the estimate $\|j \circ R(\lambda)\| \leqslant C/|\lambda|$ and our assumption $f \in \mathfrak{F}$ that the integral exists (i.e. the integral is summable). Since we have a similar estimate on $R(\lambda) \circ j$ it is easily seen that $f(A)$ maps E_1 to E_1, and that $f(A): E_1 \to E_1$ is bounded. Thus $f(A)$ is a map of Banach couples: $f(A) \in \mathfrak{L}(E)$.

THEOREM 5.7. *The mapping $(f,A) \mapsto f(A)$ from $\mathfrak{F} \times \mathcal{Q}$ to $\mathfrak{L}(E)$ is analytic.*

PROOF. Since $f(A)$ is linear in f it is sufficient to prove that $A \mapsto f(A)$ is analytic, for fixed $f \in \mathfrak{F}$. Let $B \in \mathfrak{L}(E_1,E_0)$ be small. Then we have the convergent expansion

$$j \circ R(\lambda, A + B) = j \circ R(\lambda,A) \sum_{k=0}^{\infty} (BR(\lambda,A))^k,$$

and therefore

$$f(A + B) = \sum_{k=0}^{\infty} \frac{1}{k!} (f^{(k)}(A)B), \tag{*}$$

where $f^{(k)}(A)B$ is defined by

$$f^{(k)}(A)B = \frac{k!}{2\pi i} \int_{\partial G} f(\lambda) j \circ R(\lambda,A)(BR(\lambda,A))^k d\lambda.$$

Using the uniform boundedness of $R(\lambda,A)$ on ∂G, and the fact that $f \in \mathfrak{F}$, one can verify that (*) is a convergent series expansion of $f(A+B)$ for small B. □

We have already remarked that for fixed $A \in \mathcal{A}$ the map $f \mapsto f(A)$ is linear. We shall now show that this map is also compatible with the multiplications on $\mathcal{F}$ and $\mathcal{L}(E)$. In order to do this we begin with two lemma's.

LEMMA 5.8. *Let $A \in \mathcal{A}$ be fixed, and let γ be a contour in G which goes around $\sigma(A)$. Then*

$$f(A) = \frac{1}{2\pi i} \int_\gamma f(\lambda) j \circ R(\lambda, A) d\lambda$$

REMARK. We leave it to the reader to give a more precise way of saying that γ should "go around $\sigma(A)$".

PROOF. This is just Cauchy's Theorem. The integrand $f(\lambda) j \circ R(\lambda)$ is analytic in the region between γ and ∂G, and tends to zero as $|\lambda| \to \infty$. □

LEMMA 5.9. *If $\lambda, \mu \in \rho(A)$ then*

$$(\lambda - \mu) R(\lambda) \circ j \circ R(\mu) = R(\mu) - R(\lambda).$$

REMARK. This identity is known as the *resolvent equation.* It can be easily verified by multiplying both sides with $\lambda j - A$ from the left and $\mu j - A$ from the right.

THEOREM 5.10. *For fixed $A \in \mathcal{A}$ the map $f \mapsto f(A)$ from $\mathcal{F}$ to $\mathcal{L}(E)$ is an algebra homomorphism.*

PROOF. We already know that $f(A)$ is linear in f, so we only have to verify that for any $f, g \in \mathcal{F}$ one has $(fg)(A) = f(A) g(A)$. We expand the product $f(A) g(A)$:

$$f(A) g(A) = \left(\frac{1}{2\pi i}\right)^2 \int_{\partial G} \int_{\partial G} f(\lambda) g(\mu) j \circ R(\lambda) \circ j \circ R(\mu) d\mu d\lambda$$

$$= \left(\frac{1}{2\pi i}\right)^2 \int_{\partial G} \int_{\gamma} f(\lambda) g(\mu) j \circ R(\lambda) \circ j \circ R(\mu) d\mu d\lambda$$

(by Lemma 5.8)

$$= \left(\frac{1}{2\pi i}\right)^2 \int_{\partial G} \int_{\gamma} f(\lambda) g(\mu) \frac{j \circ R(\lambda) - j \circ R(\mu)}{\mu - \lambda} d\mu d\lambda.$$

This integral consists of two terms, the first of which is

$$\left(\frac{1}{2\pi i}\right)^2 \int_{\partial G} f(\lambda) j \circ R(\lambda) \left(\int_\gamma \frac{g(\mu)}{\mu - \lambda} d\mu\right) d\lambda.$$

Since $g \in \mathcal{F}$ there is a sequence of numbers $r_n \uparrow \infty$, such that $r_n \omega_g(r_n) \downarrow 0$. Hence, by Cauchy's Theorem

$$\int_\gamma \frac{g(\mu)}{\mu - \lambda} d\mu = 0$$

for all $\lambda \in \partial G$, so that the first term in our expansion of $f(A)g(A)$ vanishes. Thus we are left with the second term which equals

$$\left(\frac{1}{2\pi i}\right)^2 \int_{\partial G}\int_{\gamma} f(\lambda)g(\mu)\frac{j\circ R(\mu)}{\lambda-\mu}\,d\mu d\lambda$$

$$=\left(\frac{1}{2\pi i}\right)^2 \int_{\gamma} g(\mu)j\circ R(\mu)\left(\int_{\partial G}\frac{f(\lambda)}{\lambda-\mu}d\lambda\right)d\mu,$$

by Fubini's Theorem. Again applying Cauchy's Theorem we get

$$\int_{\partial G}\frac{f(\lambda)}{\lambda-\mu}d\lambda = 2\pi i f(\mu),$$

whenever $\mu \in \gamma$. Therefore

$$f(A)g(A) = \frac{1}{2\pi i}\int_{\gamma} g(\mu)f(\mu)j\circ R(\mu)d\mu$$

$$= (fg)(A). \quad \square$$

We conclude this section with the following result. We define $\mathcal{L}_s(E)$ to be the vector space $\mathcal{L}(E)$ with the topology generated by the strong topologies on $\mathcal{L}(E_0)$ and $\mathcal{L}(E_1)$. A family of operators $T_\lambda \in \mathcal{L}_s(E)$ converges to a $T \in \mathcal{L}_s(E)$ if for any $x \in E_i$ $(i=0,1)$ $T_\lambda x$ converges in E_i to Tx.

THEOREM 5.11. *Let $A \in \mathcal{Q}$, $f \in \mathcal{F}$ be given. Then the family of operators $\{f(tA): 0<t\leq 1\}$ is uniformly bounded in $\mathcal{L}(E)$. Furthermore, as $t\downarrow 0$, $f(tA)$ converges in $\mathcal{L}_s(E)$ to $f(0)$.*

PROOF. We have the following expression for $f(tA)$:

$$f(tA) = \frac{1}{2\pi i}\int_{\partial G} f(\lambda)j\circ(\lambda j - tA)^{-1}d\lambda = \frac{1}{2\pi i}\int_{\partial G} f(\lambda)j\circ R\left(\frac{\lambda}{t}\right)\frac{d\lambda}{t}.$$

It follows from the estimates

$$\|j\circ R(\lambda)\|,\ \|R(\lambda)\circ j\| \leq \frac{C}{|\lambda|}$$

that $\|j\circ R(\lambda)\|_{\mathcal{L}(E)} \leq \frac{C}{|\lambda|}$, so that

$$\|f(tA)\|_{\mathcal{L}(E)} \leq \frac{1}{2\pi}\int_{\partial G}|f(\lambda)|\frac{C}{|\lambda/t|}\frac{|d\lambda|}{t} \leq \frac{C}{2\pi}\int_{\partial G}\omega_f(|\lambda|)\frac{|d\lambda|}{|\lambda|}.$$

This means that the $f(tA)$ are uniformly bounded in $\mathcal{L}(E)$, and hence in $\mathcal{L}(E_1)$ and $\mathcal{L}(E_0)$. In order to prove strong convergence it is sufficient (by the uniform boundedness) to show that $f(tA)x \to f(0)x$ in E_i for all x in a dense subset of E_i $(i=0,1)$. First we show that $f(tA)j(x) \to f(0)j(x)$ in E_0 for all $x \in E_1$. Choose a $\mu_0 \in \mathbb{C}\setminus G$, so that $\mu_0 \in \rho(A)$. Then any $x \in E_1$ may be written as $R(\mu_0)y$ for some $y \in E_0$. The resolvent equation tells us that

$$j\circ R(\lambda)\circ j\circ R(\mu_0)y = \frac{j\circ R(\lambda)y}{\mu_0-\lambda} + \frac{j\circ R(\mu_0)y}{\lambda-\mu_0}.$$

Using the following representation of $f(tA)$,

$$f(tA)=\frac{1}{2\pi i}\int_{\partial G} f(\lambda')j\circ R(\frac{\lambda'}{t})\frac{d\lambda'}{t}=\frac{1}{2\pi i}\int_{t^{-1}\partial G} f(t\lambda)j\circ R(\lambda)d\lambda$$

$$=\frac{1}{2\pi i}\int_{\partial G} f(t\lambda)j\circ R(\lambda)d\lambda$$

(by Cauchy's Theorem),
we can write $f(tA)x=f(tA)\circ j\circ R(\mu_0)y$ as:

$$f(tA)\circ j\circ R(\mu_0)y=\frac{1}{2\pi i}\int_{\partial G}\frac{j\circ R(\lambda)y}{\mu_0-\lambda}f(t\lambda)d\lambda+\frac{1}{2\pi i}\int_{\partial G}\frac{j\circ R(\mu_0)y}{\lambda-\mu_0}f(t\lambda)d\lambda.$$

The second term in this sum vanishes by Cauchy's Theorem. The integrand in the first term is dominated by $|\mu_0-\lambda|^{-1}\|j\circ R(\lambda)y\|\cdot\|f\|_{\mathcal{F}}$ which is $O(|\lambda|^{-2})$. Therefore we can apply the dominated convergence theorem as $t\downarrow 0$. We get:

$$f(tA)x\to\frac{1}{2\pi i}f(0)\int_{\partial G}\frac{j\circ R(\lambda)y}{\mu_0-\lambda}d\lambda.$$

The integrand of this integral is analytic on $\mathbb{C}\setminus(G\cup\{\mu_0\})$ and is $O(|\lambda|^{-2})$ at $\lambda=\infty$ so that we have

$$f(tA)x\to f(0)j\circ R(\mu_0)y=f(0)x$$

in E_0. This proves that $f(tA)\to f(0)$ in $\mathfrak{L}_s(E_0)$ as $t\downarrow 0$. We still have to prove this in $\mathfrak{L}_s(E_1)$. Let $y\in E_0$ be given, then $R(\mu_0)y\in E_1$. Since $R(\mu_0)$ commutes with the $f(tA)$ we get:

$$f(tA)R(\mu_0)y=R(\mu_0)f(tA)y,$$

and, using the fact that $f(tA)y$ converges to $f(0)y$ in E_0 as $t\downarrow 0$, we see that

$$f(tA)R(\mu_0)y\to f(0)R(\mu_0)y \text{ in } E_1 \quad (t\downarrow 0).$$

However, any $x\in E_1$ can be written as $R(\mu_0)y$ for some $y\in E_0$, so that $f(tA)\to f(0)$ in $\mathfrak{L}_s(E_1)$, which completes the proof. □

6. The exponential mapping

Using the functional calculus developed in the preceding section we shall construct a semigroup $\{\exp(tA)\}_{t\geq 0}$ for $A\in Hol(E)$. As the notation already suggests, A will turn out to be the generator of $\{\exp(tA)\}_{t\geq 0}$. In particular we obtain the inclusion $Hol(E)\subset Gen(E)$.

Let $E=(E_1,E_0)$ be a Banach couple, and let $A\in Hol(E)$ be given. As in Section 5.5 we assume that $A\in\mathcal{Q}$, i.e.

$$\{\lambda:|\arg(\lambda)|\leq\pi-\theta\}\subset\rho(A)$$

and

$$\|R(\lambda,A)\|\leq C \text{ for all } \lambda\in\mathbb{C} \text{ with } |\arg(\lambda)|\leq\pi-\theta,$$

hold for certain constants C and θ $(0<\theta<\frac{1}{2}\pi)$. For all t in the sector

$$S = \{t \in \mathbb{C} : t \neq 0, |\arg(t)| < \frac{\pi}{2} - \theta\}$$

we consider the function $e_t(\lambda)$ defined by

$$e_t(\lambda) = e^{t\lambda}.$$

A straightforward argument shows that for any $t \in S$ the function $e_t(\lambda)$ belongs to the class $\mathfrak{F}$ defined in Section 5.5. Thus we can define

$$e^{tA} = e_t(A) \quad (t \in S).$$

The family of operators $e_t(A)$ has the following properties:

THEOREM 5.12.
a) *The mapping $t \in S \to e_t(A) \in \mathfrak{L}(E)$ is holomorphic.*
b) *For any $\phi < \pi/2 - \theta$ one has $e_t(A) \xrightarrow{s} 1$ as $t \to 0$ with $|\arg(t)| \leqslant \phi$.*
c) *$e_{t+s}(A) = e_t(A)e_s(A)$, for all $t, s \in S$.*

PROOF.
a) This follows from the fact that the map $t \in S \to e_t \in \mathfrak{F}$ is holomorphic. Indeed it's derivative is

$$\frac{d}{dt} e_t(\lambda) = \lambda e_t(\lambda),$$

which belongs to $\mathfrak{F}$.
b) It follows from Theorem 5.11 that $e_t(A)$ converges strongly to 1 as t tends to zero along a straight line $\arg(t) =$ constant. The proof we gave of Theorem 5.11 can easily be modified to show that $e_t(A) \xrightarrow{s} 1$ as $t \to 0$ in the sector $|\arg(t)| \leqslant \phi$.
c) This follows directly from the identity

$$e_t(\lambda)e_s(\lambda) = e_{t+s}(\lambda),$$

and the fact that $f \mapsto f(A)$ is an algebra homomorphism from $\mathfrak{F}$ to $\mathfrak{L}(E)$. □

COROLLARY. *The family $\{e_t(A)\}_{t \geqslant 0}$ is a C_0-semigroup.*

REMARK. Part (a) of Theorem 5.12 can be strengthened in the following way: *the exponential mapping*

$$A \in \mathfrak{A} \mapsto e^A \in \mathfrak{L}(E)$$

is holomorphic.
This follows immediately from Theorem 5.7.

Up to this point we know that $\{e^{tA}\}_{t \geqslant 0}$ is a C_0-semigroup which extends holomorphically to $t \in S$. It is not entirely obvious, however, that A is the generator of $\{e^{tA}\}_{t \geqslant 0}$. The next lemmas will show that this is actually true.

LEMMA 5.13. *Let $A \in \mathfrak{A}$, $f \in \mathfrak{F}$ and $\lambda_0 \in \mathbf{C} \setminus G$ be given. Then*

$$j \circ R(\lambda_0, A) f(A) = g(A)$$

where $g \in \mathfrak{F}$ is defined by

$$g(\lambda) = (\lambda_0 - \lambda)^{-1} f(\lambda).$$

(Recall that G was defined in Section 5.5.)

PROOF. We have

$$j \circ R(\lambda_0, A) f(A) = (2\pi i)^{-1} \int_{\partial G} f(\lambda) j \circ R(\lambda_0, A) \circ j \circ R(\lambda, A) d\lambda$$

$$= (2\pi i)^{-1} \int_{\partial G} \{ f(\lambda) \frac{j \circ R(\lambda, A)}{\lambda_0 - \lambda} + f(\lambda) \frac{j \circ R(\lambda_0, A)}{\lambda - \lambda_0} \} d\lambda,$$

by the resolvent equation. The second term in the integral vanishes by Cauchy's Theorem. What remains is then, by definition, equal to $g(A)$. □

REMARK. The identity $j \circ R(\lambda_0, A) f(A) = g(A)$ holds in $\mathfrak{L}(E)$. In $\mathfrak{L}(E_0)$ this means that $j \circ R(\lambda_0, A) f(A) = g(A)$, whereas in $\mathfrak{L}(E_1)$ one has the identity $R(\lambda_0, A) \circ j \circ f(A) = g(A)$.

LEMMA 5.14. *For $x_0 \in E_1$ consider*

$$x(t) = e_t(A) x_0, \quad 0 \leqslant t < \infty.$$

Then $x \in C^0([0,\infty), E_1)$ and $j \circ x \in C^1([0,\infty); E_0)$. The following holds:

$$(j \circ x)'(t) = A x(t) \quad (0 \leqslant t < \infty).$$

PROOF. Define $e'_t(\lambda) = \lambda e_t(\lambda)$. We know that on $(0, \infty)$, $x(t)$ is analytic, and that $x'(t) = e'_t(A) x_0$. Therefore

$$A^{-1} (j \circ x)'(t) = R(0, A) \circ j \circ e'_t(A) x_0 = e_t(A) x_0.$$

In Theorem 5.12 we saw that $e_t(A) x_0$ belongs to $C^0([0,\infty); E_1)$. Therefore we have

$$(j \circ x)'(t) = A e_t(A) x_0 = A x(t),$$

so that $j \circ x \in C^1([0,\infty); E_0)$. □

THEOREM 5.15. *For $A \in \mathfrak{A}$, A is the generator of the C_0 semigroup $\{e_t(A)\}_{0 \leqslant t < \infty}$.*

PROOF. By the Hille-Yosida theorem $\{e_t(A)\}_{t \geqslant 0}$ has a generator $B : \mathfrak{D}(B) \to E_0$, where $\mathfrak{D}(B)$ is dense in E_0. We have to show that $E_1 = \mathfrak{D}(B)$, and that on $\mathfrak{D}(B)$, $A = B$ holds.

The domain of B is defined by

$$\mathfrak{D}(B) = \{x \in E_0 : (t \to e_t(A)x) \in C^1([0,\infty);E_1)\},$$

and on $\mathfrak{D}(B)$ we have

$$Bx = \lim_{t \downarrow 0} \frac{e_t(A)x - x}{t}.$$

The previous lemma immediately implies that $E_1 \subset \mathfrak{D}(B)$, and that $B|_{E_1}$ and A coincide. It remains to show $\mathfrak{D}(B) \subset E_1$. Let $x_0 \in \mathfrak{D}(B)$ be given. Then the derivative of $x(t) = e_t(A)x_0$,

$$x'(t) = e_t'(A)x_0,$$

is continuous up to $t=0$. Thus

$$A^{-1}x'(t) = R(0,A)e'_t(A)x_0 = e_t(A)x_0$$

belongs to $C^0([0,\infty);E_1)$, so that

$$x_0 = \lim_{t \downarrow 0} e_t(A)x_0$$

belongs to E_1, which completes the proof. □

7. Some estimates on e^{tA}

Let $A \in Hol(E)$ be fixed, and define

$$E_k(A) = \{x \in E_{k-1}(A) : Ax \in E_{k-1}(A)\} \quad (k \geq 2)$$

by induction with $E_1(A) = E_1, E_0(A) = E_0$. The spaces E_k $(k \geq 2)$ are *not* independent of the operator A. One easily verifies the following characterization of E_k:

$$E_k = \mathfrak{R}((j \circ R(\lambda_0, A))^k) \text{ for any } \lambda_0 \in \rho(A),$$

where $\mathfrak{R}(\cdot)$ denotes the *range*. It follows that the E_k are all dense in E_0. The spaces E_k are defined in such a way that whenever $x \in E_k$, $A^k x = A \circ A \circ \cdots \circ Ax$ is well defined, and $A^k x \in E_0$.

Theorem 5.16. *If $A \in \mathfrak{A}$ then for any $t > 0$ and any $x \in E_0$, one has $e_t(A) \in E_k$ $(k = 1, 2, \cdots)$ Furthermore*

$$\|A^k e_t(A)\|_{\mathfrak{L}(E)} \leq \frac{C}{t^k} \quad (k \geq 1, 0 < t \leq 1).$$

Proof. Define $e_t^{(k)}(\lambda) = \lambda^k e^{t\lambda}$. Then $e_t^{(k)} \in \mathfrak{F}$ for all $t > 0$, $k = 0, 1, 2, \cdots$. Since $A \in \mathfrak{A}$ we have $0 \in \rho(A)$ so that

$$(j \circ A^{-1})^k e_t^{(k)}(A)x = e_t(A)x.$$

In other words $e_t(A)x \in E_k$, and

$$A^k e_t(A)x = e_t^{(k)}(A)x.$$

(Strictly speaking we should write $(A \circ j^{-1})^k$ instead of A^k). Thus

$$t^k A^k e_t(A)x = e_1^{(k)}(tA)x,$$

which is uniformly bounded in E_0 for $0<t\leqslant 1$, by Theorem 5.11. This implies the claimed estimate. □

REMARK. Theorem 5.11 allows one to sharpen the previous result a little. Namely, for any $x\in E_0$ one has

$$\|A^k e_t(A)x\|_{E_0} = o(t^{-k}) \quad (t\downarrow 0),$$

instead of just $O(t^{-k})$. Indeed, $t^k A^k e_t(A)$ equals $e_1^{(k)}(tA)$ which converges strongly to $e_1^{(k)}(0)1=0$ as $t\downarrow 0$.

We see from this theorem that $\cap_{k\geqslant 1}E_k$ is dense in E_0. In fact a much stronger result holds. Define an $x\in E_\infty=\cap_{k\geqslant 1}E_k$ to be *entire* if

$$\limsup_{k\to\infty}\|\frac{1}{k!}A^k x\|_{E_0}^{1/k} = 0.$$

Denote the set of entire vectors by E_ω. Then we have:

THEOREM 5.17. *Let $x_0\in E_0$ be given, and define $x(t)=e_t(A)x_0$ $(0\leqslant t<\infty)$. Then*
(a) *$x_0\in E_k$ iff $x(\cdot)\in C^k([0,\infty);E_0)$*
(b) *$x_0\in E_\omega$ iff $x(\cdot)$ extends to an entire function with values in E_0*
(c) *E_ω is dense in E_0.*

PROOF.
(a) If $x\in C^k([0,\infty);E_0)$ then

$$x^{(k)}(t) = e_t^k(A)x_0$$

is continuous in E_0 up to $t=0$, and thus so is

$$(j\circ A^{-1})^k x^{(k)}(t) = e_t(A)x_0.$$

Therefore

$$x_0=(j\circ A^{-1})^k x^{(k)}(0)\in E_k.$$

Conversely, if $x_0\in E_k$, then $x_0=(j\circ A^{-1})^k y_0$ for some $y_0\in E_0$, and

$$x^{(k)}(t) = e_t^k(A)(j\circ A^{-1})^k y_0 = e_t(A)y_0,$$

which is continuous up to $t=0$. Hence $x\in C^k([0,\infty);E_0)$.
(b) If $x(\cdot)$ extends to an entire function, then $x(t)=\Sigma_{k=0}^{\infty}x_k t^k$, and $x\in C^\infty([0,\infty);E_0)$. Hence $x_0\in E_k$ for all $k\geqslant 0$, and

$$\frac{1}{k!}A^{-k}x_0 = x_k$$

holds for $k=0,1,2,\cdots$. The condition $x_0\in E_\omega$ is then equivalent to $\limsup_{k\to\infty}\|x_k\|^{1/k}=0$, which is actually true because $x(t)$ is an entire function. To prove the converse, suppose $x_0\in E_\omega$ is given. Then $x_0\in E_k$ for all

$k \geqslant 0$, and we have the estimate $\limsup_{k\to\infty} \|\frac{1}{k!}A^k x_0\|_{E_0}^{1/k} = 0$. As $j \circ R(\lambda, A)$ and $e_t(A)$ commute the spaces E_k $(k \geqslant 1)$ are invariant under $e_t(A)$. Therefore

$$\sup_{0\leqslant t\leqslant 1} \|x^{(k)}(t)\|_{E_0} \leqslant \sup_{0\leqslant t\leqslant 1} \|e_t(A)\| \cdot \|x^{(k)}(0)\|_{E_0} = M\|A^k x_0\|_{E_0},$$

and as a consequence of this:

$$\limsup_{k\to\infty} \{\sup_{0\leqslant t\leqslant 1} \|\frac{1}{k!}x^{(k)}(t)\|_{E_0}\}^{1/k} = 0.$$

This means that the Taylor expansion of $x(\cdot)$ at $t=0$ converges and represents the function $x(\cdot)$ on the interval $0 \leqslant t \leqslant 1$. To conclude the proof of (b) we observe that the Taylor series of $x(\cdot)$ at $t=0$ is an entire function, which must coincide with $x(t)$ on $0 \leqslant t < \infty$, since $x(t)$ is analytic for $t > 0$.
(c) Recall that $A \in \mathcal{Q}$ means that

$$\{\lambda : |\arg(\lambda)| \leqslant \pi - \theta\} \subset \rho(A)$$

and that $R(\lambda, A)$ is uniformly bounded for these values of λ. For any $A \in \mathcal{Q}$ we had an algebra homomorphism $f \in \mathfrak{F} \mapsto f(A) \in \mathfrak{L}(E)$, where $\mathfrak{F}$ contained functions $f(\lambda)$ defined on

$$G = \{\lambda \in \mathbf{C} : |\arg(-\lambda)| \leqslant \theta \text{ or } |\lambda| \leqslant \epsilon\}.$$

Now let $\alpha > 1$ be any number such that $2\alpha\theta < \pi$ holds. Then

$$f(\lambda) = \exp(a - (a - \lambda)^\alpha)$$

belongs to $\mathfrak{F}$, if $a > \epsilon$ (so that the branch point $\lambda = a$ of f lies outside of G). Indeed one has the following bound on $f(\lambda)$:

$$|f(\lambda)| \leqslant \exp(a - |a - \lambda|^\alpha \cos\alpha\theta).$$

(We urge the reader to draw some pictures in order to verify this estimate). Because of the rapid decay of $f(\lambda)$ as $|\lambda| \to \infty$ the product

$$e_t(\lambda) f(\epsilon\lambda) \quad (t \in \mathbf{C}, \epsilon > 0)$$

defines an entire function of t with values in $\mathfrak{F}$. Using the functional calculus we obtain an entire function $e_t(A) f(\epsilon A)$ of $t \in \mathbf{C}$ with values in $\mathfrak{L}(E)$. If $x_0 \in E_0$ is a given vector, then for any $\epsilon > 0$, $x_\epsilon(t) = e_t(A) f(\epsilon A) x_0$ is an entire function of t, so that $x_\epsilon(0)$ belongs to E_ω. But $x_\epsilon(0) = f(\epsilon A) x_0$ converges to x_0 as $\epsilon \downarrow 0$ (by Theorem 5.11), so that E_ω is dense in E_0. □

Remark. If, in the proof given here, we had taken $x_0 \in E_1$, then $f(\epsilon A) x_0$ would have converged to x_0 in the E_1 -norm. Hence we obtain the density of E_ω in E_1 as well.

8. EXAMPLES AND EXERCISES

Let $L_2^s(\mathbb{R})$ be the subspace of $L_2(\mathbb{R})$ defined by

$$L_2^s(\mathbb{R}) = \{f \in L_2(\mathbb{R}) : |x|^s \cdot f(x) \in L_2\}$$

with norm

$$\|f\|_s^2 = \int_{\mathbb{R}} (1 + |x|^{2s}) |f(x)|^2 dx.$$

Then $L_2^s \hookrightarrow L_2$ is a dense inclusion, so that for any $s > 0$, $L^{(s)} = (L_2^s(\mathbb{R}), L_2(\mathbb{R}))$ is a Banach couple. Define $A_s : L_2^s \to L_2$ by

$$(A_s f)(x) = -|x|^s \cdot f(x).$$

EXERCISE 5.18. Prove that $A_s \in Hol(L^{(s)})$ for all $s > 0$, and show that $\sigma(A_s) = (-\infty, 0]$.

Next we consider the operator B_s $(s > 0)$, which is defined by:

$$(B_s f)(x) = (-|x|^s + ix) f(x).$$

EXERCISE 5.19.

(a) Prove that $B_s \in \mathfrak{L}(L_2^{\bar{s}}, L_2)$, where $\bar{s} = \min(s, 1)$.

(b) Prove that $B_s \in Hol(L_2^{\bar{s}})$ if and only if $s \geqslant 1$.

(c) Determine $\sigma(B_s)$ for $s > 0$.

Thus B_s does not generate an analytic semigroup if $s \in (0,1)$. Nevertheless, if $s \in (0,1)$ then B_s, with domain L_2^1 generates a strongly continuous semigroup, which is given by

$$\exp(tB_s) f(x) = e^{(-|x|^s + ix)t} f(x) \quad (0 \leqslant t < \infty).$$

EXERCISE 5.20.

(a) Show that the operator valued function $t \mapsto \exp(tB_s)$ is C^∞ from $(0, \infty)$ to $\mathfrak{L}(L_2(\mathbb{R}))$, whenever $s \in (0,1)$.

(b) Prove that the range of $\exp(tB_s)$ is contained in $\mathfrak{D}(B_s) = L_2^1(\mathbb{R})$, and that

$$\|B_s \exp(tB_s)\| \leqslant C \cdot t^{-1/s} \quad (0 < t \leqslant 1)$$

holds for some constant $C > 0$, for $0 < s < 1$.

The point of this exercise is to show that there exist C^∞-semigroups which are not analytic, and that the estimate given in Theorem 5.16 fails for such semigroups (as it should). In fact it can be shown that, if $\{e^{tA}\}_{t \geqslant 0}$ is a C_0-semigroup on a Banach space X for which $e^{tA}(X) \subset \mathfrak{D}(A)$ holds for all $t > 0$, then $A \in Hol(\mathfrak{D}(A), X)$ if and only if the operators $\{tAe^{tA}\}_{0 < t \leqslant 1}$ are uniformly bounded on X. The proof can be found in most textbooks on linear semigroup theory (e.g. DAVIES [1980]).

EXERCISE 5.21. Let $E=(E_1,E_0)$ be a Banach couple, and let $A\in Hol(E)$ be given. By Theorem 5.16 we know that $\|Ae^{tA}\|_{\mathfrak{L}(E_0)}$ is $O(t^{-1})$ as $t\downarrow 0$. Prove that for any $x\in E_0$ one actually has

$$\|Ae^{tA}x\|_{E_0}=o(t^{-1})\quad (t\downarrow 0)$$

without using Theorem 5.11.

Hint: Since E_1 is dense in E_0 one can write $x=y_\epsilon+z_\epsilon$ with $y_\epsilon\in E_1$ and $z_\epsilon\in E_0, \|z_\epsilon\|_{E_0}\leqslant\epsilon$, for any $\epsilon>0$. Now write $Ae^{tA}x=Ae^{tA}z_\epsilon+e^{tA}Ay_\epsilon$, and estimate both terms separately.

EXERCISE 5.22. Consider the operator $C:L_2^{(1)}(\mathbb{R})\to L_2(\mathbb{R})$ given by

$$(Cf)(x)=[-\log(1+|x|)+ix]\cdot f(x).$$

(a) Show that C generates a strongly continuous semigroup e^{tC} on $L_2(\mathbb{R})$.
(b) Show that the $\mathfrak{L}(L_2(\mathbb{R}))$ -valued function $t\mapsto e^{tC}$ on $\mathbb{R}_+$ has exactly k continuous derivatives when $k<t\leqslant k+1$ (for $k=0,1,2,\cdots$).

EXERCISE 5.23. Let E be the Banach couple

$$E_0=\{u\in C([0,1]):u(1)=0\},$$

$$E_1=\{u\in C^1([0,1]):u \text{ and } u'\in E_0\},$$

and let $D\in\mathfrak{L}(E_1,E_0)$ be given by

$$Df(x)=f'(x).$$

(a) Show that D generates a strongly continuous semigroup e^{tD} on E_0.
(b) Compute the resolvent of D and show that D has empty spectrum.
(c) Verify that $D\notin Hol(E)$ (*Hint:* for $t\geqslant 1$ one has $e^{tD}=0$).

This last exercise shows that even if the spectrum of a $D:E_1\to E_0$ is contained in some sector $\{\lambda:|\arg(-\lambda)|<\theta\}$, D need not generate a holomorphic semigroup. One really needs the estimates on the resolvent of D for this.

EXERCISE 5.24 : *fractional powers*
Let $E=(E_1,E_0)$ and $A\in\mathcal{Q}$ be as in Section 5.5 (on the functional calculus). For each $z\in\mathbb{C}$ with $\operatorname{Re} z>0$ define

$$f_z(\lambda)=(1-\lambda)^{-z}=e^{-z\log(1-\lambda)},$$

where $\log(1-\lambda)$ is defined on the complex plane cut along the half-line $[1,\infty)$, and normalised by $\log(1)=0$.
(a) Show that $f_z\in\mathfrak{F}$ (notation as in Section 5.5).
The functional calculus can therefore be used to define

$$(1-A)^{-z}=f_z(A)\quad (\operatorname{Re} z>0).$$

(b) Verify that for $\operatorname{Re} z, \operatorname{Re} w>0$ one has $(1-A)^{-z}\circ(1-A)^{-w}=(1-A)^{-(z+w)}$.

(c) Show that the following identity holds:

$$(1-A)^{-z} = \frac{1}{\Gamma(z)} \int_0^\infty t^{z-1} e^{tA} e^{-t} dt,$$

where $\Gamma(z)$ is Euler's gamma function.

Hint: The integral should be interpreted as an $\mathfrak{L}(E)$ -valued Bochner integral. Use the functional calculus definition of e^{tA}.

(d) Verify that for $n=1,2,3 \cdots, (1-A)^{-n}$ actually is the n-th power of the inverse of $1-A$.

EXERCISE 5.25 : *Gaussian semigroups*

For $1 \leqslant p < \infty$ define $E_0 = L_p(\mathbb{R})$ and

$$E_1 = W_p^2(\mathbb{R}) = \{f \in L_p(\mathbb{R}) : f', f'' \in L_p(\mathbb{R})\}.$$

Let $A \in \mathfrak{L}(E_1, E_0)$ be given by $Af(x) = \frac{1}{2} f''(x)$.

(a) Determine the resolvent of A.

Answer: $\sigma(A) = (-\infty, 0]$. For $\lambda \in \rho(A)$ the operator $(\lambda j - A)^{-1}$ is given by convolution with

$$G_\lambda(x) = (2\lambda)^{-1/2} \exp(-(2\lambda)^{\frac{1}{2}} |x|).$$

(b) Verify that $A \in Hol(E)$.

Hint: use Young's inequality

$$\|f \star g\|_{L_p} \leqslant \|f\|_{L_1} \cdot \|g\|_{L_p}.$$

(c) Use the functional calculus to show that e^{tA} is given by convolution with $(2\pi t)^{-1/2} \exp(-x^2/2t)$.

Chapter 6

Interpolation and Maximal Regularity

1. Interpolation methods

An *interpolation method* is a map I from the class of Banach couples $\mathcal{C}$ to the class of Banach spaces $\mathcal{B}$, which satisfies the following two axioms:

(1) If $X = I(E)$ then $E_1 \subset X \subset E_0$

(2) If $E, F \in \mathcal{C}$ are couples and $T \in \mathcal{L}(E,F)$ is a map of pairs then T maps $I(E)$ into $I(F)$

$$\begin{array}{ccc} & T & \\ E_0 & \longrightarrow & F_0 \\ \cup & & \cup \\ I(E) & \dashrightarrow T \dashrightarrow & I(F) \\ \cup & & \cup \\ E_1 & \longrightarrow & F_1 \\ & T & \end{array}$$

The closed graph theorem implies that T is bounded from $I(E)$ to $I(F)$.

The interpolation method I is said to be *of exponent* $\theta \in (0,1)$ if there is a constant C such that

$$\|T\|_{I(E),I(F)} \leqslant C \|T\|^{\theta}_{E_1,F_1} \cdot \|T\|^{1-\theta}_{E_0,F_0}.$$

The method is called exact of exponent θ if the given constant can be chosen to be $C = 1$. The following lemma illustrates how property (2) can be used. Recall that intermediate spaces were defined in Chapter 5.

LEMMA 6.1. *If I is of exponent θ, then for any couple E, $I(E)$ is an intermediate space of exponent θ for E.*

PROOF. Let $x \in E_1$ be given, and let F be the couple $(\mathbb{C},\mathbb{C})$. Define $T \in \mathfrak{L}(F,E)$ by $Ts = sx$ ($\forall s \in \mathbb{C}$). Then

$$\|T\|_{\mathbb{C},E_k} = \|x\|_{E_k} \quad (k = 0,1),$$

and $\|T\|_{\mathbb{C},X} = \|x\|_X$, so that property (2) of the interpolation method implies

$$\|x\|_X \leqslant C\|x\|_{E_1}^{\theta} \cdot \|x\|_{E_0}^{1-\theta}. \quad \square$$

2. THE CONTINUOUS INTERPOLATION METHOD

In this section we construct a one parameter family of methods I_θ $(0<\theta<1)$, and prove that the given method I_θ is exact of exponent θ. For any couple E we define a Banach space $V_\theta(E)$ of functions $u:\mathbb{R}_+ \to E_1$. More precisely:

$V_\theta(E) =$ $\{u \in C(\mathbb{R}_+;E_1) \cap C^1(\mathbb{R}_+;E_0)$: the functions $t \to t^{1-\theta}\|u(t)\|_{E_1}$ and $t \to t^{1-\theta}\|u'(t)\|_{E_0}$ are bounded and tend to zero as $t \downarrow 0\}$.

Clearly $V_\theta(E)$ is a vector space, and it is a Banach space if we give it the norm

$$\|u\|_{V_\theta(E)} = \sup_{t>0} t^{1-\theta}\max(\|u(t)\|_{E_1}, \|u'(t)\|_{E_0}).$$

Since $u'(t)$ is $o(t^{\theta-1})$ $(t \downarrow 0)$ in E_0, whenever $u \in V_\theta(E)$, $u(t)$ is uniformly continuous and

$$u(0) = \lim_{t \downarrow 0} u(t) = u(1) - \int_0^1 u'(s)ds$$

exists in E_0. Thus we obtain a map

$$r_\theta : V_\theta(E) \to E_0$$

$$r_\theta(u(\cdot)) = u(0),$$

which is bounded and linear.

DEFINITION. $I_\theta(E) = \mathfrak{R}(r_\theta)$, and the norm on $I_\theta(E)$ is given by:

$$\|x\|_\theta = \inf\{\|u\|_{V_\theta(E)} : r_\theta u = x\}.$$

If we identify $\mathfrak{R}(r_\theta)$ with $V_\theta(E)/\mathfrak{N}(r_\theta)$, (here $\mathfrak{R}(\cdot)$ and $\mathfrak{N}(\cdot)$ denote the range and null space respectively) then the norm on $I_\theta(E)$ is the same as the quotient norm on $V_\theta(E)/\mathfrak{N}(r_\theta)$. Hence $I_\theta(E)$ is a Banach space.

THEOREM 6.2. *I_θ is an interpolation method, which is exact of exponent θ $(0<\theta<1)$.*

PROOF. It is evident that $I_\theta(E) \subset E_0$. Also, if $x \in E_1$ is given then $u(t) = \zeta(t)\cdot x$ with $\zeta \in C_c^\infty(\mathbb{R})$, $\zeta(0) = 1$, defines an element of $V_\theta(E)$ which satisfies $u(0) = x$, so that $E_1 \subset I_\theta(E)$. It remains to verify property (2). Let $E, F \in \mathcal{C}$ be two Banach couples, and let $T \in \mathfrak{L}(E, F)$ be given. For any positive α define $T^{(\alpha)}: V_\theta(E) \to V_\theta(F)$ by

$$(T^{(\alpha)}u)(t) = T(u(\alpha t)) \quad (0 < t < \infty).$$

Then $T^{(\alpha)}$ is bounded and linear. If $x \in I_\theta(E)$ is given then we can write $x = u(0)$ for some $u \in I_\theta(E)$, and thus $Tx = (T^{(\alpha)}u)(0)$, so that $Tx \in I_\theta(F)$. Therefore I_θ is an interpolation method.

Finally we compute the norm of T. Since $\|Tx\|_{I_\theta(F)} \leqslant \|T^{(\alpha)}u\|_{V_\theta(F)}$ we have;

$$\|Tx\|_{I_\theta(F)} \leqslant \|T^{(\alpha)}\| \cdot \|u\|_{V_\theta(E)}$$

for any $u \in V_\theta(E)$ with $u(0) = x$. Hence

$$\|Tx\|_{I_\theta(F)} \leqslant \|T^{(\alpha)}\| \cdot \|x\|_{I_\theta(E)}.$$

The norm of $T^{(\alpha)}$ may be estimated as follows:

$$\begin{aligned}\sup t^{1-\theta} \|(T^{(\alpha)}u)(t)\|_{F_1} &\leqslant \|T\|_1 \cdot \sup t^{1-\theta} \|u(\alpha t)\|_{E_1} \\ &\leqslant \alpha^{\theta-1} \cdot \|T\|_1 \cdot \|u\|_{V_\theta(E)}\end{aligned}$$

and

$$\begin{aligned}\sup t^{1-\theta} \|(T^{(\alpha)}u)'(t)\|_{F_0} &\leqslant \|T\|_0 \cdot \sup \alpha t^{1-\theta} \|u'(\alpha t)\|_{E_0} \\ &\leqslant \alpha^\theta \|T\|_0 \cdot \|u\|_{V_\theta(E)},\end{aligned}$$

so that

$$\|T^{(\alpha)}\| \leqslant \max(\alpha^\theta \|T\|_0, \alpha^{\theta-1} \|T\|_1).$$

If we take $\alpha = \|T\|_1 / \|T\|_0$ then

$$\|T^{(\alpha)}\| \leqslant \|T\|_1^\theta \cdot \|T\|_0^{1-\theta}.$$

Hence

$$\|T\|_{I_\theta(E), I_\theta(F)} \leqslant \|T\|_{E_1, F_1}^\theta \cdot \|T\|_{E_0, F_0}^{1-\theta},$$

which completes the proof. □

3. Alternative descriptions of I_θ

If one is given a Banach couple E then it may be quite hard to determine $I_\theta(E)$ from first principles. Fortunately there are several other characterizations of the space $I_\theta(E) \subset E_0$. Our first alternative is the following: in order to determine for a given $x \in E_0$ whether or not x belongs to $I_\theta(E)$ we investigate how good x can be approximated by elements of E_1. More precisely, define the subspace X of E_0 by:

$x \in X$ if and only if there is a sequence $\{a_n, b_n\} \in E_1 \times E_0$ such that:
(1) $x = a_n + b_n$
(2) $\|a_n\|_{E_1} = o(2^{n(1-\theta)})$ $(n \to \infty)$
(3) $\|b_n\|_{E_0} = o(2^{-n\theta})$ $(n \to \infty)$.

Then the following holds:

THEOREM 6.3. $I_\theta(E) = X$.

PROOF. "$\subset$": Let $x \in I_\theta(E)$ be given. Then $x = u(0)$ for some $u \in V_\theta(E)$. Consider

$$a_n = u(2^{-n}),$$
$$b_n = x - u(2^{-n})$$

(so that $a_n + b_n = x$). Since $u \in V_\theta(E)$ we have

$$\|a_n\|_{E_1} = 2^{n(1-\theta)} \cdot \|(2^{-n})^{1-\theta} u(2^{-n})\|_{E_1} = o(2^{n(1-\theta)}),$$

and

$$-b_n = \int_0^{2^{-n}} u'(s) ds,$$

i.e.

$$\|b_n\|_{E_0} \leqslant \sup_{s \leqslant 2^{-n}} \|s^{1-\theta} u'(s)\|_{E_0} \cdot \int_0^{2^{-n}} s^{\theta-1} ds$$

so that $\|b_n\|_{E_0} = o(2^{-n\theta})$ $(n \to \infty)$. Hence $x \in X$.
"$\supset$". Let $x \in X$ be given. Then there are $a_n \in E_1$ such that $\|a_n\|_{E_1} = o(2^{n(1-\theta)})$, and $\|x - a_n\|_{E_0} = o(2^{-n\theta})$ $(n \to \infty)$. For $n \leqslant 0$ we define $a_n = 0$. Next, construct a positive C^∞ function $\phi(x)$ with support in $(1/4, 1)$ such that $\Sigma_{n \in \mathbf{Z}} \phi(2^n x) = 1$. One might take

$$\phi(x) = \psi(x) \cdot (\Sigma_{n \in \mathbf{Z}} \psi(2^n x))^{-1},$$

where

$$\psi(x) = \begin{cases} 0, & x \leqslant 1/4 \text{ or } x \geqslant 1 \\ \exp[-((1-x)(4x-1))^{-1}], & 1/4 < x < 1. \end{cases}$$

Given this function $\phi(x)$ we put

$$u(t) = \sum_{n=-\infty}^{\infty} a_n \phi(2^n \cdot t) \quad (0 < t < \infty),$$

and proceed to verify that $u \in V_\theta(E)$. Consider

$$t^{1-\theta} u(t) = \sum_{u \in \mathbf{Z}} a_n \cdot t^{1-\theta} \phi(2^n t).$$

In this sum only three terms are non zero, say those with $n=N-1$, N and $n=N+1$. Since $\phi(2^n t)\neq 0$ for those values of n we have $1/16<2^N t<4$, so that $t^{1-\theta}\leqslant C.2^{-N(1-\theta)}$. Thus

$$\|t^{1-\theta}u(t)\|_{E_1}\leqslant \text{const}\cdot 2^{-N(1-\theta)}\cdot(\|a_N\|+\|a_{N-1}\|+\|a_{N+1}\|)$$
$$=o(1) \quad (N\to\infty, \text{ or } t\downarrow 0).$$

Next, consider

$$u(t)=x+(u(t)-x)$$
$$=x+\sum_{n\in\mathbf{Z}}(a_n-x)\cdot\phi(2^n t).$$

After differentiating this, one gets

$$u'(t)=\sum_{n\in\mathbf{Z}}(a_n-x)\cdot 2^n\phi'(2^n t),$$

so that a similar argument as above gives

$$t^{1-\theta}\|u'(t)\|_{E_0}\leqslant\sum_{n\in\mathbf{Z}}t^{1-\theta}\|a_n-x\|_{E_0}\cdot 2^n|\phi'(2^n t)|$$
$$\leqslant \text{const}.2^N 2^{-N(1-\theta)}(\|a_{N-1}-x\|_{E_0}+\|a_N-x\|_{E_0}+\|a_{N+1}-x\|_{E_0})$$
$$=o(1) \quad (N\to\infty, \text{ or } t\downarrow 0).$$

These estimates show that $u\in V_\theta(E)$. Since it is easily seen that $u(0)=x$ we have proved that $x\in I_\theta(E)$, and so $X=I_\theta(E)$. □

REMARK. The quantity

$$\inf\{\sup_{n\geqslant 1}(2^{-n(1-\theta)}\|a_n\|_{E_1},\, 2^{n\theta}\|b_n\|_{E_0})\},$$

where the infimum is taken over all sequences $\{a_n,b_n\}$ with $a_n+b_n=x$, defines a norm on X equivalent to the norm of $I_\theta(E)$. We leave it to the reader to verify this (hint: show that the given expression defines a complete norm on X and apply the closed graph theorem to the identity $id:X\to I_\theta(E)$).

As the reader may have guessed the number "2" in the definition of X may be replaced by an arbitrary real number $a>1$.

A slight variant of the space X is the space Y defined by $x\in Y$ iff there are $c_n\in E_1$ $(n\geqslant 1)$ such that

(1) $x=\Sigma_{n\geqslant 1}c_n$

(2) $\|c_n\|_{E_1}=o(2^{n(1-\theta)})$, $\|c_n\|_{E_0}=o(2^{-n\theta})$.

THEOREM 6.4. $Y=I_\theta(E)$.

PROOF. We show that $Y=X$. Let $x=\Sigma c_n\in Y$ be given, with c_n as above. Define

$$a_n = \sum_{k=1}^{n} c_k, \quad b_n = \sum_{k=n+1}^{\infty} c_k,$$

then $a_n + b_n = x$ and

$$\|a_n\|_{E_1} \leqslant \sum_{1}^{n} \|c_k\|_{E_1} = o(2^{n(1-\theta)}),$$

since $\|c_n\|_{E_1} = 0(2^{n(1-\theta)})$, and

$$\|b_n\|_{E_0} \leqslant \sum_{n+1}^{\infty} \|c_k\|_{E_0} = o(2^{-n\theta}),$$

for similar reasons. So $x \in X$.

Conversely, if $x \in X$ is given, then there exist a_n, b_n such that $x = a_n + b_n$ and

$$\|a_n\|_{E_1} = 0(2^{n(1-\theta)}), \ \|b_n\|_{E_0} = o(2^{-n\theta}).$$

It is then easily seen that $c_n = a_n - a_{n-1}$ satisfies $\|c_n\|_{E_1} = o(2^{n(1-\theta)})$ and $\|c_n\|_{E_0} = o(2^{-n\theta})$ (note that $a_n - a_{n-1} = -b_n + b_{n-1}$). Therefore we have $x \in Y$. □

The two characterizations of the space $I_\theta(E)$ we have up to now (the spaces X and Y) hold for any Banach couple $E \in \mathcal{C}$. In the following we shall assume the existence of an operator $A \in \mathcal{L}(E_1, E_0)$ of *positive type* and show that such an operator can be used to describe $I_\theta(E)$.

An operator $A \in \mathcal{L}(E_1, E_0)$ is said to be of *positive type* if for some $\lambda_0 > 0$ we have $[\lambda_0, \infty) \subset \rho(A)$, and if the resolvent $R(\lambda, A) = (\lambda j - A)^{-1}$ of A satisfies the following condition:

$$R(\lambda, A) \text{ is uniformly bounded in } \mathcal{L}(E_0, E_1) \text{ for } \lambda_0 \leqslant \lambda < \infty.$$

In particular, any $A \in Hol(E)$ is of positive type. In fact, if A generates a C_0-semigroup, then A is of positive type. From here on we shall consider one fixed operator A of positive type. Given this operator we define the space $Z \subset E_0$ by

$$Z = \{x \in E_0 : \lim_{t \to \infty} t^\theta \|R(t, A)x\|_{E_1} = 0\}.$$

On Z we have the norm

$$\|x\|_Z = \sup_{\lambda_0 \leqslant t < \infty} t^\theta \|R(t, A)x\|_{E_1},$$

which turns Z into a Banach space.

THEOREM 6.5. *$I_\theta(E) = Z$ with equivalent norms.*

PROOF. If $x \in Z$, then consider

$$u(t) = \zeta(t) t^{-1} R(t^{-1}, A)x,$$

where $\zeta \in C^\infty(\mathbb{R})$, with $\zeta(t) \equiv 1$ near $t=0$, and $\mathrm{supp}(\zeta) \subset (-\infty, \lambda_0^{-1})$. We claim that $u \in V_\theta(E)$. Indeed as $t \downarrow 0$ we have

$$\|u(t)\|_{E_1} = t^{-1}\|R(t^{-1},A)x\|_{E_1} = o(t^{-1+\theta}),$$

and, since

$$u'(t) = \frac{d}{dt}(j - tA)^{-1}x$$
$$= t^{-1}R(\frac{1}{t},A)At^{-1}R(\frac{1}{t},A)x$$

holds, we also have

$$\|j \circ u'(t)\|_{E_0} \leq \|t^{-1} j \circ R(\frac{1}{t},A)\| \cdot \|A\| \cdot \|u(t)\|_{E_1}.$$

However $t^{-1} j \circ R(t^{-1},A)$ can be written as

$$t^{-1} j \circ R(\frac{1}{t},A) = (\frac{1}{t}j - A + A)R(\frac{1}{t},A)$$
$$= 1 + AR(\frac{1}{t},A),$$

and is therefore uniformly bounded in $\mathfrak{L}(E_0)$ as $t \downarrow 0$. It follows that $\|j \circ u'(t)\|_{E_0} = o(t^{\theta-1})$, so that $u \in V_\theta(E)$. Finally, note that $u(0)=x$. This is a consequence of the identity

$$u(t) = x + AR(\frac{1}{t},A)x,$$

and the fact that the second term tends to zero as $t \downarrow 0$ (which may be shown in the same way as Lemma 5.5 was proven). These considerations indicate that $x \in I_\theta(E)$, i.e. we have shown that $Z \subset I_\theta(E)$.

To prove the converse let $x \in I_\theta(E)$ be given. Then $x = u(0)$ for some $u \in V_\theta(E)$, and we can estimate $R(t,A)x$ as follows:

$$R(t,A)x = R(t,A)u(t^{-1}) - \int_0^{t^{-1}} R(t,A)u'(s)ds.$$

For the first term we have:

$$\|R(t,A)u(t^{-1})\|_{E_1} \leq \text{const.} t^{-1}\|u(t^{-1})\|_{E_1} = o(t^{\theta-1}),$$

and for the second term

$$\|R(t,A)\int_0^{t^{-1}} u'(s)ds\|_{E_1} \leq \text{const.} \|\int_0^{t^{-1}} u'(s)ds\|_{E_0}$$
$$\leq \text{const.} \int_0^{t^{-1}} \|u'(s)\|_{E_0} ds = o(t^\theta)$$

holds. Adding these estimates we see that $\|R(t,A)x\|_{E_1}$ is $o(t^\theta)$ as $t \to \infty$, so that x belongs to Z. We may conclude that Z and $I_\theta(E)$ coincide. □

Equivalence of the norms on Z and $I_\theta(E)$ follows from the closed graph theorem, or, alternatively, from a more precise version of the proof given above.

We conclude this section with one other description of $I_\theta(E)$. Let $A \in \mathfrak{L}(E_1, E_0)$ generate a C_0 semigroup. Then, by the Hille-Yosida theorem, A is of positive type. Denote the semigroup generated by A by $\{T(t)\}_{t\geqslant 0}$. For any $x \in E_0$ we have a function $x(t)=T(t)x$ from $[0,1]$ to E_0 which is always continuous. Applying the Hille-Yosida theorem again, we have the following description of E_1:

$$x \in E_1 \Leftrightarrow x(\cdot) \in C^1([0,1];E_0).$$

Thus the smaller space E_1 is determined by the smoothness properties of $x(\cdot)$ $(0 \leqslant t \leqslant 1)$. A similar characterization of the interpolation spaces $I_\theta(E)$ will now be given.

Define the *little-Hölder space* $h^\theta(E_0)$ to be the set of $u \in C([0,1];E_0)$ for which

$$\lim_{\epsilon \downarrow 0} \sup_{|t-s| \leqslant \epsilon} \epsilon^{-\theta} \|u(t)-u(s)\|_{E_0} = 0$$

holds. Give $h^\theta(E_0)$ the norm

$$\|u\|_{h^\theta} = \|u(0)\|_{E_0} + \sup_{\epsilon > 0} \epsilon^{-\theta} \cdot \sup_{|t-s| \leqslant \epsilon} \|u(t)-u(s)\|_{E_0}.$$

With this norm $h^\theta(E_0)$ is a Banach space, and we have the following result:

THEOREM 6.6. *For any* $x \in E_0$, x *belongs to* $I_\theta(E)$ *if and only if* $x(\cdot)=T(\cdot)x \in h^\theta(E_0)$. *The norm*

$$x \to \|T(\cdot)x\|_{h^\theta(E)}$$

is equivalent to the $I_\theta(E)$ *norm.*

Before we prove this theorem we give a lemma which says that the smoothness of $x(\cdot)$ is determined by its behaviour at $t=0$.

LEMMA 6.7. *Let* $x \in E_0$ *be given. Then*

$$x(\cdot) \in h^\theta \Leftrightarrow \lim_{t \downarrow 0} t^{-\theta} \|x(t)-x\|_{E_0} = 0.$$

PROOF. The implication "⇒" being obvious, we shall prove the converse: "⇐". For any $t,s \in [0,1]$ with $t \geqslant s$ we have

$$\|x(t)-x(s)\|_{E_0} = \|T(t)(x(t-s)-x)\|_{E_0}$$

$$\|x(t)-x(s)\|_{E_0} \leqslant \text{const. } \|x(t-s)-x\|_{E_0} = o(|t-s|^\theta)$$

as $|t-s|\downarrow 0$. Hence $x(\cdot)\in h^\theta(E_0)$ if $\|x(t)-x\|_{E_0}=o(t^\theta)$. □

PROOF OF THEOREM 6.6.
Let $x\in I_\theta(E)$, say $x=u(0), u\in V_\theta(E)$, be given. Then

$$x = u(t) - \int_0^t u'(s)ds,$$

and so we can estimate $\|T(t)x-x\|_{E_0}$ as follows:

$$\|T(t)u(t)-u(t)\|_{E_0} \leqslant Ct\cdot\|u(t)\|_{E_1} = o(t^\theta),$$

and

$$\|\int_0^t (T(t)u'(s)-u'(s))ds\|_{E_0} \leqslant C\int_0^t \|u'(s)\|_{E_0} ds = o(t^\theta).$$

Therefore $\|T(t)x-x\|_{E_0}=o(t^\theta)$, and by the previous lemma we have $T(\cdot)x\in h^\theta(E_0)$.

Conversely, let $T(\cdot)x$ belong to $h^\theta(E_0)$ for some $x\in E_0$. Consider

$$u(t) = \frac{1}{t}\int_0^t T(s)xds = S(t)x,$$

so $S(t)$ is the average of $T(s)$ over $0\leqslant s\leqslant t$. We claim that $u\in V_\theta(E)$. In fact, $u'(t)$ is given by

$$u'(t) = \frac{1}{t}T(t)x - \frac{1}{t^2}\int_0^t T(s)xds = \frac{1}{t^2}\int_0^t (T(t)-T(s))xds.$$

Since $T(\cdot)x$ belongs to $h^\theta(E_0)$ this expression may be estimated by

$$\|u'(t)\|_{E_0} = o(t^{\theta-1}).$$

To see that $u(t)\in E_1$ for $t>0$ we note that $S(t)$ is bounded from E_0 to E_1, and that

$$AS(t)=t^{-1}(T(t)-1).$$

Indeed, if $x\in E_1$ then $T(t)x\in E_1$ and

$$AS(t)x = \frac{1}{t}\int_0^t AT(s)xds = t^{-1}(T(t)x-x),$$

since $x(\cdot)$ is in $C^1([0,1];E_0)$ and $x'(t)=Ax(t)$. Now using the density of E_1 in

E_0 and the boundedness of $t^{-1}(T(t)-1)$ on E_0, one easily proves our statements concerning $S(t)$. Returning to $u(\cdot)$ again, we note that the E_1-norm of $u(t)$ is equivalent to $\|u(t)\|_{E_0}+\|Au(t)\|_{E_0}$. Here, the first term is bounded as $t\downarrow 0$, and the second term is

$$\begin{aligned}\|Au(t)\|_{E_0} &= \|AS(t)x\|_{E_0}\\ &= t^{-1}\|T(t)x - x\|_{E_0} = o(t^{\theta-1}).\end{aligned}$$

So we see that u indeed belongs to $V_\theta(E)$, and, because $u(0)=x$, x belongs to $I_\theta(E)$. Therefore $x\in I_\theta(E)$ iff $x(\cdot)\in h^\theta(E_0)$. As before, we leave it to the reader to prove the equivalence of the norms in question. □

4. Inhomogeneous initial value problems

In Chapter 5 we saw that the initial value problem

$$x'(t) = Ax,\ x(0) = x_0,$$

with $A\in Hol(E)$ and $x_0\in E_1$ for some Banach couple $E=(E_1,E_0)$ can always be solved by using the functional calculus. The solution is of course $e^{tA}x_0$. In this section we shall be concerned with the inhomogeneous initial value problem

$$\begin{cases} x'(t)=Ax(t)+f(t), & 0<t\leqslant T\\ x(0)=x_0. \end{cases} \tag{IVP}$$

Formally this problem has the following solution:

$$x(t) = e^{tA}x_0 + \int_0^t e^{(t-s)A}f(s)ds.$$

This equation is known as the *variation-of-constants formula.* The question is, can we make sense of this expression and is it a solution of (IVP)? In view of the theory in the previous chapter we shall consider the first term, $e^{tA}x_0$, to be under control. Therefore we shall assume that $x_0=0$ from here on, and concentrate on the expression

$$Sf(t) = \int_0^t e^{(t-s)A}f(s)ds,$$

where, as usual, we are given a Banach couple E and an $A\in Hol(E)$. Our first result says that all is well if we have strong bounds on $f(t)$.

LEMMA 6.8. *If $f\in C([0,T];E_1)$ then $x(t)=Sf(t)$ satisfies*

$$x\in C([0,T];E_1)\cap C^1([0,T];E_0)$$

and

$$x'(t) = Ax(t) + f(t), x(0) = 0.$$

PROOF. Since the operators e^{tA} $(0 \leqslant t \leqslant T)$ are uniformly bounded and strongly continuous in $\mathfrak{L}(E_1)$, the integral defining $Sf(t)$ exists (as a Bochner integral) and by the dominated convergence theorem it represents a continuous E_1-valued function. It follows from Lemma 5.14 that

$$jx(t) = \int_0^t j \circ e^{(t-s)A} f(s)ds$$

$$= \iint_{s+r \leqslant t} Ae^{rA} f(s)drds + \int_0^t f(s)ds.$$

Now write $s+r=\tau$, $r=\rho$ in the first integral:

$$jx(t) = \int_0^t \int_0^\tau Ae^{\rho A} f(\tau-\rho)d\rho d\tau + \int_0^t f(s)ds = \int_0^t (Ax(\tau)+f(\tau))d\tau.$$

Therefore we see that $jx(\cdot)$ is C^1 and that x is a solution of $x' - Ax = f$. Obviously $x(0)=0$ also holds. □

Thus if f is E_1-valued, then we have a good solution for IVP. The next lemma shows us what can be expected if f takes values in E_0.

LEMMA 6.9. *Let $f \in C([0,T]);E_0)$ be given, and define $x(t)=Sf(t)$ as before. Then, for any $\theta \in [0,1)$ we have $x \in C^\theta([0,T];E_0)$ and $x \in C([0,T];E_\theta)$, where E_θ denotes the continuous interpolation space $I_\theta(E)$.*

PROOF. Again, using the uniform boundedness and strong continuity of the operators e^{tA} in $\mathfrak{L}(E_0)$, one easily sees that $x = Sf \in C([0,T];E_0)$. In order to see that $x(t) \in E_\theta$ we use the criterion of Theorem 6.6, i.e. we check if $e^{\tau A}x(t)$ is Hölder continuous in τ, at $\tau=0$. The following holds:

$$e^{\tau A}x(t) - x(t) = \int_0^t (e^{(\tau+s)A} - e^{sA})f(t-s)ds$$

$$= \int_0^t \int_0^\tau Ae^{(\sigma+s)A} f(t-s)d\sigma ds,$$

so that

$$\|e^{\tau A}x(t) - x(t)\|_{E_0} \leqslant \max_{0 \leqslant s \leqslant T} \|f(s)\|_{E_0} \cdot \int_0^t \int_0^\tau \|Ae^{(\sigma+s)A}\|_{\mathfrak{L}(E_0)} d\sigma ds.$$

Using the estimate in Theorem 5.16 we get

$$\|e^{\tau A}x(t) - x(t)\|_{E_0} \leqslant C \max_{0 \leqslant s \leqslant T} \|f(s)\|_{E_0} \cdot \int_0^t \int_0^\tau \frac{d\sigma ds}{\sigma+s}$$

$$\leqslant C \max_{0 \leqslant s \leqslant T} \|f(s)\|_{E_0} \cdot \tau \log(\frac{1}{\tau}).$$

Thus $\|e^{\tau A}x(t)-x(t)\|_{E_0}=o(\tau^\theta)$ for any $\theta<1$, uniformly in $t\in[0,T]$, so that $x(t)\in E_\theta$ and $\|x(t)\|_{E_\theta}$ is uniformly bounded. We leave it to the reader to verify that $t\mapsto x(t)$ is continuous from $[0,T]$ to E_θ.

We now check that $x\in C^\theta([0,T];E_0)$. For any $\tau>0$ we have

$$x(t+\tau)-x(t)=\int_t^{t+\tau} e^{(t+\tau-s)A}f(s)ds+\int_0^t(e^{(t+\tau-s)A}-e^{(t-s)A})f(s)ds,$$

so that

$$\|x(t+\tau)-x(t)\|_{E_0}\leqslant C\max\|f(s)\|_{E_0}\{\tau+\int_0^t\|e^{(s+\tau)A}-e^{sA}\|ds\}.$$

We have just seen how we can estimate the remaining integral. What we get is this:

$$\|x(t+\tau)-x(t)\|_{E_0}\leqslant C\max_{0\leqslant s\leqslant T}\|f(s)\|_{E_0}\cdot\tau\log\frac{1}{\tau}.$$

Clearly this implies that $x\in C^\theta([0,T];E_0)$ for all $\theta<1$. □

This lemma should be a bit of a disappointment since we *almost* get $x\in C([0,T];E_1)$ and $x\in C^1([0,T];E_0)$ if $f\in C([0,T];E_0)$. Moreover, the lemma does *not* allow us to conclude that x is a solution of (IVP). Unfortunately examples show that in general Lemma 6.9 is optimal in the sense that one cannot require θ to be 1 (see Exercises 6.21 and 6.19). In fact, in BAILLON [1980] it is shown that if $E_1\neq E_0$, and $f\in C([0,T];E_0)$ implies $x\in C([0,T];E_1)$, then the Banach space E_0 contains a closed subspace isomorphic to (c_0). One way out of this problem is to assume extra regularity on f, e.g. one could assume that $f\in C^\alpha([0,T];E_0)$. In that case it can be shown that $x=Sf$ belongs to $C^{1+\alpha}([0,T];E_0)\cap C^\alpha([0,T];E_1)$, and also is a solution of the initial value problem. This is described in the book of PAZY [1983]. In the next section we shall present another solution which is due to DA PRATO and GRISVARD ([1975, 1979]).

5. MAXIMAL REGULARITY

For $k=2,3,4,\cdots$ we define the spaces $E_k(=$"$\mathfrak{D}(A^k)$") as in Section 5.7. For $0<\theta<1$ we define $E_{1+\theta}=I_\theta((E_2,E_1))$. One easily verifies that $A\in Hol(E_2,E_1)$ and by interpolation (axiom (2) in Section 6.1) also that $A\in Hol(E_{1+\theta},E_\theta)$. Our next result is due to DA PRATO and GRISVARD [1979].

THEOREM 6.10. *Let $f\in C([0,T];E_\theta)$ be given and define*

$$x(t)=Sf(t)=\int_0^t e^{(t-s)A}f(s)ds.$$

Then $x\in C([0,T];E_{1+\theta})\cap C^1([0,T];E_\theta)$, and satisfies $x'(t)=Ax(t)+f(t)$.

In order to convince the reader that this is a remarkable result we define F to be the Banach couple $(E_{1+\theta}, E_\theta)$. Now read Lemma 6.9 again and substitute an "F" whenever it says "E". What we couldn't prove then (i.e. the case $\theta=1$) turns out to be true under some condition on the Banach couple F and the operator $A \in Hol(F)$ (the condition is that F should be of the form $(E_{1+\theta}, E_\theta)$ and that $A \in Hol(F)$ should extend to an $A \in Hol(E)$).

PROOF OF THEOREM 6.10.
The proof consists of two parts. In the first part we use a density argument to show that we only have to prove an a priori estimate on $x = Sf$ for "nice" f. Then we prove the a priori estimate. Since the density argument is standard we shall only sketch it, and pay more attention to the a priori estimate.

The estimate we mean is the following. Let $f \in C([0,T];E_1)$ be given and define

$$x(t) = Sf(t) = \int_0^t e^{(t-s)A} f(s) ds.$$

Since $A \in Hol((E_2, E_1))$ we know by Lemma 6.9 that $x \in C([0,T];E_{1+\theta})$, and by Lemma 6.8 that $x' \in C([0,T];E_0)$. Suppose that there exists a constant $C<\infty$, independent of f, such that

$$\max_{0 \leq t \leq T} \|x(t)\|_{E_{1+\theta}} \leq C \cdot \max_{0 \leq t \leq T} \|f(t)\|_{E_\theta} \tag{EST}$$

holds. Given this estimate we shall now prove the theorem. Let the function $f \in C([0,T];E_\theta)$ be given. For small $\epsilon>0$ define

$$f_\epsilon(t) = \exp(\epsilon A) f(t).$$

Then $f_\epsilon \in C([0,T];E_1)$ and f_ϵ converges uniformly (in the E_θ norm) to f, as $\epsilon \downarrow 0$. Consider the corresponding functions $x_\epsilon = Sf_\epsilon$. It follows from our a priori estimate that the x_ϵ converge uniformly (in $E_{1+\theta}$) to some $x \in C([0,T];E_{1+\theta})$. Since the x_ϵ all satisfy $x'_\epsilon(t) = Ax_\epsilon(t) + f_\epsilon(t)$, they also converge in $C^1([0,T];E_\theta)$, so that x belongs to this space. Finally one can deduce from the estimates given in the proof of Lemma 6.9 that the x_ϵ converge in $C([0,T];E_0)$ to Sf. Hence $x = Sf$.

It remains to prove the a priori estimate. Before we start the computation we make some extra hypothesis on A. To be precise, we assume that for some $\theta \in (0, \frac{\pi}{2})$ we have

(a) $\lambda j - A$ is invertible if $|\arg(\lambda)| \leq \frac{\pi}{2} + \theta$

(b) $C = \sup\{\|(\lambda j - A)^{-1}\|_{E_0, E_1} : |\arg(\lambda)| \leq \frac{\pi}{2} + \theta\}$ is finite.

Since $A \in Hol(E)$ there is an $\omega \in \mathbb{R}$ such that $A + \omega j$ satisfies these conditions. Given such an ω we substitute $x(t) = e^{\omega t} y(t)$. The new unknown satisfies $y'(t) = (A + \omega j) y(t) + e^{-\omega t} f(t)$. Clearly an estimate like (EST) for y implies a similar estimate for x so that we may as well suppose that A satisfies the conditions (a) and (b). These conditions imply that $A: E_1 \to E_0$ is an isomorphism. In particular the norms $\|x\|_{E_{1+\theta}}$ and $\|Ax\|_{E_\theta}$ are equivalent. In order to estimate $\|Ax\|_{E_\theta}$ we use the description of the E_θ norm given in Theorem 6.5. Thus for any $\alpha>0$ we compute

$$\alpha^\theta\|(\alpha j-A)^{-1}Ax\|_{E_1}$$

and take the supremum over all $\alpha\geqslant 0$. Using the functional calculus of Section 5.5 we get the following identity

$$\alpha^\theta(\alpha j-A)^{-1}Ax(t)=\alpha^\theta\int_0^t(\alpha j-A)^{-1}Ae^{(t-s)A}f(s)ds$$

$$=\frac{\alpha^\theta}{2\pi i}\int_0^t\int_\Gamma\frac{\lambda e^{(t-s)\lambda}}{\alpha-\lambda}(\lambda j-A)^{-1}\cdot f(s)d\lambda ds.$$

Here Γ is the contour $\Gamma=\{z\in\mathbf{C}:\arg(z)=\pm(\pi/2+\theta)\}$ oriented "downwards". Using Theorem 6.5 again we see that

$$\|(\lambda j-A)^{-1}f(s)\|_{E_1}\leqslant C|\lambda|^{-\theta}\|f(s)\|_{E_\theta}\leqslant C\cdot M|\lambda|^{-\theta}$$

where we have written

$$M=\max_{0\leqslant t\leqslant T}\|f(t)\|_{E_\theta}.$$

So we get:

$$\alpha^\theta\|(\alpha j-A)^{-1}Ax(t)\|_{E_1}\leqslant\frac{C\cdot\alpha^\theta}{2\pi}M\int_\Gamma\int_0^t\frac{|e^{(t-s)\lambda}||\lambda|^{1-\theta}}{|\alpha-\lambda|}ds|d\lambda|$$

$$\leqslant\frac{CM\alpha^\theta}{\sin\theta}\cdot\int_\Gamma\frac{|d\lambda|}{|\alpha-\lambda||\lambda|^\theta},$$

where we have evaluated the integral with respect to ds. The remaining integral can be estimated by substituting $\alpha\mu=\lambda$ and noting that the contour Γ is invariant under dilations. The result is:

$$\alpha^\theta\|(\alpha j-A)^{-1}Ax(t)\|_{E_1}\leqslant\frac{CM}{\sin\theta}\cdot\int_\Gamma\frac{|d\lambda|}{|1-\lambda||\lambda|^\theta}\leqslant C'\cdot M.$$

If we now take the supremum over α we see that this implies our estimate (EST). Thus the theorem is proved. □

6. Examples and exercises

The first two exercises illustrate the use of interpolation inequalities.

Let $E=(E_1,E_0)$ be a Banach couple, and let F be an *intermediate Banach space* of exponent θ $(0<\theta<1)$ for E. Recall that this means that $E_1\subset F\subset E_0$, and for all $x\in E_1$ we have

$$\|x\|_F\leqslant C\cdot\|x\|_{E_1}^\theta\|x\|_{E_0}^{1-\theta},$$

where C is independent of x.

Exercise 6.11. Let $f:[0,1]\to E_1$ be a bounded function such that $j\circ f:[0,1]\to E_0$

is continuous. Then $j_F \circ f : [0,1] \to F$ is also continuous (where $j_F : E_1 \to F$ denotes the inclusion).

EXERCISE 6.12. Let G be a Banach space, and let $T:G \to E_1$ be a bounded linear operator, such that $j_1 \circ T : G \to E_0$ is compact. Then $j_F \circ T : G \to F$ is also compact.

In general it is quite hard to describe the space $I_\theta(E)$ for a given couple E. The following is perhaps the easiest example.

Let E_0 be the sequence space

$$(c_0) = \{\{x_n\} \in \mathbb{C}^{\mathbb{N}} : \lim_{n\to\infty} x_n = 0\}$$

with norm $\|\{x_n\}\|_{E_0} = \sup\{|x_n| : n = 1,2, \cdots\}$. Define E_1 to be the space

$$E_1 = \{\{x_n\} \in E_0 : \lim_{n\to\infty} nx_n = 0\}$$

with norm $\|\{x_n\}\|_{E_1} = \sup\{|nx_n| : n = 1,2,3, \cdots\}$. The operator $A : E_1 \to E_0$ given by

$$(Ax)_n = -nx_n \quad (n = 1,2, \cdots)$$

belongs to $Hol(E)$.

EXERCISE 6.13. The continuous interpolation space $I_\theta(E)$ consists of those sequences $\{x_n\}$ such that $|x_n| = o(n^{-\theta})$ $(n \to \infty)$. The norm on $I_\theta(E)$ is equivalent to

$$\|\{x_n\}\|_\theta = \sup\{n^\theta |x_n| : n = 1,2, \cdots\}.$$

Hint: use Theorem 6.6.

Another example where it is not so hard to compute $I_\theta(E)$ is the following. Let T be the "1-torus" $\mathbb{R}(\mathbb{Z})$, and define

$$E_0 = C(T), E_1 = C^1(T).$$

For any $\theta \in (0,1)$ we define the so-called little Hölder space $h^\theta(T)$ by

$$f \in h^\theta(T) \Leftrightarrow \lim_{\epsilon \downarrow 0} \sup_{|x-y| \leq \epsilon} \epsilon^{-\theta} |f(x) - f(y)| = 0.$$

On $h^\theta(T)$ we have the norm

$$\|f\|_{h^\theta(T)} = |f(0)| + \sup_{x \neq y} \frac{|f(x) - f(y)|}{|x-y|^\theta}.$$

EXERCISE 6.14. For $0 < \theta < 1$ one has

$$I_\theta(E) = h^\theta(T)$$

Hint: The operator $D : E_1 \to E_0$ defined by $Df(x) = f'(x)$ generates a strongly continuous semigroup $(e^{tD}f)(x)=f(x+t)$. Use theorem 6.6.

In the following exercises the reader is asked to take a closer look at the Banach space $h^{\theta}(T)$. The first exercise shows that $h^{\theta}(T)$ may be obtained by interpolating between different Banach couples.

Let $W^1_\infty(T)$ be the space of $f\in C(T)$ whose distributional derivative f' belongs to $L_\infty(T)$. One can show that $W^1_\infty(T)$ coincides with the space of Lipschitz continuous functions on T. With the norm

$$\|f\|_{W^1_\infty} = |f(0)| + \operatorname*{ess\,sup}_{x\in T}|f'(x)|,$$

W^1_∞ is a Banach space, and $C^1(T)$ is a closed subspace of W^1_∞.

EXERCISE 6.15. Show that $I_\theta(E)=I_\theta(F)$, where $E=(C^1(T),C(T))$ and $F=(W^1_\infty(T),C(T))$.

Hint: Obviously $I_\theta(E)\subset I_\theta(F)$. Using a "mollifier" one shows that any $f\in W^1_\infty(T)$ can be decomposed as $f=g_\epsilon+h_\epsilon$, where $g_\epsilon\in C^1(T)$ and $h_\epsilon\in C(T)$ satisfy $\|g_\epsilon\|_{C^1}\leqslant 2.\|f\|_{W^1_\infty}$ and $\|h_\epsilon\|_{C(T)}\leqslant\epsilon$. Here the number $\epsilon>0$ can be chosen arbitrarily small. Now suppose $f\in I_\theta(F)$, then by Theorem 6.3, we can write $f=a_n+b_n$, with $\|a_n\|_{W^1_\infty}=o(2^{n(1-\theta)})$, $\|b_n\|_C=o(2^{-n\theta})$. By the "mollifier"-argument we can write $a_n=a_n'+a_n''$, with $\|a_n'\|_{C^1}=o(2^{n(1-\theta)})$ and $\|a_n''\|_C\leqslant 10^{-n}$. After defining $b_n'=b_n+a_n''$ we see (by Theorem 6.3 again) that $f=a_n'+b_n'$ belongs to $I_\theta(E)$.

The next exercise will reveal that the space $h^{\theta}(T)$ is isomorphic (as a topological vector space) to the sequence space (c_0). First we introduce some notation. For $n=0,1,2,3,\cdots$ and $k=1,2,3,\ldots,2^n$ define the function

$$g_{k,n}(x)=\max(0,1-|2^{n+1}x-2k+1|)$$

for $0\leqslant x\leqslant 1$. Since $g_{k,n}(0)=g_{k,n}(1)$ we may regard $g_{k,n}$ as an element of $C(T)$, in fact of $W^1_\infty(T)$. The function $g_{k,n}$ is a piecewise linear peak of height one, width 2^{-n} which is centered at $(2k-1)\cdot 2^{-(n+1)}$ (draw a picture). Define the subspace $V_n\subset W^1_\infty$ to be the span of the $g_{k,n}$ $(1\leqslant k\leqslant 2^n)$ for $n=0,1,2,\cdots$. If we put $V_{-1}=\{c\cdot 1 : c\in\mathbb{C}\}$ the spaces V_n $(n\geqslant -1)$ generate a dense subspace of $C(T)$. For each n we define the projection p_n in $C(T)$ onto V_n as

$$p_{-1}(f)=f(0)\quad (n=-1)$$

$$(p_nf)(x)=-\sum_{k=1}^{2^n}\tfrac{1}{2}\{f(\frac{k-1}{2^n})-2f(\frac{k-1/2}{2^n})+f(\frac{k}{2^n})\}\cdot g_{k,n}(x)\quad (n\geqslant 0).$$

EXERCISE 6.16.

(a) Verify that $p_n\circ p_m=0$ if $n\neq m$.

(b) For any $f\in h^{\theta}(T)$ the following estimate holds:

$$\|p_n(f)\|_{C(T)}\leqslant 2^{-n\theta}\|f\|_{h^{\theta}(T)}.$$

Furthermore $\lim_{n\to\infty} 2^{n\theta}\|p_n(f)\|_{C(T)}=0$.
(*Hint*: $a-2b+c=(a-b)-(b-c)$).

(c) On the subspace V_n the norms of $C(T)$ and of $W^1_\infty(T)$ are equivalent since V_n is finite dimensional. Show that for $f\in V_n$

$$\|f\|_{W^1_\infty} = 2^{-n-1}\|f\|_{C(T)}.$$

(d) Show that there is a constant c independent of $\theta\in(0,1)$, such that for all $f\in W^1_\infty(T)$ one has the estimate

$$c\cdot\theta(1-\theta)\cdot\|f\|_{h^\theta(T)} \leqslant \sup_{n\geqslant -1} 2^{n\theta}\|p_n(f)\|_{C(T)}.$$

Hint: For given $x,y\in T$ one has to estimate $f(x)-f(y)$. Fix an integer $N\geqslant 0$ such that $2^{-N}\leqslant|x-y|<2^{-N+1}$. Write $f=g+h$, where $g=p_{-1}(f)+\cdots+p_N(f)$, and $h=p_{N+1}(f)+\cdots$. Now one can estimate $|h(x)-h(y)|\leqslant 2\|h\|_{C(T)}$ via the triangle inequality: $|h(x)-h(y)|\leqslant 2\Sigma_{k>N}\|p_n(f)\|_{C(T)}$. In order to estimate $|g(x)-g(y)|$ use $g\in W^1_\infty$. One gets $|g(x)-g(y)|\leqslant \Sigma_{k=-1}^N\|p_n(f)\|_{W^1_\infty}\cdot|x-y|$. Apply part (c) of this exercise to compute the sum. Adding both terms leads to the desired estimate for $|f(x)-f(y)|$; in the end one should compare $2^{-N\theta}$ with $|x-y|^\theta$.

(e) For any $f\in C(T)$ one has $f\in h^\theta(T)$ iff $\lim_{n\to\infty} 2^{n\theta}\|p_n(f)\|_{C(T)}=0$. Show this, and conclude that $h^\theta(T)$ is isomorphic with the sequence space (c_0).

As a corollary we see that the interpolation space $I_\theta(E)$ of a Banach couple E need not be isomorphic with either of the spaces E_0 or E_1 (since $h^\theta(T)\cong(c_0)$, $h^\theta(T)$ has a separable dual, isomorphic to l_1, whereas $C(T)$ does not).

The next few exercises explore the relations between the interpolation spaces $I_\theta(E)$ and the so called *"fractional power spaces"*. Let $E=(E_1,E_0)$ be a Banach couple, and let $A\in Hol(E)$ be given. In Section 5.7 we defined the spaces E_k $(k=2,3,\cdots)$. Loosely speaking $E_k=\mathfrak{D}(A^k)$. For any θ with $0<\theta<1$ we define $E_{k+\theta}=I_\theta((E_{k+1},E_k))$.

EXERCISE 6.17. (notation as in Exercise 5.24)

(a) For all $n=1,2,3,\cdots$ show that

$$(1-A)^{-n}: E_s\to E_{s+n}$$

is an isomorphism, for any $s\geqslant 0$.
Hint: use the interpolation property.

(b) Show that the operator $(1-A)^{-\sigma}$ is bounded from E_s to $E_{s+\mathrm{Re}\,\sigma}$, provided s and $s+\mathrm{Re}\,\sigma$ are not integers.
Hint: mimic the proof of Theorem 6.10. First reduce the problem to the case $0<s<1$ and $0<\mathrm{Re}\,\sigma<1$ by using part (a). Next, let $x\in E_1$ be given. Then $(1-A)^{-\sigma}x\in E_1$ is given by

$$(1-A)^{-\sigma}x = \frac{1}{2\pi i}\int_\Gamma (1-\lambda)^{-\sigma}(\lambda j-A)^{-1}x\,d\lambda,$$

where Γ is the same contour as in the proof of Theorem 6.10. If $\mathrm{Re}\,\sigma+s<1$ then use Theorem 6.5 to estimate $\|(1-A)^{-\sigma}x\|_{E_{s+\mathrm{Re}\,\sigma}}$. If $1<\mathrm{Re}\,\sigma+s<2$, one estimates $\|A(1-A)^{-\sigma}x\|_{E_{s+\mathrm{Re}\,\sigma-1}}$ in the same way.

(c) Under the conditions of part (b), $(1-A)^{-\sigma}$ is even an isomorphism from E_s onto $E_{s+\mathrm{Re}\,\sigma}$.
Hint: if $N>\mathrm{Re}\,\sigma$ is an integer then $(1-A)^{\sigma-N}(1-A)^{-\sigma}=(1-A)^{-N}$ by Exercise 5.24. Now remember part (a) again.

This exercise shows us that the fractional powers $(1-A)^{-z}$ are well-behaved provided we avoid the spaces E_s with integral s. We shall now introduce a new scale of Banach spaces, usually known as the *fractional power spaces* of the operator A.

For real $s>0$ define

$$D_s=(1-A)^{-s}(E_0)=\{(1-A)^{-s}x : x\in E_0\}.$$

If $s\in\mathbb{Z}^+$ then D_s and E_s coincide. For any $s>0$, $(1-A)^{-s}$ is a bijection from E_0 to D_s. The inverse will be denoted by $(1-A)^s : D_s\to E_0$. If we give D_s the norm

$$\|x\|_{D_s}=\|(1-A)^s x\|_{E_0},$$

then D_s is a Banach space and $(1-A)^{-s}$ trivially becomes an isomorphism from E_0 onto D_s.

EXERCISE 6.18. Show that for $0<\theta<1$ one has the inclusion $D_\theta\subset I_\theta(E)=E_0$.
Hint: if $x\in D_\theta$ then $x=(1-A)^{-\theta}y$ and by Exercise 5.24 we have

$$x=\frac{1}{\Gamma(\theta)}\int_0^\infty e^{tA}ye^{-t}t^{\theta-1}\,dt.$$

Now define

$$a_n=\frac{1}{\Gamma(\theta)}\int_{2^{-n}}^{2^{1-n}} e^{tA}ye^{-t}t^{\theta-1}\,dt$$

for $n\geqslant 1$ and

$$a_0=\frac{1}{\Gamma(\theta)}\int_1^\infty e^{tA}ye^{-t}t^{\theta-1}\,dt$$

for $n=0$, and use Exercise 5.21 combined with Theorem 6.4 to show that $x\in E_\theta$.

In the last few exercises the reader is invited to think about the maximal regularity result of Section 6.5.

Let $E=(E_1,E_0)$ be a Banach couple and assume $A\in Hol(E)$ is given. We shall say that A has the *property MR* if for any $f\in C([0,1];E_0)$ the function

$$Sf(t)=\int_0^t e^{(t-s)A}f(s)\,ds$$

belongs to $C([0,1];E_1)$. Using the closed graph theorem (and Theorem 6.9) one concludes that if A has the property MR then the operator S is bounded from $C([0,1];E_0)$ to $C([0,1];E_1)$.

EXERCISE 6.19.
(a) Let E be the Banach couple of Exercise 6.13, and let A be the operator $(ax)_n = -nx_n$. Then A has the property MR.
(b) Let E_0 be the sequence space l_p $(1 \leqslant p < \infty)$ and let E_1 be the dense subspace given by

$$E_1 = \{\{x_n\} \in l_p : \sum_1^{\infty} |nx_n|^p < \infty\}$$

with the obvious norm. Again consider $A : E_1 \to E_0$ defined by $(Ax)_n = -nx_n$. Then $A \in Hol(E)$, but A fails to have the maximal regularity property.

Hint: let e_n be the n-th unit vector in l_p. Choose a nonnegative continuous function $g(t)$ on $[0,\infty)$ with $\mathrm{supp}(g) \subset [1/2,1]$ and $g(3/4) \neq 0$. Define $f_n(t)$ to be $f_n(t) = g(2^n(1-t)) \cdot e_{2^n}$ $(n=1,2,\cdots)$. Then $f_n \in C([0,1];E_0)$ and $\|f_n(t)\|_{E_0} \leqslant 1$. Since the f_n have disjoint supports we also have $h_N = f_1 + \cdots + f_N \in C([0,1];E_0)$ with $\|h_N(t)\|_{E_0} \leqslant 1$. A short computation shows that, if $c = \int_0^\infty e^{-\tau} g(\tau) d\tau$, then $Sf_n(1) = c \cdot 2^{-n} e_{2^n}$. Hence $\|h_N(1)\|_{E_1} = cN^{1/p}$, so that S cannot be bounded from $C([0,1];E_0)$ to $C([0,1];E_1)$.

We now turn to the *Gaussian semigroup* on $C(T)$ (see Exercise 5.25). Define $E_0 = C(T)$ and $E_1 = C^2(T)$. Then the operator A given by

$$Af(x) = f''(x)$$

belongs to $Hol(E)$. In fact e^{tA} is given by

$$(e^{tA}f)(x) = \int_T G(t,x-y) f(y) dy$$

where

$$G(t,z) = (4\pi t)^{-\frac{1}{2}} \sum_{n=-\infty}^{+\infty} \exp\left(-\frac{(z+n)^2}{4t}\right).$$

It can be shown that for $0 < \theta < 1$ and $\theta \neq 1/2$ the space $I_\theta(E)$ is again a little Hölder space (the reader may wish to try to prove this by hand). We have

$$I_\theta(E) = h^{2\theta}(T) \quad (0 < \theta < 1,\ \theta \neq 1/2),$$

where $h^{1+\sigma}(T)$ contains those $C^1(T)$ functions whose derivative belongs to $h^\sigma(T)$ $(0 < \sigma < 1)$. The inhomogeneous initial value problem of Section 6.5 can be written as

$$u_t - u_{xx} = f(x,t) \quad x \in T,\ 0 \leqslant t \leqslant 1 \qquad \text{(HEQ)}$$
$$u(x,0) = 0,$$

i.e. the *inhomogeneous heat equation.*

EXERCISE 6.20. Use the maximal regularity result to establish the following a priori estimate. Let $f \in C^\infty(T \times [0,1])$ be given and let $u \in C^\infty(T \times [0,1])$ be the classical solution of (HEQ). Then

$$[u_{xx}]_\theta + [u_t]_\theta \leqslant c \cdot [f]_\theta$$

where c does not depend on f, and $[\cdot]_\theta$ denotes the norm

$$[v]_\theta = \max_{T \times [0,1]} |v| + \sup_{\substack{x \neq y \\ 0 \leqslant t \leqslant 1}} \frac{|v(x,t) - v(y,t)|}{|x-y|^\theta},$$

with $0 < \theta < 1$.

In our last exercise we show that the operator A on the pair $E = (C^2(T), C(T))$ does not have the maximal regularity property.

EXERCISE 6.21.

(a) Let $\phi(x,t)$ be a C^∞ function on $\mathbb{R}^2$ such that

$$\phi(x,t) = \begin{cases} 0, & \text{if } x^2 + |t| \geqslant 1, \\ 1, & \text{if } x^2 + |t| \leqslant \frac{3}{4}. \end{cases}$$

Consider the function

$$u_N(x,t) = \sum_{k=1}^{N} \phi(2^k x, 2^{2k} t)\,(x^2 + 2t).$$

Show that $f_N = (u_N)_t - (u_N)_{xx}$ satisfies

$$\max_{T \times [0,1]} |f_N(x,t)| \leqslant M$$

(with M independent of N), where as

$$\max_{T \times [0,1]} |(u_N)_{xx}| \geqslant (u_N)_{xx}(0,0) = N.$$

(b) Prove that $A \in Hol(E)$ does not have the maximal regularity property.

Part III

Positive Semigroups

Chapter 7

Generators of Positive Semigroups

1. Elementary properties of positive semigroups

Let $(X, \|\cdot\|)$ be an ordered Banach space (real or complex) with positive cone X^+. Suppose that $\{T(t)\}_{t\geqslant 0}$ is a C_0-semigroup of linear operators in X, with generator A. As usual, we denote the type (growth bound) of $\{T(t)\}_{t\geqslant 0}$ by ω_0 (see also Section 3.7), i.e.,

$$\omega_0 = \lim_{t\to\infty} \frac{\log\|T(t)\|}{t} = \inf_{t>0} \frac{\log\|T(t)\|}{t}.$$

The semigroup $\{T(t)\}_{t\geqslant 0}$ is called *positive* if all the operators $T(t)$, $t\geqslant 0$, are positive, i.e., if $T(t)X^+ \subset X^+$ for all $t\geqslant 0$. The following characterization of positivity will be used frequently.

PROPOSITION 7.1. *The C_0-semigroup $\{T(t)\}_{t\geqslant 0}$ is positive if and only if the resolvent $R(\lambda,A)$ is positive for all $\lambda>\omega_0$ (equivalently, $R(\lambda,A)$ is positive for all $\lambda>\lambda_0$ for some $\lambda_0\geqslant\omega_0$).*

PROOF. First assume that $\{T(t)\}_{t\geqslant 0}$ is positive. If $\lambda>\omega_0$, then

$$R(\lambda,A)x = \int_0^\infty e^{-\lambda t}T(t)x\,dt$$

for all $x\in X$ (see Theorems 3.5 and 3.31), and since X^+ is closed, this implies that $R(\lambda,A)x\geqslant 0$ whenever $x\geqslant 0$.

Conversely, if $R(\lambda,A)x \geq 0$ for all $\lambda > \lambda_0$ ($\lambda_0 \geq \omega_0$), then it follows from

$$T(t)x = \lim_{n\to\infty} \{\frac{n}{t}R(\frac{n}{t},A)\}^n x, \quad x \in X,\ t>0,$$

(see Theorem 3.6) that $T(t)x \geq 0$ for all $0 \leq x \in X$. □

A closed linear operator A for which there exists λ_0 such that $(\lambda_0,\infty) \subset \rho(A)$ and $R(\lambda,A) \geq 0$ for all $\lambda > \lambda_0$ is called *resolvent positive.*
The next proposition shows that if $\{T(t)\}_{t\geq 0}$ is a positive C_0-semigroup, then both the domain $\mathfrak{D}(A)$ of the generator A, and the domain $\mathfrak{D}(A^*)$ of the adjoint A^* contain sufficiently many positive elements.

PROPOSITION 7.2. *If $\{T(t)\}_{t\geq 0}$ is a positive C_0-semigroup, then $\mathfrak{D}(A) \cap X^+$ is norm dense in X^+ and $\mathfrak{D}(A^*) \cap (X^*)^+$ is $\sigma(X^*,X)$-dense in $(X^*)^+$.*

PROOF. Take any $0 \leq x \in X$. Then $t^{-1}\int_0^t T(s)xds \geq 0$ for $t>0$. Moreover, $t^{-1}\int_0^t T(s)xds \to x$ in norm as $t\downarrow 0$ and $t^{-1}\int_0^t T(s)xds \in \mathfrak{D}(A)$ for all $t>0$ (see Lemma 3.2). Hence $\mathfrak{D}(A) \cap X^+$ is norm dense in X^+.
For the second part of the statement, note that, by the above proposition, $R(\lambda,A)$ is a positive operator for all $\lambda > \omega_0$, and that $\lambda R(\lambda,A)x \to x$ (norm) as $\lambda \to \infty$ for all $x \in X$ (see the discussion preceeding Theorem 2.2). Furthermore, since $R(\lambda,A)X = \mathfrak{D}(A)$, it is clear that $R(\lambda,A)^* X^* = \mathfrak{D}(A^*)$. Therefore, if $0 \leq \phi \in X^*$, then $0 \leq \lambda R(\lambda,A)^* \phi \in \mathfrak{D}(A^*)$ for all $\lambda > \omega_0$ and $\lambda R(\lambda,A)^* \phi \to \phi$ in the weak * topology as $\lambda \to \infty$. □

For a C_0-semigroup $\{T(t)\}_{t\geq 0}$ with generator A, the *spectral bound* is defined by

$$s(A) = \sup\{\mathrm{Re}\,\lambda : \lambda \in \sigma(A)\},$$

where $s(A) = -\infty$ if $\sigma(A) = \varnothing$.
A complex number λ with $\mathrm{Re}\,\lambda > \omega_0$ belongs to the resolvent set $\rho(A)$ and

$$R(\lambda,A)x = \int_0^\infty e^{-\lambda t} T(t)x dt$$

for all $x \in X$ (the integral being absolutely norm convergent). This implies that $s(A) \leq \omega_0$, and in general the inequality may be strict (it is possible that $-\infty = s(A) < \omega_0$; examples will be presented in Sections 8.1 and 9.1). If $\{T(t)\}_{t\geq 0}$ is a positive C_0-semigroup, then the resolvent $R(\lambda,A)$ has the above integral representation for all λ with $\mathrm{Re}\,\lambda > s(A)$ (although the integral need not be absolutely convergent). For the proof we need a lemma.

LEMMA 7.3. *Let X be a Banach space and $\{T(t)\}_{t\geq 0}$ a C_0-semigroup in X with generator A. If $\lambda \in \mathbb{R}$ satisfies*

$$\int_0^\infty e^{-\lambda t} |\langle T(t)x,\phi \rangle| dt < \infty$$

for all $x \in X$ *and all* $\phi \in X^*$, *then* $\lambda > s(A)$. *Moreover, if* $\mathrm{Re}\,\mu > \lambda$, *then for all* $x \in X$,

$$R(\mu, A)x = \int_0^\infty e^{-\mu t} T(t)x\,dt,$$

the integral being norm convergent as an improper Riemann-integral.

PROOF. For $\mu \in \mathbb{C}$ with $\mathrm{Re}\,\mu \geqslant \lambda$ define

$$R_t(\mu)x = \int_0^t e^{-\mu s} T(s)x\,ds, \quad x \in X$$

for all $t \geqslant 0$. Clearly $R_t(\mu) \in \mathcal{L}(X)$ for all such μ and t.
If $x \in X$ and $\phi \in X^*$, then

$$|\langle R_t(\mu)x, \phi\rangle| \leqslant \int_0^t e^{-s\mathrm{Re}\,\mu} |\langle T(s)x, \phi\rangle| ds \leqslant$$

$$\leqslant \int_0^\infty e^{-\lambda s} |\langle T(s)x, \phi\rangle| ds < \infty,$$

and so it follows from the uniform boundedness theorem that there exists a constant M such that $\|R_t(\mu)\| \leqslant M$ for all $t \geqslant 0$ and all μ with $\mathrm{Re}\,\mu \geqslant \lambda$. Using integration by parts we find that

$$R_t(\mu)x = e^{-(\mu-\lambda)t} R_t(\lambda)x + (\mu - \lambda)\int_0^t e^{-(\mu-\lambda)s} R_s(\lambda)x\,ds,$$

for all $x \in X$ and $\mathrm{Re}\,\mu > \lambda$. Now, since $\|e^{-(\mu-\lambda)t} R_t(\lambda)\| \to 0$ as $t \to \infty$, and since $\int_0^\infty \|e^{-(\mu-\lambda)s} R_s(\lambda)x\| ds \leqslant M\|x\| \int_0^\infty e^{-s\mathrm{Re}(\mu-\lambda)} ds < \infty$, it follows that

$$R(\mu)x = \int_0^\infty e^{-\mu s} T(s)x\,ds = \lim_{t\to\infty} \int_0^t e^{-\mu s} T(s)x\,ds$$

$$= (\mu - \lambda)\int_0^\infty e^{-(\mu-\lambda)s} R_s(\lambda)x\,ds$$

exists. Note that the convergence is uniform on $\{\mu \in \mathbb{C}; \mathrm{Re}\,\mu \geqslant \lambda + \delta\}$ for all $\delta > 0$, and so $R(\mu)x$ is an analytic function of μ in $\{\mu \in \mathbb{C} : \mathrm{Re}\,\mu > \lambda\}$. Furthermore, $R(\mu) \in \mathcal{L}(X)$ and $\|R(\mu)\| \leqslant M$ for all μ with $\mathrm{Re}\,\mu > \lambda$. Since $R(\mu) = R(\mu, A)$ whenever $\mathrm{Re}\,\mu > \omega_0$, it follows from the analyticity of $R(\mu)$ that $\{\mu \in \mathbb{C} : \mathrm{Re}\,\mu > \lambda\} \subset \rho(A)$ and $R(\mu) = R(\mu, A)$ whenever $\mathrm{Re}\,\mu > \lambda$. Finally, since $\|R(\mu, A)\| \leqslant M$ for all μ with $\mathrm{Re}\,\mu > \lambda$, it is clear that $s(A) < \lambda$, and by this the lemma is completely proved. □

Before stating the next theorem we recall some facts concerning the Laplace transform. Let $f(t)$ be a locally integrable function on $[0, \infty)$. The Laplace transform of f is defined by

$$\mathfrak{L}(f;z) = \int_0^\infty e^{-zt} f(t)dt$$

for all $z \in \mathbf{C}$ for which this integral converges (as an improper integral). If the integral converges for some $z_0 \in \mathbf{C}$, then it exists for all $z \in \mathbf{C}$ with $\mathrm{Re}\, z > \mathrm{Re}\, z_0$, and so the region of convergence is a half-plane. Define $\sigma_c = \inf\{\mathrm{Re}\, z : \mathfrak{L}(f;z)$ exists$\}$, which is called the abscissa of convergence. The function $\mathfrak{L}(f;z)$ is analytic in $\{z \in \mathbf{C} : \mathrm{Re}\, z > \sigma_c\}$. In general, $\mathfrak{L}(f;z)$ need not have any singularities on the axis of convergence, i.e., it is possible that $\mathfrak{L}(f;z)$ has an analytic continuation defined on a half-plane that contains $\{z \in \mathbf{C} : \mathrm{Re}\, z > \sigma_c\}$ properly. However, if f is a positive function this cannot happen, as a consequence of the Pringsheim-Landau theorem: if f is a positive, locally integrable function on $[0, \infty)$, then the abscissa of convergence σ_c is a singularity of the Laplace transform $\mathfrak{L}(f;z)$.

We now return to positive semigroups on an ordered Banach space, which is tacitly assumed to be complex.

THEOREM 7.4. *Let X be an ordered Banach space with normal and generating positive cone X^+. Suppose that $\{T(t)\}_{t \geq 0}$ is a positive C_0-semigroup in X with generator A. Then $s(A) = \sup\{\lambda \in \mathbb{R} : \lambda \in \sigma(A)\}$ and*

$$R(\mu, A)x = \int_0^\infty e^{-\mu t} T(t)x dt$$

for all $x \in X$ and all $\mu \in \mathbf{C}$ with $\mathrm{Re}\, \mu > s(A)$, where the integral exists as a norm convergent improper integral.

PROOF. It is sufficient to show that for any $\lambda \in \mathbb{R}$ for which $[\lambda, \infty) \subset \rho(A)$ we have $s(A) < \lambda$, and that the integral representation of $R(\mu, A)$ holds for all $\mu \in \mathbf{C}$ with $\mathrm{Re}\, \mu > \lambda$. To this end, fix $\lambda \in \mathbb{R}$ such that $[\lambda, \infty) \subset \rho(A)$. Take $0 \leq x \in X$ and $0 \leq \phi \in X^*$, and consider the positive continuous function $f(t) = \langle T(t)x, \phi \rangle$ on $[0, \infty)$. The abscissa of convergence σ_c of $\mathfrak{L}(f;\mu)$ clearly satisfies $\sigma_c \leq \omega_0$, and $\mathfrak{L}(f,\mu) = \langle R(\mu, A)x, \phi \rangle$ for all μ with $\mathrm{Re}\, \mu > \omega_0$. Furthermore, $\langle R(\mu, A)x, \phi \rangle$ is an analytic function of μ on some open set containing $[\lambda, \infty)$, whereas, by the Pringsheim-Landau theorem, σ_c is a singularity of $\mathfrak{L}(f,\mu)$. Therefore $\sigma_c < \lambda$, and so

$$\int_0^\infty e^{-\lambda t} \langle T(t)x, \phi \rangle dt < \infty.$$

Since X^+ and $(X^*)^+$ are both generating (as X^+ is normal), it follows that

$$\int_0^\infty e^{-\lambda t} |\langle T(t)x, \phi \rangle| dt < \infty$$

for all $x \in X$ and all $\phi \in X^*$. An application of Lemma 7.3 now yields the desired result. □

Since $\sigma(A)$ is closed, it is an immediate consequence of the present result that, under the above assumptions, $s(A)\in\sigma(A)$ whenever $\sigma(A)\neq\varnothing$. The nature of the singularity of $R(\mu,A)$ at $s(A)$ will be discussed in Sections 8.2 and 8.3. For future reference we collect two other consequences of the above theorem.

COROLLARY 7.5. *If $\{T(t)\}_{t\geqslant 0}$ is a positive C_0-semigroup with generator A in the ordered Banach space X with generating normal positive cone, then $R(\lambda,A)\geqslant 0$ for all $\lambda>s(A)$. Moreover, if X is a Banach lattice, then*

$$|R(\mu,A)x|\leqslant R(\mathrm{Re}\,\mu,A)|x|,$$

for all $x\in X$ and all $\mu\in\mathbb{C}$ with $\mathrm{Re}\,\mu>s(A)$.

PROOF. The first statement is clear from the integral representation of $R(\lambda,A)$ in the above theorem. Now assume that X is a Banach lattice, and take μ with $\mathrm{Re}\,\mu>s(A)$. Then

$$\left|\int_0^t e^{-\mu s}T(s)x\,ds\right|\leqslant\int_0^t e^{-s\mathrm{Re}\,\mu}T(s)|x|\,ds$$

for all $t\geqslant 0$, and letting $t\to\infty$ we obtain $|R(\mu,A)x|\leqslant R(\mathrm{Re}\,\mu,A)|x|$. □

Next we mention some results concerning irreducible semigroups. For the sake of simplicity we restrict ourselves to Banach lattices. First we recall some of the relevant facts. Let X be a Banach lattice. The element $0\leqslant u\in X$ is called a quasi-interior point if $\overline{X_u}=X$ (where X_u denotes the principal ideal generated by u). Note that $0\leqslant u\in X$ is a quasi-interior point if and only if $\langle u,\phi\rangle>0$ for all $0<\phi\in X^*$. Given the positive linear operator T from X into itself, the closed ideal J in X is called T-invariant if $T(J)\subset J$. The operator T is called *irreducible* if $\{0\}$ and X are the only T-invariant closed ideals. Furthermore, T is called *strongly irreducible* if Tu is a quasi-interior point for all $0<u\in X$. Clearly, a strongly irreducible operator is irreducible.

Now suppose that $\{T(t)\}_{t\geqslant 0}$ is a positive C_0-semigroup in X. A closed ideal J in X is called $\{T(t)\}$- invariant if J is $T(t)$-invariant for all $t\geqslant 0$, and $\{T(t)\}$ is called irreducible if $\{0\}$ and X are the only $\{T(t)\}$-invariant closed ideals. Furthermore, $\{T(t)\}$ is called strongly irreducible if all operators $T(t)$, $t>0$, are strongly irreducible.

PROPOSITION 7.6. *For a positive C_0-semigroup $\{T(t)\}_{t\geqslant 0}$, with generator A, in a Banach lattice X the following statements are equivalent.*

(i) *$\{T(t)\}$ is irreducible.*

(ii) *For every $0<x\in X$ and $0<\phi\in X^*$ there exists $t\geqslant 0$ such that $\langle T(t)x,\phi\rangle>0$.*

(iii) *$R(\lambda,A)$ is strongly irreducible for all (for some) $\lambda>s(A)$.*

(iv) *$R(\lambda,A)$ is irreducible for all (for some) $\lambda>s(A)$.*

PROOF. (i)⇒(ii). Take $0<\phi\in X^*$ and define

$$J = \{x\in X : <T(t)|x|,\phi> = 0 \text{ for all } t\geqslant 0\},$$

which is a closed $\{T(t)\}$-invariant ideal in X. Since $\{T(t)\}$ is irreducible and $J\neq X$ we conclude that $J=\{0\}$, which implies (ii).
(ii)⇒(iii). Suppose that $0<u\in X$ and $\lambda>s(A)$. It follows from Theorem 7.4 that

$$<R(\lambda,A)u,\phi> = \int_0^\infty e^{-\lambda t}<T(t)u,\phi>dt>0$$

for all $0<\phi\in X^*$, which implies that $R(\lambda,A)u$ is a quasi-interior point in X.
(iii)⇒(iv). Trivial.
(iv)⇒(i). This implication is an immediate consequence of the fact that, by Theorem 7.4, any closed $\{T(t)\}$-invariant ideal in X is $R(\lambda,A)$-invariant for all $\lambda>s(A)$. □

In relation to statement (ii) above, note that $\{T(t)\}$ is strongly irreducible if and only if $<T(t)x,\phi>>0$ for all $t>0$ whenever $0<x\in X$ and $0<\phi\in X^*$. In general, an irreducible semigroup is not strongly irreducible (e.g. consider translation semigroups). However, for positive analytic semigroups the two notions of irreducibility coincide, as is shown by the following interesting result.

PROPOSITION 7.7. *Suppose that $\{T(t)\}_{t\geqslant 0}$ is a positive analytic semigroup on the Banach lattice X. If $0\leqslant u\in X$ and $0\leqslant\phi\in X^*$ are such that $<T(t_0)u,\phi>=0$ for some $t_0>0$, then $<T(t)u,\phi>=0$ for all $t\geqslant 0$.*

PROOF. First observe that, if $0\leqslant x\in X$ and $0\leqslant x_n\in X$ $(n=1,2,...)$ such that $\Sigma_{n=1}^\infty\|x-x_n\|<\infty$, then there exists $y\in X$ such that $0\leqslant y\leqslant x_n$ and $\|x-y\|\leqslant\Sigma_{n=1}^\infty\|x-x_n\|$. Indeed, replacing x_n by $x_n\wedge x$, we may assume $0\leqslant x_n\leqslant x$ for all n. Now let $y_n=x_1\wedge\cdots\wedge x_n$, then $0\leqslant y_n-y_{n+1}\leqslant x-x_{n+1}$, so $\|y_n-y_{n+1}\|\leqslant\|x-x_{n+1}\|$ for all n, and hence $\Sigma_{n=1}^\infty\|y_n-y_{n+1}\|<\infty$. This shows that $\{y_n : n=1,2,...\}$ is a Cauchy sequence in X, so $y_n\to y$ for some $y\in X$. Note that $0\leqslant y\leqslant y_n\leqslant x_n$ for all n. Furthermore,

$$0\leqslant x-y_n = x-x_1\wedge\cdots\wedge x_n\leqslant\Sigma_{k=1}^n(x-x_k),$$

so $\|x-y_n\|\leqslant\Sigma_{k=1}^n\|x-x_k\|\leqslant\Sigma_{k=1}^\infty\|x-x_k\|$ for all n, which implies that $\|x-y\|\leqslant\Sigma_{k=1}^\infty\|x-x_k\|$.

Now consider the situation of the proposition. Take a sequence $t_n\downarrow 0$ such that $\Sigma_{n=1}^\infty\|T(t_n)u-u\|<\infty$. By the above observation, for each $m=1,2,...$ there exists $v_m\in X$ such that $0\leqslant v_m\leqslant T(t_n)u$ for all $n\geqslant m$ and $\|u-v_m\|\leqslant$ $\leqslant\Sigma_{n=m}^\infty\|T(t_n)u-u\|$. In particular, $\|u-v_m\|\to 0$ as $m\to\infty$. For $n\geqslant m$ we have

$$0\leqslant<T(t_0-t_n)v_m,\phi>\leqslant<T(t_0-t_n)T(t_n)u,\phi> = <T(t_0)u,\phi> = 0,$$

so the function $f_m(t)=<T(t)v_m,\phi>$ has zeros at the points t_0-t_n $(n\geqslant m)$. By hypothesis, f_m is the restriction of an analytic function to $(0,\infty)$, and since $t_0-t_n\uparrow t_0$, this implies that $f_m(t)=0$ for all $t\geq 0$. Finally, since $v_m\to u$ as $m\to\infty$, we may conclude that $<T(t)u,\phi>=0$ for all $t\geq 0$. □

COROLLARY 7.8. *An irreducible positive analytic semigroup in a Banach lattice is strongly irreducible.*

Note that Proposition 7.7 implies in particular that for a positive analytic semigroup $\{T(t)\}_{t\geq 0}$ in a Banach function space the support of $T(t)f$ is independent of t for all $t>0$, for any function $f\geq 0$.

2. HALF-NORMS AND DISSIPATIVE OPERATORS

In this section we discuss some properties of certain sublinear functionals and their relation to positive cones in ordered Banach spaces. Furthermore, we consider operators dissipative with respect to such sublinear functionals, which will be used for the characterization of positive (contraction) semigroups in ordered Banach spaces.

Let X be a real vector space. Recall that a functional $p:X\to\mathbb{R}$ is called sublinear if $p(x+y)\leq p(x)+p(y)$ and $p(\lambda x)=\lambda p(x)$ for all $x,y\in X$ and all $0\leq\lambda\in\mathbb{R}$. Note that $p(x)+p(-x)\geq p(0)=0$, and hence $\max(p(x),p(-x))\geq 0$ for all $x\in X$. The sublinear functional p is called a *half-norm* if $\max(p(x),p(-x))>0$ for all $x\neq 0$ in X. If p is a half-norm, then $\|x\|_p=\max(p(x),p(-x))$ defines a norm on X, which is called *the norm induced by p*. Note that $|p(x)|\leq\|x\|_p$.

Given the sublinear functional $p:X\to\mathbb{R}$, define $X^p=\{x\in X:p(-x)\leq 0\}$, which is a cone in X, i.e., if $x,y\in X^p$ then $x+y\in X^p$ and $\lambda x\in X^p$ for all $\lambda\geq 0$. If p is a half-norm, then X^p is proper, i.e., $X^p\cap(-X^p)=\{0\}$, so X^p induces a partial ordering in X by setting $x\leq y$ whenever $y-x\in X^p$. In this situation X is a partially ordered vector space with positive cone $X^+=X^p$.

From now on we assume that X is a Banach space, and that any sublinear functional $p:X\to\mathbb{R}$ is continuous, i.e., $|p(x)|\leq C\|x\|$ for all $x\in X$ and some constant $C\geq 0$. Clearly, the cone X^p is closed, and if p is a half-norm, then X is an ordered Banach space with respect to the induced ordering, and $\|x\|_p\leq C\|x\|$ for all $x\in X$. Furthermore, note that $x\leq y$ implies that $p(x)\leq p(y)$, and that $\|x\|_p=p(x)$ for all $x\in X^p$. In particular, $\|\cdot\|_p$ is a monotone norm with respect to the induced ordering, i.e., $\|x\|_p\leq\|y\|_p$ whenever $0\leq x\leq y$ in X.

Conversely, suppose that X is an ordered Banach space with positive cone X^+. For $x\in X$ define

$$p(x)=\mathrm{dist}(-x,X^+)=\inf\{\|x+y\|:y\in X^+\},$$

then p is a continuous sublinear functional in X. Clearly p is a half-norm in X and $X^p=X^+$. This half-norm p is called the *canonical half-norm* with respect to the given ordering. Note that $0\leq p(x)\leq\|x\|$ and $\|x\|_p\leq\|x\|$ for all $x\in X$. In general the norms $\|\cdot\|$ and $\|\cdot\|_p$ are not equivalent. Recall that the positive cone X^+ is normal if there exists a constant $M>0$ such that $\|\cdot\|$ is M-monotone, i.e., $\|x\|\leqslant M\|y\|$ whenever $0\leqslant x\leqslant y$ in X. Using the canonical half-norm we get the following characterization of normal cones.

PROPOSITION 7.9. *The positive cone X^+ in the ordered Banach space X is normal if and only if the norm induced by the canonical half-norm is equivalent to the given norm in X.*

PROOF. Since $\|\cdot\|_p$ is a monotone norm, it suffices to show that $\|\cdot\|_p$ is equivalent to the given norm if X^+ is normal. To this end observe that

$$\|x\|_p = \inf\{\max(\|y\|,\|z\|): y,z\in X,\, z\leqslant x\leqslant y\}$$

for all $x\in X$. Now assume that $\|\cdot\|$ is M-monotone and $z\leqslant x\leqslant y$, then

$$\max(\|y\|,\|z\|)\geqslant\frac{1}{2}\|y-z\|\geqslant\frac{1}{2M}\|x-z\|\geqslant\frac{1}{2M}\|x\|-\frac{1}{2M}\max(\|y\|,\|z\|),$$

from which we obtain that $\max(\|y\|,\|z\|)\geqslant(2M+1)^{-1}\|x\|$. This shows that $\|x\|_p\geqslant(2M+1)^{-1}\|x\|$ for all $x\in X$. □

Suppose that p is a sublinear functional on the Banach space X. The subdifferential $\partial p(x)$ of p in $x\in X$ consists of all $\phi\in X^*$ with $<x,\phi>=p(x)$ and $<y,\phi>\leqslant p(y)$ for all $y\in X$ (see Appendix 1, Remark A.1.26). If p is a half-norm and X is an ordered Banach space with $X^+=X^p$ and with dual positive cone $(X^*)^+=\{\phi\in X^*:<x,\phi>\geqslant 0 \text{ for all } x\in X^+\}$, then $\partial p(x)\subset(X^*)^+$ for all $x\in X$. Indeed, if $\phi\in\partial p(x)$ and $y\in X^+$, then $<-y,\phi>\leqslant p(-y)\leqslant 0$, and so $<y,\phi>\geqslant 0$.

EXAMPLES 7.10.

(i) If $(X,\|\cdot\|)$ is a Banach space, then $p(x)=\|x\|$ is a half-norm with $X^p=\{0\}$. Note that $\partial p(x)=\{\phi\in X^*:\|\phi\|=1 \text{ and } <x,\phi>=\|x\|\}$ for all $x\neq 0$ in X, and that $\partial p(0)=\{\phi\in X^*:\|\phi\|\leqslant 1\}$.

(ii) Let X be an ordered Banach space with canonical half-norm p. If $\phi\in X^*$, then $<y,\phi>\leqslant p(y)$ for all $y\in X$ if and only if $\phi\geqslant 0$ and $\|\phi\|\leqslant 1$. Indeed, if $<y,\phi>\leqslant p(y)$ for all $y\in X$, then, as observed above, $\phi\in(X^*)^+$, and $|<y,\phi>|\leqslant\|y\|_p\leqslant\|y\|$ so $\|\phi\|\leqslant 1$. For the converse, take $y,z\in X$ with $z\geqslant 0$, then $<y,\phi>\leqslant<y+z,\phi>\leqslant\|y+z\|$, which shows that $<y,\phi>\leqslant p(y)$. From this observation it follows that

$$\partial p(x)=\{\phi\in X^*:\phi\geqslant 0,\|\phi\|\leqslant 1 \text{ and } <x,\phi>=p(x)\}$$

for all $x\in X$. Furthermore, if $p(x)>0$, then it is easy to see that $\|\phi\|=1$ for all $\phi\in\partial p(x)$.

(iii) Let X be a Banach lattice. We assert that the canonical half-norm p in X is given by $p(x)=\|x^+\|$ for all $x\in X$. Indeed, since p obeys $p(x)=\inf\{\|x+y\|:0\leqslant y\in X\}$ and $x^+=x+x^-$, it is clear that $p(x)\leqslant\|x^+\|$. For any $0\leqslant y\in X$ we have $|x+y|\geqslant(x+y)^+\geqslant x^+$, so $\|x+y\|\geqslant\|x^+\|$, which shows that $p(x) \geqslant \|x^+\|$. The norm induced by p is given by $\|x\|_p = \max(\|x^+\|, \|x^-\|)$ for all $x \in X$, which is a monotone norm, equivalent to $\|\cdot\|$. We note, however, that in general $\|\cdot\|_p$ is not a lattice norm. It follows from (ii) that

$$\partial p(x) = \{\phi \in X^* : \phi \geqslant 0, \|\phi\| \leqslant 1 \text{ and } <x, \phi> = \|x^+\|\},$$

for all $x \in X$. Furthermore, $<x^-, \phi> = 0$ for all $\phi \in \partial p(x)$. Indeed, $\|x^+\| = <x^+, \phi> - <x^- \phi>$, so $<x^-, \phi> + \|x^+\| = <x^+, \phi> \leqslant \|x^+\|$ and hence $<x^-, \phi> = 0$. This shows that $<x^+, \phi> = <x, \phi> = \|x^+\|$ for all $\phi \in \partial p(x)$, and therefore $\partial p(x) \subset \partial p(x^+) \subset \partial N(x^+)$, where $N(x) = \|x\|$.

(iv) Let X be a Banach lattice and fix a positive linear functional $0 < \phi \in X^*$. For $x \in X$ define $p(x) = <x^+, \phi>$, then p is a sublinear functional, which is, in general, not a half-norm. Note that $\partial p(x) = \{\phi \in X^* : 0 \leqslant \psi \leqslant \phi$ and $<x, \psi> = <x^+, \phi>\}$, and that $<x^-, \psi> = 0$ whenever $\psi \in \partial p(x)$.

(v) Let Ω be a compact Hausdorff space and consider the space $X = C(\Omega)$ of all real continuous functions on Ω, which is a Banach lattice with respect to the sup-norm and pointwise ordering. The canonical half-norm on $C(\Omega)$ is given by

$$p(f) = \|f^+\|_\infty = \inf\{\lambda \geqslant 0 : f \leqslant \lambda \mathbf{1}\},$$

(where the function $\mathbf{1}$ is given by: $\mathbf{1}(\omega) = 1$ for all $\omega \in \Omega$), and the induced norm $\|f\|_p = \max(\|f^+\|_\infty, \|f^-\|_\infty) = \|f\|_\infty$. Another natural half-norm in $C(\Omega)$ is defined by

$$\hat{p}(f) = \inf\{\lambda \in \mathbb{R} : f \leqslant \lambda \mathbf{1}\}.$$

The cone defined by $\hat{p}$ is equal to the positive cone in $C(\Omega)$, and the norm induced by $\hat{p}$ is the sup-norm. Finally, $p(f) = \max(\hat{p}(f), 0)$ for all $f \in C(\Omega)$.

Let p be a continuous sublinear functional on the real Banach space X. Suppose that A is a linear operator from the linear subspace $\mathfrak{D}(A)$ of X into X. Recall that A is called *p-dissipative* if $p(x) \leqslant p(x - tAx)$ for all $t \geqslant 0$ and all $x \in \mathfrak{D}(A)$, which is equivalent to saying that

$$p'(x; -Ax) = \lim_{t \downarrow 0} t^{-1}\{p(x - tAx) - p(x)\} \geqslant 0$$

for all $x \in \mathfrak{D}(A)$ (see Section 3.3). Further, A is called *strictly p-dissipative* if $p'(x; Ax) \leqslant 0$ for all $x \in \mathfrak{D}(A)$. Since $-p'(x; -Ax) \leqslant p'(x; Ax)$, it is clear that any strictly p-dissipative operator is p-dissipative. The following equivalences follow from Proposition A.1.24:

(1) A is p-dissipative if and only if for each $x \in \mathfrak{D}(A)$ there exists $\phi \in \partial p(x)$ such that $<Ax, \phi> \leqslant 0$

(2) A is strictly p-dissipative if and only if $<Ax, \phi> \leqslant 0$ for all $\phi \in \partial p(x)$ and all $x \in \mathfrak{D}(A)$.

From the results concerning dissipativity obtained in Theorem 3.14, we recall in particular that for densely defined operators the notions of p-dissipativity and strict p-dissipativity coincide. Moreover, if p is a half-norm, then any densely defined p-dissipative operator A is closable, and the closure $\tilde{A}$ is strictly p-dissipative.

EXAMPLE 7.11. If X is a Banach lattice and p is the canonical half-norm, i.e., $p(x)=\|x^+\|$, then p-dissipative operators are called *dispersive*. It follows from the above observations combined with Example 7.10-(iii), that an operator $A:\mathfrak{D}(A)\to X$ is dispersive if and only if for every $x\in\mathfrak{D}(A)$ there exists $0\leqslant\phi\in X^*$ with $\|\phi\|\leqslant 1$ and $<x,\phi>=\|x^+\|$ such that $<Ax,\phi>\leqslant 0$. Note that the latter condition is trivially satisfied for all $x\in\mathfrak{D}(A)$ with $x^+=0$.

Now consider the special case that $X=L_2(\Omega,\mu)$. Identifying, as usual, X^* with $L_2=L_2(\Omega,\mu)$ and setting $N(f)=\|f\|_2$, we have $\partial N(f)=\{f/\|f\|\}$ for all $f\neq 0$ in L_2. As observed in Example 7.10-(iii), $\partial p(f)\subset\partial N(f^+)$ and so $\partial p(f)=\partial N(f^+)=\{f^+/\|f^+\|\}$ for all $f\in L_2$ with $f^+\neq 0$. This shows that an operator A in L_2 is dispersive if and only if $<Af,f^+>\leqslant 0$ for all $f\in\mathfrak{D}(A)$.

Let X be a real Banach space and let $p:X\to\mathbb{R}$ be a sublinear functional. The linear mapping T in X is called a p-contraction if $p(Tx)\leqslant p(x)$ for all $x\in X$. It is clear that the cone $X^p=\{x\in X:p(-x)\leqslant 0\}$ is invariant under every p-contraction in X. In particular, if X is an ordered Banach space with canonical half-norm p, then any p-contraction in X is positive. Now assume that $\{T(t)\}_{t\geqslant 0}$ is a C_0-semigroup in X with generator A. If $T(t)$ is a p-contraction for all $t\geqslant 0$, then $\{T(t)\}_{t\geqslant 0}$ is called a p-contraction semigroup. The following proposition is proved in Chapter 3 (Theorem 3.15).

PROPOSITION 7.12. *The C_0-semigroup $\{T(t)\}_{t\geqslant 0}$ is a p-contraction semigroup if and only if the generator A is p-dissipative.*

In the above proposition it is given in advance that the operator A is the generator of a C_0-semigroup. However, in many situations, p-dissipativity combined with a range condition already implies that A is the generator of a C_0-semigroup. For the sake of convenience we consider the case that p is the canonical half-norm in an ordered Banach space.

PROPOSITION 7.13. *Let X be an ordered Banach space with normal positive cone, and let p be the canonical half-norm in X. For a linear operator $A:\mathfrak{D}(A)\to X$ the following two statements are equivalent.*
(i) *A is the generator of a p-contraction C_0-semigroup.*
(ii) *A is densely defined, p-dissipative and $\mathfrak{R}(\lambda I-A)=X$ for some $\lambda>0$.*

PROOF. As before, let $\|x\|_p=\max(p(x),p(-x))$ be the norm induced by p. Since X^+ is normal, it follows from Proposition 7.9 that $\|\cdot\|_p$ is equivalent to the given norm in X. Furthermore, it is evident that any p-dissipative operator in X is dissipative with respect to $\|\cdot\|_p$. The equivalence of (i) and (ii) now follows from a combination of the Hille-Yosida theorem (see Theorem 3.4) and Proposition 7.12. □

Note that in an ordered Banach space X with canonical half-norm p, any positive contraction T is a p-contraction. Indeed, if $x \in X$ then

$$p(Tx) = \inf\{\|y\| : y \in X, Tx \leqslant y\} \leqslant \inf\{\|Ty\| : y \in X, x \leqslant y\} \leqslant$$
$$\leqslant \inf\{\|y\| : y \in X, x \leqslant y\} = p(x).$$

In order that the converse holds we need a stronger relationship between the norm and the order structure in X. Recall that the norm in X is called monotone if $\|x\| \leqslant \|y\|$ whenever $0 \leqslant x \leqslant y$. Furthermore, the operator norm in $\mathfrak{L}(X)$ is called *positively attained* if

$$\|T\| = \sup\{\|Tx\| : 0 \leqslant x \in X, \|x\| \leqslant 1\}$$

for all positive T in $\mathfrak{L}(X)$ (see Appendix A.2.3 for some further remarks concerning these properties). In an ordered Banach space with monotone norm and positively attained operator norm, especially in a Banach lattice, any p-contraction is a positive contraction. Indeed, observe first that $p(x) = \|x\|$ for all $0 \leqslant x \in X$, as the norm is monotone. Therefore, if T is a p-contraction in X, then T is positive, and so $\|Tx\| = p(Tx) \leqslant p(x) = \|x\|$ for all $0 \leqslant x \in X$. Since the operator norm is positively attained, this implies that $\|T\| \leqslant 1$. Combining these remarks with the previous proposition we obtain the following result.

PROPOSITION 7.14. *Let X be an ordered Banach space with monotone norm and suppose that the operator norm in $\mathfrak{L}(X)$ is positively attained. Let p be the canonical half-norm in X. For a linear operator $A : \mathfrak{D}(A) \to X$ the following statements are equivalent.*
(i) *A is the generator of a positive contraction semigroup.*
(ii) *A is densely defined, p-dissipative and $\mathfrak{R}(\lambda I - A) = X$ for some $\lambda > 0$.*

The above proposition can be applied in particular in a Banach lattice.

COROLLARY 7.15. *Let X be a Banach lattice and $A : \mathfrak{D}(A) \to X$ linear. The following statements are equivalent.*
(i) *A is the generator of a positive contraction semigroup.*
(ii) *A is densely defined, $\mathfrak{R}(\lambda I - A) = X$ for some $\lambda > 0$, and A is dispersive (i.e., for every $x \in \mathfrak{D}(A)$ there exists $0 \leqslant \phi \in X^*$ with $\|\phi\| \leqslant 1$ and $<x, \phi> = \|x^+\|$ such that $<Ax, \phi> \leqslant 0$).*

If, in statement (ii) of the above proposition, instead of $\mathfrak{R}(\lambda I - A) = X$ for some $\lambda > 0$, we only assume that $\mathfrak{R}(\lambda I - A)$ is dense in X for some $\lambda > 0$, then we may conclude that the closure of A is the generator of a positive contraction semigroup.

The next proposition characterizes the generators of positive C_0-semigroups which are not necessarily contractive, and can be considered as an analogue of the Hille-Yosida-Phillips theorem.

PROPOSITION 7.16. *Let X be an ordered Banach space with monotone norm and suppose that the operator norm in $\mathfrak{L}(X)$ is positively attained. The canonical half-norm in X is denoted by p. The linear operator $A:\mathfrak{D}(A)\to X$ is the generator of a positive C_0-semigroup $\{T(t)\}_{t\geq 0}$ satisfying $\|T(t)\|\leq Me^{\omega t}$, if and only if*
(i) *A is closed and densely defined,*
(ii) *$(\omega,\infty)\subset\rho(A)$ and*

$$p[R(\lambda,A)^n x]\leq\frac{M}{(\lambda-\omega)^n}p(x),\ x\in X,$$

for all $\lambda>\omega$ and all $n=1,2,\ldots$.

PROOF. The result follows from an application of the Hille-Yosida-Phillips theorem (see Theorems 3.5 and 3.6) with respect to the norm $\|x\|_p=\max(p(x),p(-x))$, which is equivalent to the given norm in X. □

COROLLARY 7.17. *Let X be a Banach lattice. The linear operator $A:\mathfrak{D}(A)\to X$ is the generator of a positive C_0-semigroup $\{T(t)\}_{t\geq 0}$ with $\|T(t)\|\leq Me^{\omega t}$, if and only if*
(i) *A is closed and densely defined.*
(ii) *$(\omega,\infty)\subset\rho(A)$ and*

$$\|[R(\lambda,A)^n x]^+\|\leq\frac{M}{(\lambda-\omega)^n}\|x^+\|,\ x\in X,$$

for all $\lambda>\omega$ and all $n=1,2,\ldots$.

Dispersiveness can be viewed as an abstract 'maximum principle'. In fact, let $X=C(\Omega)$, the Banach lattice of all real continuous functions on the compact Hausdorff space Ω, and suppose that A is a densely defined linear operator in $C(\Omega)$. Then A is dispersive if and only if $Af(\omega)\leq 0$ whenever $f\in\mathfrak{D}(A)$ and $\omega\in\Omega$ such that $f(\omega)=\|f^+\|_\infty$. This follows immediately from the observation that for any $f\in C(\Omega)$ and $\omega\in\Omega$ with $f(\omega)=\|f^+\|_\infty$, the point evaluation ϕ_ω at ω belongs to $\partial p(f)$ (where, as before, p denotes the canonical half-norm in $C(\Omega)$). In the next section we will discuss the case in which X is a $C(\Omega)$ space in more detail.

3. THE POSITIVE-OFF-DIAGONAL PROPERTY

In the previous section we have characterized generators of positive contraction semigroups by means of dissipativity with respect to appropriate sublinear functionals. In contrast to dissipativity, positivity is not a metric condition. To be more precise, if A is the generator of a positive C_0-semigroup in an ordered Banach space, then the same holds for $A-\lambda I$ for all $\lambda\in\mathbb{R}$. Consequently, characterization of positivity in terms of the generator is, in general, a more complicated problem. In this and the next section such characterizations will be presented. In particular in spaces with an order unit the situation turns out to be quite satisfactory.

Let $(X, \|\cdot\|)$ be a real ordered Banach space with positive cone X^+. As before, the canonical half-norm is denoted by p. We start with a definition.

DEFINITION 7.18. The linear operator $A:\mathcal{D}(A)\to X$ is said to have the *positive off-diagonal property (POD)* if $<Au,\phi>\geqslant 0$ whenever $0\leqslant u\in\mathcal{D}(A)$ and $0\leqslant\phi\in X^*$ with $<u,\phi>=0$.

First of all, notice that if A is the generator of a positive C_0-semigroup $\{T(t)\}_{t\geqslant 0}$, then A has POD. Indeed, take $0\leqslant u\in\mathcal{D}(A)$ and $0\leqslant\phi\in X^*$ such that $<u,\phi>=0$, then

$$<Au,\phi> = \lim_{t\downarrow 0}\frac{<T(t)u,\phi>-<u,\phi>}{t} = \lim_{t\downarrow 0}\frac{<T(t)u,\phi>}{t}\geqslant 0.$$

Furthermore, the condition POD for an arbitrary operator A is non-trivial only if $\mathcal{D}(A)$ contains sufficiently many positive elements. In order to motivate the name 'positive off-diagonal property', we consider the finite dimensional situation.

EXAMPLE 7.19. Suppose that $X=\mathbb{R}^n$ with the coordinatewise ordering and let $A:\mathbb{R}^n\to\mathbb{R}^n$ be a linear operator with matrix (a_{ij}) with respect to the standard basis. We assert that A has POD if and only if $a_{ij}\geqslant 0$ for all $i\neq j$. Indeed, assume that A has POD and $i\neq j$, then $<e_j,e_i>=0$, $e_j\geqslant 0$, $e_i\geqslant 0$, and hence $a_{ij}=<Ae_j,e_i>\geqslant 0$. Conversely, assume that $a_{ij}\geqslant 0$ for all $i\neq j$, and take $0\leqslant x=(x_1,\ldots,x_n)\in\mathbb{R}^n$ and $0\leqslant y=(y_1,\ldots,y_n)\in(\mathbb{R}^n)^*=\mathbb{R}^n$ such that $<x,y>=\sum_{i=1}^n y_ix_i=0$. Then $y_ix_i=0$ $(i=1,\ldots,n)$ and so

$$<Ax,y> = \sum_{i\neq j}a_{ij}y_ix_j + \sum_i a_{ii}y_ix_i = \sum_{i\neq j}a_{ij}y_ix_j\geqslant 0,$$

which shows that A has POD.
Now consider the semigroup $\{e^{tA}\}_{t\geqslant 0}$ generated by A, which is norm continuous (with respect to any norm in $\mathbb{R}^n$). If $\{e^{tA}\}_{t\geqslant 0}$ is positive, then it follows from the above observations that A has POD. Conversely, assume that A has POD. Then $a_{ij}\geqslant 0$ for all $i\neq j$, and so there exists $\lambda\in\mathbb{R}$ such that $B=A+\lambda I\geqslant 0$, which implies that $e^{tA}=e^{-\lambda t}e^{tB}\geqslant 0$ for all $t\geqslant 0$. We may conclude, therefore, that the following statements are equivalent:

(1) A has POD
(2) $a_{ij}\geqslant 0$ whenever $i\neq j$, i.e., the off-diagonal elements of the matrix (a_{ij}) are positive
(3) the semigroup $\{e^{tA}\}_{t\geqslant 0}$ generated by A consists of positive operators.

The above example shows that it is a natural question whether, in general, POD of the generator implies positivity of the semigroup. Before we will show that the answer is affirmative in spaces with an order unit, we make some general remarks.

REMARK 7.20.

(i) It follows immediately from the definition that, if the linear operator $A:\mathfrak{D}(A)\to X$ has POD, then $A-\lambda I$ has POD as well for all $\lambda\in\mathbb{R}$.

(ii) If the operator $A:\mathfrak{D}(A)\to X$ is strictly p-dissipative, then A has POD. Indeed, first recall that, by Example 7.10-(ii),

$$\partial p(x)=\{\phi\in X^*:\phi\geqslant 0,\ \|\phi\|\leqslant 1 \text{ and } <x,\phi>=p(x)\}$$

for all $x\in X$. Now suppose that $0\leqslant u\in\mathfrak{D}(A)$ and $0<\phi\in X^*$ such that $<u,\phi>=0$. It follows from $p(-u)=0$ that $\phi/\|\phi\|\in\partial p(-u)$, and since A is strictly p-dissipative this implies that $<A(-u),\phi>\leqslant 0$, i.e., $<Au,\phi>\geqslant 0$.

(iii) To some extent the POD property asserts that the non-local part of the operator is positive, as is illustrated by Example 7.19. To make this statement more precise, we restrict ourselves to the Banach lattice situation. Assume that X is a Banach lattice and $A:\mathfrak{D}(A)\to X$ a linear operator. Recall that X_u denotes the principal ideal generated by the element $0\leqslant u\in X$, i.e., $X_u=\{x\in X:|x|\leqslant nu \text{ for some } n\in\mathbb{N}\}$. The following statements are equivalent:
(1) A has POD,
(2) the negative part $(Au)^-$ of Au is contained in $\overline{X}_u$ for all $0\leqslant u\in\mathfrak{D}(A)$.
Indeed, first assume that A has POD and suppose that there exists $0\leqslant u\in\mathfrak{D}(A)$ with $(Au)^-\notin\overline{X}_u$. Then there exists $0\leqslant\phi\in X^*$ such that $<x,\phi>=0$ for all $x\in\overline{X}_u$ and $<(Au)^-,\phi>>0$. Let ψ be the minimal positive extension of the restriction of ϕ to the ideal generated by $(Au)^-$, then $<u,\psi>=0$, $<(Au)^+,\psi>=0$ and $<(Au)^-,\psi>>0$. Since, by hypothesis, A has POD, this implies that $<Au,\psi>\geqslant 0$, so

$$0\leqslant <Au,\psi> = <(Au)^+,\psi> - <(Au)^-,\psi> = -<(Au)^-,\psi><0,$$

a contradiction. This shows that (1) implies (2). To prove the converse implication, assume that A satisfies (2) and that $0\leqslant u\in\mathfrak{D}(A), 0\leqslant\phi\in X^*$ with $<u,\phi>=0$. Then $<x,\phi>=0$ for all $x\in\overline{X}_u$, and so in particular $<(Au)^-,\phi>=0$, which implies that $<Au,\phi>\geqslant 0$.

We now turn our attention to spaces with an order unit. The examples we have primarily in mind are spaces of real continuous functions on some compact Hausdorff space, and the self-adjoint part of unitary C^*-algebras. Recall that the element e in the (real) ordered Banach space X is called an order unit if for every $x\in X$ there exists $0\leqslant\lambda\in\mathbb{R}$ such that $-\lambda e\leqslant x\leqslant\lambda e$, which is equivalent to $e\in\text{int}X^+$ (see Appendix A.2.2). For $e\in\text{int}X^+$ we define the functional Φ_e on X by

$$\Phi_e(x)=\inf\{\lambda\geqslant 0: x\leqslant\lambda e\},\ x\in X.$$

It should be noted that $x\leqslant\Phi_e(x)e$ for all $x\in X$. In the next lemma we collect some properties of Φ_e.

LEMMA 7.21. *Φ_e is a continuous half-norm on X which induces the given ordering in X. Furthermore, the norm induced by Φ_e in X is the order unit norm $\|\cdot\|_e$ corresponding to e, i.e.,*

$$\max(\Phi_e(x),\Phi_e(-x)) = \|x\|_e = \inf\{\lambda\geq 0: -\lambda e\leq x\leq \lambda e\}$$

for all $x\in X$, and Φ_e is the canonical half-norm corresponding to the given ordering and the order unit norm $\|\cdot\|_e$.

PROOF. The only non-trivial statement of the lemma is, perhaps, the continuity of Φ_e. Since $e\in \text{int}X^+$, there exists $\delta>0$ such that the ball $B(e,\delta)$ is contained in X^+. Hence $e-\delta\|x\|^{-1}x\in B(e,\delta)\subset X^+$, so that $\Phi_e(x)\leq\delta^{-1}\|x\|$, which shows that Φ_e is continuous. □

The next objective is to relate the POD property to Φ_e-dissipativity for appropriate order unit e. The following lemma will be useful.

LEMMA 7.22. *Let X be an ordered Banach space and $e\in\text{int}X^+$. If $x\in X$ is such that $\Phi_e(x)>0$, then $<e,\phi>=1$ for all $\phi\in\partial\Phi_e(x)$.*

PROOF. Suppose $x\in X$ with $\Phi_e(x)>0$ and take $\phi\in\partial\Phi_e(x)$, so $<x,\phi>=\Phi_e(x)$ and $<y,\phi>\leq\Phi_e(y)$ for all $y\in X$. As noted in Section 7.2 (see the remark following Proposition 7.9), this implies that $\phi\geq 0$. Therefore, it follows from $x\leq\Phi_e(x)e$ that

$$\Phi_e(x) = <x,\phi>\leq\Phi_e(x)<e,\phi>\leq\Phi_e(x)\Phi_e(e) = \Phi_e(x),$$

so $\Phi_e(x)=\Phi_e(x)<e,\phi>$. Since $\Phi_e(x)>0$, we may conclude that $<e,\phi>=1$. □

In connection to the next lemma, it should be noted that for any densely defined linear operator A in the ordered Banach space X with order unit, the intersection $\mathfrak{D}(A)\cap\text{int}X^+$ is non-empty.

LEMMA 7.23. *Let X be an ordered Banach space, $A:\mathfrak{D}(A)\to X$ a densely defined linear operator and suppose that $e\in\mathfrak{D}(A)\cap\text{int}X^+$. The following statements are equivalent.*

(i) *A is Φ_e-dissipative*
(ii) *A has POD and $Ae\leq 0$*
(iii) *$Ae\leq 0$ and for all $\lambda>0$ (or, equivalently, there exist arbitrarily large $\lambda>0$ such that) it follows from $x\in\mathfrak{D}(A)$ and $(\lambda I-A)x\geq 0$ that $x\geq 0$.*

PROOF. (i)⇒(ii). Since A is densely defined, it follows that A is strictly Φ_e-dissipative, and now Remark 7.20-(ii) implies that A has POD. It remains to show that $Ae\leq 0$, i.e., that $\Phi_e(Ae)=0$. Assume, on the contrary, that $\Phi_e(Ae)>0$, and take any $\phi\in\partial\Phi_e(Ae)$. By the above lemma, $<e,\phi>=1$ and hence $\phi\in\partial\Phi_e(e)$. Since A is strictly Φ_e-dissipative, this implies that $<Ae,\phi>\leq 0$, which is at variance with $<Ae,\phi>=\Phi_e(Ae)>0$. Hence $Ae\leq 0$.

(ii)⇒(iii). Take $x\in\mathfrak{D}(A),\lambda>0$ and assume that $(\lambda I-A)x\geqslant 0$, i.e., $Ax\leqslant\lambda x$. Suppose that $\Phi_e(-x)>0$ and take any $\phi\in\partial\Phi_e(-x)$. From the above lemma we infer $<e,\phi>=1$ and so $<\Phi_e(-x)e+x,\phi>=0$. Since $\Phi_e(-x)e+x\geqslant 0$ and $\phi\geqslant 0$, the hypothesis on A implies that $<\Phi_e(-x)Ae+Ax,\phi>\geqslant 0$. Using that $Ae\leqslant 0$ we find

$$0\leqslant\Phi_e(-x)<-Ae,\phi>\leqslant<Ax,\phi>\leqslant\lambda<x,\phi> = -\lambda\Phi_e(-x)<0,$$

which is a contradiction. Consequently, $\Phi_e(-x)=0$, i.e., $x\geqslant 0$.
(iii)⇒(i). Take $x\in\mathfrak{D}(A)$. We have to show that $\Phi_e(x)\leqslant\Phi_e(x-\alpha Ax)$ for arbitrarily small $\alpha>0$. Choose $\alpha>0$ such that $(\alpha^{-1}I-A)z\geqslant 0$, $z\in\mathfrak{D}(A)$, implies that $z\geqslant 0$. Observing that $(I-\alpha A)e\geqslant e$, we obtain

$$\begin{aligned}\Phi_e(x-\alpha Ax) &= \inf\{\lambda\geqslant 0:(I-\alpha A)x\leqslant\lambda e\}\\ &\geqslant \inf\{\lambda\geqslant 0:(I-\alpha A)x\leqslant\lambda(I-\alpha A)e\}\\ &\geqslant \inf\{\lambda\geqslant 0:x\leqslant\lambda e\}=\Phi_e(x).\end{aligned}$$

Since, by the assumptions on A, there exist arbitrarily small $\alpha>0$ with the above property, we conclude that A is Φ_e-dissipative. □

It follows easily that in (ii) of the above lemma the condition that $Ae\leqslant 0$ can be replaced by the assumption that A is dissipative with respect to the order unit norm $\|\cdot\|_e$.
The next proposition shows the precise relation between POD and Φ_e-dissipativity. It is convenient to introduce some notation. Suppose X is an ordered Banach space and $e\in\text{int}X^+$. For $x\in X$ define

$$\hat{\Phi}_e(x)=\inf\{\lambda\in\mathbb{R}:x\leqslant\lambda e\},$$

then $\hat{\Phi}_e$ is a continuous half-norm in X (note that $|\hat{\Phi}_e(x)|\leqslant\|x\|_e$), $x\leqslant\hat{\Phi}_e(x)e$ for all $x\in X$ and $\Phi_e(x)=\max(\hat{\Phi}_e(x),0)$. Furthermore, note that the ordering induced by $\hat{\Phi}_e$ is equal to the given ordering in X and that the norm induced by $\hat{\Phi}_e$ is the order unit norm $\|\cdot\|_e$. Now suppose that $A:\mathfrak{D}(A)\to X$ is densely defined, $e\in\mathfrak{D}(A)\cap\text{int}X^+$ and that $A-\lambda I$ is Φ_e-dissipative. By the above lemma ((i)⇒(ii)) this implies that $Ae\leqslant\lambda e$, and hence $\hat{\Phi}_e(Ae)\leqslant\lambda$. Conversely, suppose that A has POD and take $\lambda\geqslant\hat{\Phi}_e(Ae)$, then $A-\lambda I$ has POD as well and $(A-\lambda I)e\leqslant 0$. By the lemma, this implies that $A-\lambda I$ is Φ_e-dissipative. These remarks show that, if A is a densely defined operator in X, which has the POD property, then

$$\begin{aligned}\hat{\Phi}_e(Ae) &= \min\{\lambda\in\mathbb{R}:A-\lambda I \text{ is } \Phi_e-\text{dissipative}\}\\ &= \min\{\lambda\in\mathbb{R}:A-\lambda I \text{ is } \|\cdot\|_e-\text{dissipative}\}.\end{aligned}$$

A combination of the above observations and Lemma 7.23 yields the following result.

PROPOSITION 7.24. *If A is a densely defined linear operator in the ordered Banach space X and $e \in \mathfrak{D}(A) \cap \operatorname{int} X^+$, then the following statements are equivalent.*
(i) *A has POD*
(ii) *$A - \lambda I$ is Φ_e-dissipative for some (for all) $\lambda \geq \hat{\Phi}_e(Ae)$*
(iii) *For all $\alpha > \hat{\Phi}_e(Ae)$ (or, equivalently, there exist arbitrarily large α such that) it follows from $x \in \mathfrak{D}(A)$ and $(\alpha I - A)x \geq 0$ that $x \geq 0$.*

Since a densely defined Φ_e-dissipative operator is closable with a Φ_e-dissipative closure (see Theorem 3.14), and since the domain of a densely defined operator contains an order unit, we immediately obtain the following corollary from the above proposition.

COROLLARY 7.25. *Let X be an ordered Banach space with order unit, and let A be a densely defined linear operator in X. If A has the POD-property, then A is closable and the closure $\tilde{A}$ has POD as well.*

Condition (iii) in the above proposition is, of course, closely related to the assertion that the resolvent $R(\alpha,A)$ is positive. However, $\alpha I - A$ need not be surjective and so $R(\alpha,A)$ need not exist for $\alpha > \hat{\Phi}_e(Ae)$. We claim that, if A has the POD property and $\alpha > \hat{\Phi}_e(Ae)$ is such that $\mathfrak{R}(\alpha I - A) = X$, then $\alpha \in \rho(A)$. Indeed, if $\lambda = \hat{\Phi}_e(Ae)$, then $A - \lambda I$ is Φ_e-dissipative and, since $\alpha I - A = (\alpha - \lambda)I - (A - \lambda I)$, this implies that $\alpha I - A$ is injective. Thus $\alpha I - A$ is a bijection from $\mathfrak{D}(A)$ onto X. Therefore, it remains to show that $\alpha I - A$ is closed. By Corollary 7.25, A is closable and the closure $\tilde{A}$ satisfies the same conditions as A. In particular, $\alpha I - \tilde{A}$ is a bijection from $\mathfrak{D}(\tilde{A})$ onto X. Since $\alpha I - A \subseteq \alpha I - \tilde{A}$ and $\alpha I - A$ is a bijection, we conclude that $\mathfrak{D}(A) = \mathfrak{D}(\tilde{A})$, which shows that A is closed. Thus $\alpha \in \rho(A)$, and it is obvious from Proposition 7.24 (iii) that $R(\alpha,A) \geq 0$. These observations yield the following characterization of resolvent positive operators.

PROPOSITION 7.26. *Suppose that X is an ordered Banach space with $\operatorname{int} X^+ \neq \varnothing$ and let A be a densely defined linear operator in X. The following two statements are equivalent.*
(i) *A has POD and $\mathfrak{R}(\alpha I - A) = X$ for all sufficiently large real α*
(ii) *$\alpha \in \rho(A)$ and $R(\alpha,A) \geq 0$ for all sufficiently large real α.*

Much more can be said in the case that the positive cone in X is in addition normal. Before we consider that situation we mention another interesting corollary of Proposition 7.24.

THEOREM 7.27. *Let A be the generator of the C_0-semigroup $\{T(t)\}_{t \geq 0}$ in the ordered Banach space X with $\operatorname{int} X^+ \neq \varnothing$. The following statements are equivalent.*
(i) *$\{T(t)\}_{t \geq 0}$ is a positive semigroup*

(ii) *A has POD*
(iii) $R(\lambda,A)\geqslant 0$ *for all sufficiently large real* λ.

PROOF. The equivalence of (i) and (iii) is already observed in Proposition 7.1. Furthermore, that (i) implies (ii) is proved at the beginning of the present section, and so it remains to show that (ii) implies (i). Take any $e\in\mathfrak{D}(A)\cap\mathrm{int}X^+$ and $\lambda\geqslant\hat{\Phi}_e(Ae)$. By Proposition 7.24, $A-\lambda I$ is Φ_e-dissipative. In view of Proposition 7.12, this implies that $\{e^{-\lambda t}T(t)\}_{t\geqslant 0}$ is a Φ_e-contraction semigroup, and hence positive. Therefore $\{T(t)\}_{t\geqslant 0}$ is positive, and we are done. □

Recall that the type of the C_0-semigroup $\{T(t)\}_{t\geqslant 0}$ is defined by

$$\omega_0 = \inf_{t>0}\frac{\log\|T(t)\|}{t},$$

and that the spectral bound of the generator A is defined by $s(A)=\sup\{\mathrm{Re}\lambda:\lambda\in\sigma(A)\}$. As noted before, $-\infty\leqslant s(A)\leqslant\omega_0<\infty$.
If X is an ordered Banach space with $\mathrm{int}X^+\neq\varnothing$, then the cone X^+ is normal if and only if for any $e\in\mathrm{int}X^+$ the order unit norm $\|\cdot\|_e$ is equivalent to the given norm. Now we are in a position to prove the main result in the present section. For sake of convenience we isolate the following lemma from the proof of the theorem.

LEMMA 7.28. *Let X be an ordered Banach space with* $\mathrm{int}X^+\neq\varnothing$ *and suppose that A is a densely defined linear operator in X with the POD property. If* $\lambda\in\rho(A)\cap\mathbb{R}$ *such that* $R(\lambda,A)\geqslant 0$, *then there exists* $e\in\mathfrak{D}(A)\cap\mathrm{int}X^+$ *such that* $Ae\leqslant\lambda e$.

PROOF. Since $\mathfrak{D}(A)$ is dense, there exists $w\in\mathfrak{D}(A)\cap\mathrm{int}X^+$. Setting $e=R(\lambda,A)w$, we have $0\leqslant e\in\mathfrak{D}(A)$ and $(\lambda I-A)e=w\geqslant 0$, so $Ae\leqslant\lambda e$. To show that e is an order unit in X, it is sufficient to prove that e is a quasi-interior point (see Proposition A.2.10). For this purpose take $0<\phi\in X^*$ and suppose that $<e,\phi>=0$. Now POD implies that $<Ae,\phi>\geqslant 0$. Furthermore, $<w,\phi>>0$, as w is an order unit. Hence, it follows from $\lambda e=Ae+w$ that $0=<\lambda e,\phi>=<Ae,\phi>+<w,\phi>\geqslant<w,\phi>>0$, which is a contradiction. Thus the element e has the desired properties. □

THEOREM 7.29. *Let X be an ordered Banach space with normal positive cone and* $\mathrm{int}X^+\neq\varnothing$. *If A is a densely defined linear operator in X, then the following statements are equivalent.*
(i) *A is the generator of a positive* C_0*-semigroup in X*
(ii) *A has POD and* $\mathfrak{R}(\lambda I-A)=X$ *for all sufficiently large real* λ
(iii) $\lambda\in\rho(A)$ *and* $R(\lambda,A)\geqslant 0$ *for all sufficiently large real* λ.
Moreover, if A satisfies one of the above conditions, then

$$s(A)=\omega_0=\inf\{\lambda\in\mathbb{R}: Ae\leqslant\lambda e \textit{ for some } e\in\mathfrak{D}(A)\cap\mathrm{int}X^+\}.$$

PROOF. In view of Proposition 7.26 and Theorem 7.27, only implication (ii)$\Rightarrow$(i) and the additional statement remain to be shown. To this end, take $e \in \mathfrak{D}(A) \cap \operatorname{int} X^+$. Since X^+ is normal, the order unit norm $\|\cdot\|_e$ is equivalent to the given norm $\|\cdot\|$. Furthermore, $\|\cdot\|_e$ is absolutely monotone and absolutely dominating (i.e., $\|\cdot\|_e$ is a strong Riesz norm; see Section A.2.3), and Φ_e is the canonical half-norm with respect to $\|\cdot\|_e$. Now take $\lambda = \hat{\Phi}_e(Ae)$, then it follows from Proposition 7.24 that $A - \lambda I$ is Φ_e-dissipative, and so, by Proposition 7.14, $A - \lambda I$ generates a positive $\|\cdot\|_e$-contraction semigroup $\{S(t)\}_{t \geqslant 0}$. Therefore, A generates the positive C_0-semigroup $\{e^{\lambda t} S(t)\}_{t \geqslant 0}$, which is assertion (i). Note that the type of the semigroup generated by A satisfies $\omega_0 \leqslant \lambda = \hat{\Phi}_e(A)$, which shows that

$$\omega_0 \leqslant \inf\{\lambda \in \mathbb{R} : Ae \leqslant \lambda e \text{ for some } e \in \mathfrak{D}(A) \cap \operatorname{int} X^+\}.$$

For the converse inequality, take any $\lambda > s(A)$, then $\lambda \in \rho(A)$ and, by Corollary 7.5, $R(\lambda, A) \geqslant 0$. From Lemma 7.28 we obtain $e \in \mathfrak{D}(A) \cap \operatorname{int} X^+$ with $Ae \leqslant \lambda e$, which shows that

$$\inf\{\lambda \in \mathbb{R} : Ae \leqslant \lambda e \text{ for some } e \in \mathfrak{D}(A) \cap \operatorname{int} X^+\} \leqslant s(A).$$

A combination of the above two inequalities together with $s(A) \leqslant \omega_0$ yields the final statement of the theorem. □

REMARK 7.30. *We assume the same hypothesis as in the previous theorem.*

(i) Let us denote for a moment

$$\lambda_0 = \inf\{\hat{\Phi}_e(Ae) : e \in \mathfrak{D}(A) \cap \operatorname{int} X^+\},$$

then it is clear that

$$\lambda_0 = \inf\{\lambda \in \mathbb{R} : Ae \leqslant \lambda e \text{ for some } e \in \mathfrak{D}(A) \cap \operatorname{int} X^+\} = \omega_0 = s(A).$$

A close inspection of the proof of the above theorem shows that assertion (ii) can be replaced by
(ii)′ *A has POD and* $\mathfrak{R}(\lambda I - A) = X$ *for some* $\lambda > \lambda_0$.
Furthermore, it should be noted that, if A satisfies the conditions of the theorem and $\lambda \in \mathbb{R}$ is such that $\lambda \in \rho(A)$ and $R(\lambda, A) \geqslant 0$, then $\lambda > s(A)$.

(ii) We assert that, if A satisfies the conditions of the above theorem, then $s(A) > -\infty$. Indeed, take any $e \in \mathfrak{D}(A) \cap \operatorname{int} X^+$, then $Ae \geqslant \lambda e$ for some $\lambda \in \mathbb{R}$. Suppose that $\lambda > s(A)$. Then $R(\lambda, A) \geqslant 0$ from which we derive that $e = R(\lambda, A)(\lambda e - Ae) \leqslant 0$, which is a contradiction. Hence $s(A) \geqslant \lambda > -\infty$. This shows in particular that $\sigma(A) \neq \varnothing$.

(iii) The above theorem implies in particular that any resolvent positive operator, in an ordered Banach space with normal cone and order unit, is the generator of a (necessarily positive) C_0-semigroup.

Before illustrating, by way of some examples, that the assumption $\operatorname{int} X^+ \neq \varnothing$ in the above theorem cannot be omitted, we discuss briefly the case that X is a Banach lattice of continuous functions.

EXAMPLE 7.31. Suppose that $X=C(\Omega)$, the Banach lattice of real continuous functions on some compact Hausdorff space Ω. As usual, we consider in $C(\Omega)$ the sup-norm. Let A be a densely defined linear operator in $C(\Omega)$. We assert that A has the POD property if and only if $(Au)(\omega)\geqslant 0$ whenever $0\leqslant u\in\mathfrak{D}(A)$ and $\omega\in\Omega$ with $u(\omega)=0$. Indeed, first assume that A satisfies the latter condition, and suppose that $0\leqslant u\in\mathfrak{D}(A)$ and $0\leqslant\phi\in C(\Omega)^*$ with $<u,\phi>=0$. Let μ be the Borel measure representing ϕ, so $<f,\phi>=\int_\Omega f d\mu$ for all $f\in C(\Omega)$. Denote the support of μ by supp (μ). Then $u(\omega)=0$ for all $\omega\in\text{supp}(\mu)$, and so by hypothesis, $(Au)(\omega)\geqslant 0$ for all $\omega\in\text{supp}(\mu)$, hence $<Au,\phi>=\int_\Omega Au d\mu\geqslant 0$. This shows that A has POD. Conversely, if A has POD and $u(\omega)=0$ for some $0\leqslant u\in\mathfrak{D}(A)$ and $\omega\in\Omega$, then, denoting the point evaluation at ω by ϕ_ω, it follows from $<u,\phi_\omega>=0$ that $<Au,\phi_\omega>\geqslant 0$, i.e., $(Au)(\omega)\geqslant 0$. Assuming that A is the generator of a C_0-semigroup in $C(\Omega)$, it is a consequence of the above observations in conjuction with Theorem 7.29 (or Theorem 7.27) that the following assertions are equivalent:

(1) $(Au)(\omega)\geqslant 0$ whenever $0\leqslant u\in\mathfrak{D}(A)$ and $\omega\in\Omega$ with $u(\omega)=0$

(2) $\{T(t)\}_{t\geqslant 0}$ consists of positive operators.

It is a special feature of a positive C_0-semigroup $\{T(t)\}_{t\geqslant 0}$ in $C(\Omega)$ that $\|T(t)\|$ depends continuously on t, since $\|T(t)\|=\|T(t)\mathbf{1}\|_\infty$. In particular $\|T(t)\|\to 1$ as $t\downarrow 0$. Let A be the generator of $\{T(t)\}_{t\geqslant 0}$, and assume that $\mathbf{1}\in\mathfrak{D}(A)$, as is the case in many examples. Putting $\omega=\|A\mathbf{1}\|_\infty$, we have $(A-\omega I)\mathbf{1}\leqslant 0$, and so it follows from Lemma 7.23 that $A-\omega I$ is $\Phi_\mathbf{1}$-dissipative. Therefore $A-\omega I$ generates a $\|\cdot\|_\mathbf{1}=\|\cdot\|_\infty$-contraction semigroup, i.e., $\|e^{-\omega t}T(t)\|\leqslant 1$. This shows that $\|T(t)\|\leqslant e^{\|A\mathbf{1}\|t}$ for all $t\geqslant 0$. It should be noted, however, that if $\mathbf{1}\notin\mathfrak{D}(A)$, then an estimate of the form $\|T(t)\|\leqslant e^{\omega t}$ with $\omega\in\mathbb{R}$ need not hold. By way of example, let $\Omega=[0,\infty]$ be the one point compactification of $\mathbb{R}^+$ and suppose that $0<w\in C[0,\infty]$, bounded away from 0. For $f\in C[0,\infty]$ define

$$(T(t)f)(x)=\frac{w(x+t)}{w(x)}f(x+t),\ t\geqslant 0,\ x\in\mathbb{R}^+,$$

then $\{T(t)\}_{t\geqslant 0}$ is a positive C_0-semigroup in $C[0,\infty]$ (note that multiplication by w is an isomorphism from $C[0,\infty]$ into itself). Moreover,

$$\|T(t)\|=\|T(t)\mathbf{1}\|_\infty=\sup_{x\in\mathbb{R}^+}\frac{w(x+t)}{w(x)},$$

and so

$$\frac{\log\|T(t)\|}{t}\geqslant\frac{\log w(t)-\log w(0)}{t},\ t>0.$$

This shows that, if we take the function w such that $\log w$ has upper right-derivative at 0 equal to $+\infty$, then $\|T(t)\|\leqslant e^{\omega t}(t\geqslant 0)$ cannot hold for any $\omega\in\mathbb{R}$. Note finally that $\mathbf{1}\in\mathfrak{D}(A)$, if and only if the function w is differentiable and $w'\in C[0,\infty]$.

Next we will show how the criterion in Theorem 7.29 can be useful to detect generators of positive C_0-semigroups in $C(\Omega)$ spaces. Consider the space $X=C[0,1]$ and the operator A defined by $Af=-f'$ with maximal domain $C^{(1)}[0,1]$. It is easy to see that A is not the generator of a C_0-semigroup on this maximal domain. We impose boundary conditions of the form $f'(0)=<f,B>$ where $B\in C[0,1]^*$, ie., we consider A on the domain

$$\mathfrak{D}(A) = \{f\in C^{(1)}[0,1]: f'(0) = <f,B>\}.$$

It is not difficult to see that $\mathfrak{D}(A)$ is dense in $C[0,1]$. Write $B=m_0\delta-\mu$, where δ is the point evaluation at 0, and $\mu\perp\delta$ (i.e., μ is represented by a Borel measure with zero mass at the point 0). First assume that A is the generator of a positive C_0-semigroup $\{T(t)\}_{t\geqslant 0}$. Note that $<Af,\delta>=<f,-B>$ for all $f\in\mathfrak{D}(A)$, so $\delta\in\mathfrak{D}(A^*)$ and $A^*\delta=-B$. Take $0\leqslant u\in C[0,1]$ with $<u,\delta>=0$, then

$$<u,\mu> = <u,-B> = <u,A^*\delta> = \lim_{t\downarrow 0}<\frac{T(t)u-u}{t},\delta> =$$

$$= \lim_{t\downarrow 0}<\frac{T(t)u}{t},\delta>\geqslant 0.$$

Since $\mu\perp\delta$, this shows that $\mu\geqslant 0$. The condition that $\mu\geqslant 0$ is sufficient in order that A generates a positive C_0-semigroup. Indeed, by Theorem 7.29 it suffices to show that A has the POD property and that $\lambda I-A$ is surjective for all sufficiently large λ. To this end, suppose that $0\leqslant u\in\mathfrak{D}(A)$ and $0\leqslant\phi\in C[0,1]^*$ such that $<u,\phi>=0$. Then $u(\omega)=0$ for all $\omega\in\text{supp}(\phi)$. If $\omega\in(0,1]$ and $u(\omega)=0$, then $u'(\omega)=0$, and if $u(0)=0$, then $u'(0)=<u,B>=-<u,\mu>\leqslant 0$. Hence $<Au,\phi>\geqslant 0$, which shows that A has POD. Finally, using that $<e^{-\lambda x},\mu>\downarrow 0$ as $\lambda\to\infty$, a simple computation shows that the equation $\lambda f+f'=g$ has a solution $f\in\mathfrak{D}(A)$ for any $g\in C[0,1]$, for all sufficiently large values of λ. We may conclude therefore that A generates a positive C_0-semigroup in $C[0,1]$ if and only if $\mu\geqslant 0$.

As promised we shall give some examples to show that the existence of an order unit is essential in Theorem 7.29. The first example shows that, in general, the POD property of a generator does not imply that the semigroup is positive, and the second example shows that a resolvent positive operator need not be the generator of a C_0-semigroup (cf. Remark 7.30 (c)).

EXAMPLE 7.32.

(i) Let $X=L_2[0,1]$, a Banach lattice with respect to the pointwise ordering. Let $\{T(t)\}_{t\geqslant 0}$ be the C_0-semigroup in $L_2=L_2[0,1]$ defined by

$$T(t)f(x) = (-1)^n f(x+t-n) \text{ if } n\leqslant x+t\leqslant n+1.$$

The generator A of $\{T(t)\}_{t\geqslant 0}$ is given by $Af=f'$, with domain the space of all absolutely continuous functions f in L_2 such that $f'\in L_2$ and $f(0)=-f(1)$. Clearly, the semigroup $\{T(t)\}_{t\geqslant 0}$ is not positive, and it is not difficult to see that A has the POD property. Moreover, it should be observed that $\mathfrak{D}(A)^+$ is dense in L_2^+.

(ii) Let $X = C_0(\mathbb{R})$ be the Banach lattice of all real continuous functions on $\mathbb{R}$ vanishing at infinity. Let $v \in L_\infty + L_1$ be a positive continuously differentiable function on $\mathbb{R}$, which is bounded away from zero. The operator A in $C_0(\mathbb{R})$ is defined by

$$Af = -\frac{(fv)'}{v}$$

on the domain

$$\mathcal{D}(A) = \{f \in C_0(\mathbb{R}): f \text{ is continuously differentiable and } \frac{(fv)'}{v} \in C_0(\mathbb{R})\}.$$

Denoting by $C_{00}^{(1)}(\mathbb{R})$ the space of all continuously differentiable functions on $\mathbb{R}$ with compact support, it is clear that $C_{00}^{(1)}(\mathbb{R})$ is contained in $\mathcal{D}(A)$, and hence A is densely defined. Furthermore, a simple argument shows that A is closed. It is easy to check that any $\lambda > 0$ belongs to $\rho(A)$ and that

$$R(\lambda, A)g(x) = \frac{1}{v(x)} \int_{-\infty}^{x} e^{-\lambda(x-y)} g(y)v(y)dy$$

for all $g \in C_0(\mathbb{R})$. Consequently, A is a resolvent positive operator. We assert that the function v can be chosen in such a way that A is not the generator of a C_0-semigroup in $C_0(\mathbb{R})$. To this end, assume that A generates the semigroup $\{T(t)\}_{t \geqslant 0}$. Take any $f \in C_{00}^{(1)}(\mathbb{R})$ and for $t \geqslant 0$ define the functions

$$f_t(x) = \frac{v(x-t)}{v(x)} f(x-t), \; x \in \mathbb{R}.$$

Then $f_t \in C_{00}^{(1)}(\mathbb{R}) \subset \mathcal{D}(A)$ for all $t \geqslant 0$ and $f_0 = f$. Moreover, it is easy to verify that the mapping $t \mapsto f_t$ is differentiable (with respect to the sup-norm), with derivative Af_t. We may conclude that $f_t = T(t)f$ for all $t \geqslant 0$ (see Davies [1980], Theorem 1.7). Since $C_{00}^{(1)}(\mathbb{R})$ is dense in $C_0(\mathbb{R})$, the continuity of $T(t)$ implies that

$$T(t)f(x) = \frac{v(x-t)}{v(x)} f(x-t), \; x \in \mathbb{R},$$

for all $f \in C_0(\mathbb{R})$ and all $t \geqslant 0$. Therefore, for t fixed, the function $v(x-t)/v(x)$ multiplies $C_0(\mathbb{R})$ into itself, and so

$$\sup_{x \in \mathbb{R}} \frac{v(x-t)}{v(x)} < \infty$$

for all $t \geqslant 0$. Now we define v such that A is not the generator of a C_0-semigroup in $C_0(\mathbb{R})$. Let v_1 be a continuously differentiable positive integrable function on $\mathbb{R}$ with $v_1(-n) = n!$ for all $n \in \mathbb{N}$, and define $v(x) = 1 + v_1(x)$. Clearly, v satisfies all the conditions imposed on v at the beginning of this example, but since

$$\sup_{x \in \mathbb{R}} \frac{v(x-1)}{v(x)} = \infty,$$

it follows from the above observations that the corresponding operator A is not the generator of a C_0-semigroup in $C_0(\mathbb{R})$.

Next we say a few words about the complex situation. Suppose that X is a complex ordered Banach space, i.e., $X = X_{\mathbb{R}} + iX_{\mathbb{R}}$, the complexification of the real Banach space $X_{\mathbb{R}}$. As usual we denote $X^+ = X_{\mathbb{R}}^+$. For any $z \in X, z = x + iy$ with $x, y \in X_{\mathbb{R}}$, the conjugate is defined by $\bar{z} = x - iy$. A subset M of X is called conjugate closed if $z \in M$ implies that $\bar{z} \in M$. We omit the simple verification of the following complex version of Theorem 7.29.

THEOREM 7.33. *Let $X = X_{\mathbb{R}} + iX_{\mathbb{R}}$ be a complex ordered Banach space with normal positive cone and order unit. If A is a densely defined linear operator in X with conjugate closed domain, then the following assertions are equivalent.*
(i) *A is the generator of a positive C_0-semigroup in X*
(ii) *$\mathcal{R}(\lambda I - A) = X$ for all sufficiently large real λ, and $\mathrm{Re}\langle Ax, \phi\rangle \geq 0$ whenever $x \in \mathcal{D}(A)$ and $0 \leq \phi \in X^*$ with $\mathrm{Re}x \geq 0$ and $\mathrm{Re}\langle x, \phi\rangle = 0$*
(iii) *$\lambda \in \rho(A)$ and $R(\lambda, A) \geq 0$ for all sufficiently large real λ.*

It should be noticed that each of the statements (i) and (ii) above imply that $\mathcal{D}(A)$ is conjugate closed.

For a measure space (S, Σ, μ) the space $L_\infty(S, \mu)$ of all μ-essentially bounded (real) functions on S is a Banach lattice with order unit (with respect to the sup-norm), and so the above results apply in particular to this space. We end this section by pointing out that $L_\infty(S, \mu)$ belongs to a class of Banach spaces X which have the peculiar property that any C_0-semigroup in X is uniformly continuous (i.e., has a bounded generator). To this end, we recall that the Banach space X is called a *Grothendieck space* if every $\sigma(X^*, X)$-convergent sequence is $\sigma(X^*, X^{**})$-convergent. Clearly, reflexive spaces are Grothendieck, but not conversely. In fact, if $X = C(\Omega)$ with Ω a quasi-Stonian compact space (i.e., such that the Banach lattice $C(\Omega)$ is Dedekind σ-complete), then X is a Grothendieck space (see SCHAEFER [1974], Theorem II. 10.4). In particular, $L_\infty(S, \mu)$ is a Grothendieck space.

PROPOSITION 7.34. *If $\{T(t)\}_{t \geq 0}$ is a C_0-semigroup in the Grothendieck space X, then the adjoint semigroups $\{T^*(t)\}_{t \geq 0}$ and $\{T^{**}(t)\}_{t \geq 0}$ are strongly continuous in X^* and X^{**} respectively.*

PROOF. First we show that $(T^*(t_n) - I)\phi_n \to 0$ weakly in X^* as $n \to \infty$, whenever $t_n \downarrow 0$ and $\phi_n \in X^*$ with $\|\phi_n\| = 1$ $(n = 1, 2, ...)$. Indeed, for any $x \in X$ we have

$$|\langle x, (T^*(t_n) - I)\phi_n\rangle| = |\langle (T(t_n) - I)x, \phi_n\rangle| \leq \|T(t_n)x - x\|,$$

which shows that $(T^*(t_n) - I)\phi_n \to 0$ with respect to $\sigma(X^*, X)$. Since X is a Grothendieck space it follows that $(T^*(t_n) - I)\phi_n \to 0$ weakly in X^*. In particular, $T^*(t_n)\phi \to \phi$ weakly for any sequence $t_n \downarrow 0$ and all $\phi \in X^*$, and hence $\{T^*(t)\}_{t \geq 0}$ is a $\sigma(X^*, X^{**})$-continuous semigroup. Now it follows from DAVIES

[1980], Proposition 1.23, that $\{T^*(t)\}_{t\geq 0}$ is a C_0-semigroup in X^*. It remains to show that $\{T^{**}(t)\}_{t\geq 0}$ is strongly continuous on X^{**}. Suppose, on the contrary, that there exist $\Phi \in X^{**}$, $\epsilon>0$ and a sequence $t_n \downarrow 0$ such that $\|(T^{**}(t_n)-I)\Phi\|>\epsilon$ for all n. For each n there exists $\phi_n \in X^*$ such that $\|\phi_n\|=1$ and $|<(T^{**}(t_n)-I)\Phi, \phi_n>|>\epsilon$ for all n, i.e., $|<\Phi,(T^*(t_n)-I)\phi_n>|>\epsilon$ for all n. By the first part of the proof it follows that $(T^*(t_n)-I)\phi_n \to 0$ as $n\to\infty$ with respect to $\sigma(X^*,X^{**})$, which is a contradiction. □

The space $L_\infty(S,\mu)$ possesses another important property, the so-called Dunford-Pettis property. Recall that the Banach space X is said to have the *Dunford-Pettis property* if for every Banach space Y, any weakly compact operator $T:X\to Y$ maps weakly compact subsets of X onto compact subsets of Y (equivalently, every weakly compact operator $T:X\to Y$ maps weakly convergent sequences onto norm convergent sequences). The Banach space X has the Dunford-Pettis property if and only if for any Banach space Y and any continuous bilinear form F on $X\times Y$, it follows from $x_n\to x$ weakly in X and $y_n\to y$ weakly in Y, that $\lim_{n\to\infty} F(x_n,y_n)=F(x,y)$ (see SCHAEFER [1974], Theorem II. 9.7). In particular, if X has the Dunford-Pettis property, and if $x_n \in X, \phi_n \in X^*$ $(n=1,2,...)$ such that $x_n\to 0$ weakly in X and $\phi_n \to 0$ weakly in X^*, then $<x_n,\phi_n>\to 0$ as $n\to\infty$. Every AM-space and every AL-space has the Dunford-Pettis property (see SCHAEFER [1974], Theorem II. 9.9). In particular, the spaces $C(\Omega)$ and $L_\infty(S,\mu)$ have the Dunford-Pettis property. It should be observed that a reflexive space with the Dunford-Pettis property is finite dimensional.

LEMMA 7.35. *Let X be a Banach space with the Dunford-Pettis property. If $T_n \in \mathcal{L}(X)$ $(n=1,2,...)$ such that $T_n^*\to 0$ and $T_n^{**}\to 0$ strongly as $n\to\infty$, then $\|T_n^2\|\to 0$ $(n\to\infty)$.*

PROOF. For $n=1,2,...$ there exist $x_n \in X$ and $\phi_n \in X^*$ such that $\|x_n\|\leq 2, \|\phi_n\|=1$ and $\|T_n^2\|=<T_n^2x_n,\phi_n>=<T_nx_n,T_n^*\phi_n>$. For any $\phi \in X^*$ we have $|<T_nx_n,\phi>|\leq 2\|T_n^*\phi\|$, which shows that $T_nx_n\to 0$ weakly, since $T_n^*\to 0$ strongly. Similarly, from $T_n^{**}\to 0$ strongly, we deduce that $T_n^*\phi_n \to 0$ weakly in X^*. Now, since X has the Dunford-Pettis property, it follows from the above remarks that $\|T_n^2\|=<T_nx_n,T_n^*\phi_n>\to 0$ as $n\to\infty$. □

Now we are in the position to prove the final result of the section.

THEOREM 7.36. *In a Grothendieck space with the Dunford-Pettis property any C_0-semigroup is uniformly continuous (i.e., has bounded generator).*

PROOF. Let $\{T(t)\}_{t\geq 0}$ be a C_0-semigroup in the Grothendieck space X with the Dunford-Pettis property. By Proposition 7.34, $\{T^*(t)\}_{t\geq 0}$ and $\{T^{**}(t)\}_{t\geq 0}$ are both C_0-semigroups in X^* and X^{**} respectively. For $t>0$ define $S_t \in \mathcal{L}(X)$ by

$$S_t x = \frac{1}{t}\int_0^t T(s)x ds, \ x \in X.$$

Since $T(t) \to I$ strongly as $t \downarrow 0$, it follows that $S_t \to I$ strongly as $t \downarrow 0$. Now it is clear that the adjoints S_t^* and S_t^{**} are given by similar integrals with $T(s)$ replaced by $T^*(s)$ and $T^{**}(s)$ respectively, and so in particular, $S_t^* \to I$ and $S_t^{**} \to I$ strongly as $t \downarrow 0$. It is now a consequence of Lemma 7.35 that $\|(S_t - I)^2\| \to 0$, and hence $r(S_t - I) \to 0$ as $t \downarrow 0$. Fix $t_0 > 0$ such that $r(S_{t_0} - I) < 1$, then S_{t_0} is invertible in $\mathfrak{L}(X)$. Now define $B = t_0 S_{t_0}$ and $A = (T(t_0) - I)B^{-1}$. A simple calculation shows that

$$(T(t) - I)Bx = (T(t_0) - I)\int_0^t T(s)x ds, \ x \in X$$

and so $(T(t) - I)x = \int_0^t T(s)Ax ds$. Hence $T(t)x$ is a differentiable function of t and

$$\frac{d}{dt}\{T(t)x\} = AT(t)x$$

for all $t \geqslant 0$ and all $x \in X$. This implies that A is the generator of $\{T(t)\}_{t \geqslant 0}$. Since A is bounded, we may conclude that $\{T(t)\}_{t \geqslant 0}$ is uniformly continuous. □

COROLLARY 7.37. *Let (S, Σ, μ) be a measure space. Every C_0-semigroup in the Banach space $L_\infty(S, \mu)$ is uniformly continuous.*

4. THE KATO INEQUALITY

In the previous section a characterization was given of generators of positive C_0-semigroups in ordered Banach spaces with order unit. The presence of an order unit is essential for these results, as is illustrated by Example 7.32. In the present section we will discuss characterizations of such generators, in terms of a so-called abstract *Kato inequality*, in Banach lattices which need not possess an order unit. This abstract Kato inequality was inspired by the Kato inequality for the Laplacian Δ on $\mathbb{R}^n$: the distributional inequality

$$\operatorname{Re}(\operatorname{sign} f \cdot \Delta f) \leqslant \Delta |f|$$

holds for all $f \in L_1^{loc}(\mathbb{R}^n)$ with $\Delta f \in L_1^{loc}(\mathbb{R}^n)$ (where, as usual, $\operatorname{sign} f = \bar{f}/|f|$). This inequality is related to the positivity of the semigroup generated by Δ in $L_2(\mathbb{R}^n)$. For more historical background of the Kato inequality and its relation to positivity of semigroups, we refer to the bibliographical notes at end of this book.

Let X be a Banach lattice. We restrict ourselves to real spaces (although most of the results carry over without difficulties to complex spaces via the complexification). Moreover, for the sake of simplicity, we will assume first that X is Dedekind σ-complete. For any $x \in X$ we denote by P_x the band projection in X onto the principal band $B_x=\{x\}^{dd}$ generated by x (see Appendix A.2.4). Our first objective is to give an appropriate definition of 'sign x' for $x\in X$. To this end, for any $x\in X$ we define the *sign-mapping* $\sigma_x=P_{x^+}-P_{x^-}$. It is easy to see that σ_x has the following properties:

(i) $\sigma_x(x)=|x|$

(ii) $|\sigma_x(y)|\leqslant|y|$ for all $y\in X$

(iii) $\sigma_x(y)=0$ for all $y\in X$ with $y\perp x$.

Furthermore, the operator σ_x is uniquely determined by the properties (i), (ii) and (iii). Note that $|\sigma_x(y)|=|\sigma_x(|y|)|$ for all $y\in X$ (cf. Appendix A.2.4). In view of property (i), the operator σ_x will be called 'sign x'. If, for instance, $X=L_p(S,\mu)$, $1\leqslant p\leqslant\infty$, then σ_f is the multiplication by the function sign$f$, as defined at the beginning of the present section.

Let $A:\mathfrak{D}(A)\to X$ be a densely defined closed linear operator. A possible version of an abstract Kato inequality would be $\sigma_x(Ax)\leqslant A|x|$ for all $x\in\mathfrak{D}(A)$. The problem is, however, that in general $x\in\mathfrak{D}(A)$ need not imply that $|x|\in\mathfrak{D}(A)$. Therefore, the above inequality has to be interpreted in the weak sense. As usual, $A^*:\mathfrak{D}(A^*)\to X^*$ denotes the adjoint of A.

DEFINITION. The operator A obeys the (abstract) Kato inequality if

$$<\sigma_x(Ax),\phi>\leqslant<|x|,A^*\phi>$$

for all $x\in\mathfrak{D}(A)$ and all $0\leqslant\phi\in\mathfrak{D}(A^*)$.

Clearly, the Kato inequaltiy for an operator A is non-trivial only in the situation that $\mathfrak{D}(A^*)$ contains sufficiently many positive elements. It should be observed that $A-\lambda I$ satisfies the Kato inequality for all $\lambda\in\mathbb{R}$, whenever A does.

REMARK 7.38. We say a few words about the relation between the Kato inequality and the positive off-diagonal property (POD). Let X be a Dedekind σ-complete Banach lattice and assume, in addition, that the norm is σ-order continuous (hence, X is Dedekind complete with order continuous norm). Suppose that the operator A satisfies Kato's inequality, and assume that $\mathfrak{D}(A^*)^+$ is $\sigma(X^*,X)$-dense in $(X^*)^+$, so that $\mathfrak{D}(A^*)^+$ contains sufficiently many positive elements (cf. Proposition 7.2). Given any $0\leqslant x\in\mathfrak{D}(A)$, it follows from the Kato inequaltiy that $<P_x(Ax),\phi>\leqslant<x,A^*\phi>=<Ax,\phi>$ for all $0\leqslant\phi\in\mathfrak{D}(A^*)$. By the assumption on $\mathfrak{D}(A^*)^+$, this implies that $<P_x(Ax),\phi>\leqslant<Ax,\phi>$ for all $0\leqslant\phi\in X^*$, and hence $P_x(Ax)\leqslant Ax$ for all $0\leqslant x\in\mathfrak{D}(A)$. Now assume that $0\leqslant x\in\mathfrak{D}(A)$ and $0\leqslant\phi\in X^*$ such that $<x,\phi>=0$. The norm in X is σ-order continuous, and so ϕ is a σ-order continuous linear functional. Since $P_x(Ax)\in\{x\}^{dd}$, we deduce from $<x,\phi>=0$ that $<P_x(Ax),\phi>=0$. From the above observations we obtain $<Ax,\phi>\geqslant<P_x(Ax),\phi>=0$. We thus have shown that, at least if X has σ-order continuous norm and $\mathfrak{D}(A^*)$ contains sufficiently many positive elements, the Kato inequality for A implies the POD property.

Before showing that the generator of a positive C_0-semigroup satisfies Kato's inequality, we recall the following fact. If A is the generator of the C_0-semigroup $\{T(t)\}_{t\geq 0}$ in the Banach space X, then

$$\lim_{t\downarrow 0} < \frac{T(t)x-x}{t}, \phi> = <x, A^*\phi>$$

for all $x\in X$ and all $\phi\in\mathfrak{D}(A^*)$. Indeed, if $t>0$, then

$$\frac{1}{t}\int_0^t T(s)xds \in \mathfrak{D}(A) \text{ and } A(\frac{1}{t}\int_0^t T(s)xds) = \frac{T(t)x-x}{t}$$

for all $x\in X$. Therefore, if $\phi\in\mathfrak{D}(A^*)$, then

$$\lim_{t\downarrow 0} < \frac{T(t)x-x}{t}, \phi> = \lim_{t\downarrow 0} <A(\frac{1}{t}\int_0^t T(s)xds), \phi> =$$

$$\lim_{t\downarrow 0} < \frac{1}{t}\int_0^t T(s)xds, A^*\phi> = <x, A^*\phi>,$$

since $\frac{1}{t}\int_0^t T(s)xds \to x$ (norm) as $t\downarrow 0$.

PROPOSITION 7.39. *If A is the generator of the positive C_0-semigroup $\{T(t)\}_{t\geq 0}$ in the Dedekind σ-complete Banach lattice X, then A obeys the Kato inequality.*

PROOF. Suppose $x\in\mathfrak{D}(A)$ and $0\leq\phi\in\mathfrak{D}(A^*)$. Then

$$<\sigma_x(Ax), \phi> = \lim_{t\downarrow 0} <\sigma_x(\frac{T(t)x-x}{t}), \phi> =$$

$$\lim_{t\downarrow 0} < \frac{\sigma_x(T(t)x)-|x|}{t}, \phi> \leq \lim_{t\downarrow 0} < \frac{T(t)|x|-|x|}{t}, \phi> =$$

$$<|x|, A^*\phi> \quad \square$$

As observed before, in order that the Kato inequality implies positivity of the semigroup generated by A, we have to impose conditions assuring that $\mathfrak{D}(A^*)$ contains sufficiently many positive elements. It is not sufficient to assume that $\mathfrak{D}(A^*)^+$ is weak*-dense in $(X^*)^+$. In fact, the generator in Example 7.32-(i) obeys Kato's inequality, $\mathfrak{D}(A^*)^+$ is weak*-dense in $(X^*)^+$ and $\mathfrak{D}(A)^+$ is norm dense in X^+, whereas the semigroup is not positive. For the purpose of formulating sufficient conditions on $\mathfrak{D}(A^*)$, it is convenient to introduce some terminology. Let X be a Banach lattice. A subset K of $(X^*)^+$ is called *strictly positive* if it follows from $0\leq x\in X$ and $<x,\phi>=0$ for all $\phi\in K$ that $x=0$. Note that the set $(X^*)^+$ is strictly positive, and that the linear functional $\phi\in X^*$ is called strictly positive if the set $K=\{\phi\}$ is strictly positive. Furthermore, the element $u\in X$ is called a *positive subeigenvector* of the linear operator $A:\mathfrak{D}(A)\to X$ if $0<u\in\mathfrak{D}(A)$ and $Au\leq\lambda u$ for some $\lambda\in\mathbb{R}$.

PROPOSITION 7.40. *If A is the generator of the positive C_0-semigroup $\{T(t)\}_{t\geq 0}$*

in the Banach lattice X, then there exists a strictly positive subset K of $(X^*)^+$ *consisting of subeigenvectors of* A^*. *Furthermore, if there exists a strictly positive linear functional on X, then* A^* *has a strictly positive subeigenvector.*

PROOF. Let M be a strictly positive subset of $(X^*)^+$ and take $\lambda \in \mathbb{R}$ such that $\lambda \in \rho(A)$ and $R(\lambda,A) \geqslant 0$. Now define $K = R(\lambda,A)^*(M)$. Clearly, $K \subset \mathfrak{D}(A^*)^+$. If $\psi \in K$, then $\psi = R(\lambda,A)^*\phi$ for some $\phi \in M$ and so

$$A^*\psi = [AR(\lambda,A)]^*\phi = [\lambda R(\lambda,A) - I]^*\phi = \lambda\psi - \phi \leqslant \lambda\psi.$$

Hence, K consists of positive subeigenvectors of A^*. It remains to show that K is strictly positive. For this purpose, assume that $0 \leqslant x \in X$ is such that $\langle x,\psi \rangle = 0$ for all $\psi \in K$. Then $\langle R(\lambda,A)x,\phi \rangle = 0$ for all $\phi \in M$, and so $R(\lambda,A)x = 0$, as M is strictly positive. This implies that $x = 0$, and the proof is complete. □

Next it will be shown that the Kato inequality, in combination with the existence of a strictly positive set of subeigenvectors of A^*, implies the positivity of the semigroup generated by A.

Recall from Section 7.2, that the operator $A : \mathfrak{D}(A) \to X$ is called p-dissipative, with respect to some (continuous) sublinear functional p on X, whenever, $p(x) \leqslant p(x - tAx)$ for all $x \in \mathfrak{D}(A)$ and all $t \geqslant 0$. Further, the operator A is p-dissipative if and only if for each $x \in \mathfrak{D}(A)$ there exists $\phi \in \partial p(x)$ such that $\langle Ax,\phi \rangle \leqslant 0$, where $\partial p(x)$ denotes the subdifferential of p at the point x. Given $0 \leqslant \phi \in X^*$, define $p_\phi(x) = \langle x^+,\phi \rangle$ for all $x \in X$. Clearly, p_ϕ is a sublinear functional on X, and, as already observed in Example 7.10-(iv),

$$\partial p_\phi(x) = \{\psi \in X^* : 0 \leqslant \psi \leqslant \phi \text{ and } \langle x,\psi \rangle = \langle x^+,\phi \rangle\}$$

for all $x \in X$.

LEMMA 7.41. *Suppose that A is a densely defined linear operator in the Dedekind σ-complete Banach lattice X and let* $0 \leqslant \phi \in \mathfrak{D}(A^*)$ *be such that* $A^*\phi \leqslant 0$. *If* $\langle \sigma_x(Ax),\phi \rangle \leqslant \langle |x|,A^*\phi \rangle$ *for all* $x \in \mathfrak{D}(A)$, *then A is* p_ϕ*-dissipative.*

PROOF. Fix $x \in \mathfrak{D}(A)$. We have to show that there exists $\psi \in \partial p_\phi(x)$ such that $\langle Ax,\psi \rangle \leqslant 0$. Put $\pi = 1/2(\sigma_x + I)$ and define $\psi = \pi^*\phi$. Since $0 \leqslant \pi \leqslant I$, it is clear that $0 \leqslant \psi \leqslant \phi$. Moreover, $\langle x,\psi \rangle = \langle \pi x,\phi \rangle = \langle x^+,\phi \rangle = p_\phi(x)$, which shows that $\psi \in \partial p_\phi(x)$. Finally,

$$\begin{aligned}\langle Ax,\psi \rangle &= \tfrac{1}{2}\langle (\sigma_x + I)Ax,\phi \rangle = \\ &\tfrac{1}{2}\langle \sigma_x(Ax),\phi \rangle + \tfrac{1}{2}\langle Ax,\phi \rangle \leqslant \\ &\tfrac{1}{2}\langle |x|,A^*\phi \rangle + \tfrac{1}{2}\langle x,A^*\phi \rangle = \\ &\langle x^+,A^*\phi \rangle \leqslant 0,\end{aligned}$$

since $A^*\phi \leqslant 0$, and we are done. □

REMARK 7.42. The assumptions in the above lemma can be weakened slightly. In fact, it is sufficient to assume that there exists a core $\mathfrak{D}_0$ of A such that $<\sigma_x(Ax),\phi> \leqslant <|x|,A^*\phi>$ holds for all $x \in \mathfrak{D}_0$. Indeed, the proof of the lemma then shows that the restriction of A to $\mathfrak{D}_0$ is p_ϕ-dissipative, and by the continuity of p_ϕ we may conclude that A is p_ϕ-dissipative.

PROPOSITION 7.43. *If A is the generator of the C_0-semigroup $\{T(t)\}_{t\geqslant 0}$ in the Dedekind σ-complete Banach lattice X, then the following statements are equivalent.*
(i) *$\{T(t)\}_{t\geqslant 0}$ is a positive semigroup*
(ii) *There exists a strictly positive set K consisting of subeigenvectors of A^* and there exists a core $\mathfrak{D}_0$ of A such that $<\sigma_x(Ax),\phi> \leqslant <|x|,A^*\phi>$ for all $x \in \mathfrak{D}_0$ and all $\phi \in K$.*

PROOF. (i)⇒(ii). The existence of such a set K follows from Proposition 7.40, and it follows from Proposition 7.39 that $<\sigma_x(Ax),\phi> \leqslant <|x|,A^*\phi>$ for all $x \in \mathfrak{D}(A)$ and all $0 \leqslant \phi \in \mathfrak{D}(A^*)$.
(ii)⇒(i). Fix an element $\phi \in K$. By the definition of subeigenvector, there exists $\lambda \in \mathbb{R}$ such that $A^*\phi \leqslant \lambda\phi$. Put $B = A - \lambda I$, then $B^*\phi \leqslant 0$ and $<\sigma_x(Bx),\phi> \leqslant <|x|,B^*\phi>$ for all $x \in \mathfrak{D}_0$. Hence, by Lemma 7.41 (and the preceding remark), B is p_ϕ-dissipative. Furthermore, B is the generator of the C_0-semigroup $\{e^{-\lambda t}T(t)\}_{t\geqslant 0}$. From Proposition 7.12 we readily obtain that $\{e^{-\lambda t}T(t)\}_{t\geqslant 0}$ is a p_ϕ-contraction semigroup, i.e., that

$$p_\phi(e^{-\lambda t}T(t)x) \leqslant p_\phi(x),\ t \geqslant 0,$$

for all $x \in X$. Hence, by the definition of p_ϕ,

$$e^{-\lambda t} <[T(t)x]^+,\phi> \leqslant <x^+,\phi>,\ t \geqslant 0,$$

for all $x \in X$, which implies that $<[T(t)x]^+,\phi> = 0$ for all $x \leqslant 0$ in X. This holds for all $\phi \in K$. Since K is strictly positive, we may conclude that $[T(t)x]^+ = 0$ for all $x \leqslant 0$ in X, and hence $T(t)x \geqslant 0$ for all $0 \leqslant x \in X$ and all $t \geqslant 0$. By this the proposition is completely proved. □

The following theorem is an immediate corollary of the above proposition.

THEOREM 7.44. *Let $A: \mathfrak{D}(A) \to X$ be the generator of the C_0-semigroup $\{T(t)\}_{t\geqslant 0}$ in the Dedekind σ-complete Banach lattice X. Then $\{T(t)\}_{t\geqslant 0}$ is positive if and only if A satisfies the following two conditions:*
(i) *A obeys the Kato inequality*
(ii) *A^* possesses a strictly positive set of subeigenvectors.*

In many of the familiar Banach lattices X (e.g. if $X = L_p(\Omega,\mu)$, with (Ω,Σ,μ) a σ-finite measure space), there exists a strictly positive linear functional in X^*. For such spaces, the following result is evident from Propositions 7.40 and 7.43.

COROLLARY 7.45. *Suppose that X is a Dedekind σ-complete Banach lattice with strictly positive linear functional. The C_0-semigroup $\{T(t)\}_{t\geq 0}$ in X, with generator A, is positive if and only if there exists a strictly positive subeigenvector ϕ of A^* such that $<\sigma_x(Ax),\phi> \leqslant <|x|,A^*\phi>$ for all $x \in \mathfrak{D}(A)$.*

A close inspection of the proof of Proposition 7.43 shows that the Dedekind σ-completeness of X is used only to ensure the existence of the sign-mappings σ_x. If X is a Banach lattice which is not Dedekind σ-complete (e.g. if $X = C[0,1]$), then such sign-mappings need not exist in X. However, under the canonical embedding, X can be considered as a vector sublattice of the second dual X^{**} (see Appendix A.2.4). Since X^{**} is a Dedekind complete Banach lattice, the sign-mapping exists for all elements in X^{**}. Therefore, considering X as a Banach sublattice of X^{**}, for any $x \in X$ there exists a mapping σ from X into X^{**} such that $\sigma(x) = |x|$, $|\sigma(y)| \leqslant |y|$ for all $y \in X$ and $\sigma(y) = 0$ for all $y \perp x$. In contrast to the uniqueness of a sign-mapping in the space X itself, it should be noted that, in general, such a mapping $\sigma : X \to X^{**}$ is not uniquely determined by these three properties. In fact, it is possible that an element $x \in X$ has a (uniquely determined) sign-mapping σ_x in X (e.g., if X is Dedekind σ-complete), which is not equal to the restriction of the sign-mapping of x in the space X^{**}. Furthermore, in some situations it may occur that for each element $x \in X$ there is a distinguished linear operator σ_x from X into X^{**} with $\sigma_x(x) = |x|$, $|\sigma_x(y)| \leqslant |y|$ for all $y \in X$, and $\sigma_x(y) = 0$ for all $y \perp x$, but this mapping σ_x is not equal to the restriction of the sign-mapping of x in X^{**}. By way of example, suppose that $X = C(\Omega)$ for some compact Hausdorff space Ω. As usual, the Borel function sign f on Ω is defined by

$$\operatorname{sign} f(\omega) = \begin{cases} \dfrac{f(\omega)}{|f(\omega)|} & \text{if } f(\omega) \neq 0 \\ 0 & \text{otherwise.} \end{cases}$$

Multiplication by sign$f$ is an operator σ_f from $C(\Omega)$ into $\mathfrak{B}(\Omega)$, the Banach lattice of all bounded Borel functions on Ω. In a natural way, $\mathfrak{B}(\Omega)$ can be considered as a subspace of the second dual $C(\Omega)^{**}$, but, in general, σ_f is not the restriction to $C(\Omega)$ of the sign-mapping of f in $C(\Omega)^{**}$. The reader should keep this last example in mind while considering the following proposition, the proof of which is almost identical to the proof of Proposition 7.43. Observe that we do not use the property $\sigma_x(\{x\}^{dd}) = \{0\}$ in that proof.

PROPOSITION 7.46. *Let X be a Banach lattice and suppose that for each $x \in X$ there exists a linear mapping σ_x from X into X^{**} such that $\sigma_x(x) = |x|$ and $|\sigma_x(y)| \leqslant |y|$ for all $y \in X$. If A is the generator of the C_0-semigroup $\{T(t)\}_{t\geq 0}$ in X, then the following two statements are equivalent*

(i) $\{T(t)\}_{t\geq 0}$ *is positive*

(ii) *There exists a strictly positive set K consisting of subeigenvectors of A^* and there exists a core $\mathfrak{D}_0$ of A such that $<\sigma_x(Ax),\phi> \leqslant <|x|,A^*\phi>$ for all $x \in \mathfrak{D}_0$ and all $\phi \in K$.*

Using the above extension of Proposition 7.43, it is evident how to formulate the corresponding extensions of Theorem 7.44 and Corollary 7.45.

Closely related to Kato's inequality in Hilbert space is the so-called *Beurling-Deny criterion* for positivity of semigroups generated by a form-positive self-adjoint operator. For a short discussion, let $\mathcal{H}=L_2(S,\mu)$, the space of all *complex* square integrable functions on the σ-finite measure space (S,Σ,μ), with the usual order structure and inner-product. Suppose that H is a form - positive self-adjoint operator in $\mathcal{H}$. Since $<Hf,f>\geq 0$ for all $f\in\mathcal{H}$, the operator $-H$ is dissipative, and hence $-H$ is the generator of a contraction semigroup $\{T(t)\}_{t\geq 0}$ in $\mathcal{H}$. Note that all operators $T(t)$ are self-adjoint. In fact, if $\{E(\cdot)\}$ denotes the spectral measure of H, then

$$T(t)f = \int_0^\infty e^{-\lambda t} dE(\lambda) f$$

for all $f\in\mathcal{H}$ and $t\geq 0$. Furthermore, recall that the quadratic form Q_H corresponding to H is defined by

$$Q_H(f) = \begin{cases} <H^{1/2}f, H^{1/2}f> & f\in\mathcal{D}(H^{1/2}) \\ \infty & f\notin\mathcal{D}(H^{1/2}). \end{cases}$$

Equivalently,

$$Q_H(f) = \int_0^\infty \lambda d<Ef,f>$$

for all $f\in\mathcal{H}$ and the quadratic-form domain $\mathcal{Q}(H)=\mathcal{D}(H^{1/2})$ consists precisely of the elements $f\in\mathcal{H}$ for which this integral is finite. Using these notions we can formulate the Beurling-Deny theorem.

THEOREM 7.47. *Let H be a form - positive self-adjoint operator in the Hilbert space $\mathcal{H}=L_2(S,\mu)$, and let $\{T(t)\}_{t\geq 0}$ be the contraction semigroup generated by $-H$. The following assertions are equivalent.*
(i) *$\{T(t)\}_{t\geq 0}$ is a positive semigroup*
(ii) *The quadratic-form domain $\mathcal{Q}(H)$ is a sublattice of $\mathcal{H}$ and $Q_H(|f|)\leq Q_H(f)$ for all $f\in\mathcal{Q}(H)$ (equivalently $Q_H(|f|)\leq Q_H(f)$ for all $f\in\mathcal{H}$).*

PROOF. (i)⇒(ii). First note that

$$<\frac{f-T(t)f}{t}, f> = \int_{\lambda=0}^{\infty} \frac{1-e^{-\lambda t}}{t} d<Ef,f>$$

for all $f\in\mathcal{H}$, and hence, by monotone convergence,

$$\lim_{t\downarrow 0}<\frac{f-T(t)f}{t}, f> = \int_{\lambda=0}^{\infty} \lambda d<Ef,f> = Q_H(f)$$

for all $f\in\mathcal{H}$. Now assume that $T(t)\geq 0$ for all $t\geq 0$. Then $|T(t)f|\leq T(t)|f|$, and hence by the definition of the inner-product in $\mathcal{H}=L_2$,

$$<T(t)f,f> \leqslant <|T(t)f|,|f|> \leqslant <T(t)|f|,|f|>,$$

for all $f\in\mathcal{H}$. Since $<f,f>=<|f|,|f|>$, we obtain

$$<\frac{|f|-T(t)|f|}{t},|f|>\leqslant<\frac{f-T(t)f}{t},f>$$

for all $f\in\mathcal{H}$ and all $t>0$. Letting $t\downarrow 0$ we find $Q_H(|f|)\leqslant Q_H(f)$ for all $f\in\mathcal{H}$.
(ii)$\Rightarrow$(i). As already observed in Proposition 7.1, we have to show that $R(\lambda,-H)=(\lambda I+H)^{-1}\geqslant 0$ for all $\lambda>0$. To this end, fix $\lambda>0$, take $0\leqslant u\in\mathcal{H}$ and put $h=(\lambda I+H)^{-1}u$. Define the non-negative quadratic form Q on $\mathcal{H}$ by

$$Q(f)=Q_H(f)+\lambda<f,f>$$

with $\mathcal{Q}(Q)=\mathcal{Q}(H)$. A simple calculation shows that, for any $g\in\mathcal{Q}(H)$ with $\operatorname{Re} g\geqslant 0$, we have

$$\begin{aligned} Q(h+g) &= Q(h)+Q(g)+2\operatorname{Re}<(\lambda I+H)h,g> = \\ & Q(h)+Q(g)+2<u,\operatorname{Re}g>\geqslant \\ & Q(h)+Q(g)\geqslant Q(h), \end{aligned}$$

and $Q(h+g)=Q(h)$ if and only if $g=0$. By hypothesis, $Q(|h|)\leqslant Q(h)$, and so, by taking $g=|h|-h$ it follows from the above that $Q(|h|)=Q(h)$. Hence $g=0$, i.e., $h=|h|\geqslant 0$. This shows that $(\lambda I+H)^{-1}\geqslant 0$, and the proof is complete. □

Chapter 8

Perron-Frobenius Theory for Positive Semigroups

1. Spectral mapping theorems

In the beginning of this century, Perron and Frobenius discovered that a positive square matrix $L=(l_{ij})$ has some nice spectral properties. For instance, the set of eigenvalues with modulus equal to the spectral radius is multiplicatively cyclic (see Section 8.2 for a definition). Since that time, their results have been extended into several directions: there is a vast literature on positive (or order-preserving) linear operators. In the present chapter we will describe the extension of the Perron-Frobenius theory to strongly continuous semigroups of positive operators. For unknown terminology we refer to Appendices 2 and 3.

If A is a bounded linear operator and $T(t)=e^{tA}, t \geqslant 0$, then $\sigma(T(t))= e^{t\sigma(A)}=\{e^{t\lambda}:\lambda\in\sigma(A)\}$. In general such a straightforward relationship between the spectrum of a semigroup and the spectrum of its generator does not exist, as the following counterexample illustrates.

EXAMPLE 8.1.

Let l_n^2 denote the n-dimensional Hilbert space, and let H be the Hilbert space direct sum consisting of all sequences $\mathbf{x}=(x_1,x_2,...)$ with $x_n\in l_n^2$, for which $\Sigma_{n=1}^{\infty}\|x_n\|^2<\infty$. The inner product in H is defined by $(\mathbf{x},\mathbf{y})=\Sigma_{n=1}^{\infty}(x_n,y_n)$. If $L_n:l_n^2\rightarrow l_n^2$ are linear operators such that $\sup_n\|L_n\|<\infty$, then we can define the Hilbert direct sum L of these operators by $L\mathbf{x}=(L_1x_1,L_2x_2,...)$ for every $\mathbf{x}=(x_1,x_2,...)\in H$. Obviously $\|L\|\leqslant\sup_n\|L_n\|$. Let for every n the operator Q_n

on l_n^2 be defined by the matrix

$$Q_n = \begin{bmatrix} 0 & 1 & 0 & \cdots & 0 \\ 0 & 0 & 1 & \cdots & 0 \\ 0 & 0 & 0 & \cdots & 0 \\ \vdots & \vdots & \vdots & & \vdots \\ 0 & 0 & 0 & \cdots & 1 \\ 0 & 0 & 0 & \cdots & 0 \end{bmatrix}.$$

Then $Q_n^n=0$ (hence Q_n is nilpotent), and therefore $\sigma(Q_n)=\{0\}$. Obviously, $\|Q_n\|=1$. Defining $A_n=Q_n+inI_n$, where I_n is the identity operator on l_n^2, we have $\sigma(A_n)=\{in\}$. Let $\{T_n(t)\}_{t\geqslant 0}$ be the semigroup generated by A_n,

$$T_n(t) = e^{tA_n} = e^{int}.e^{tQ_n}\ , t\geqslant 0,$$

then $\|T_n(t)\|\leqslant e^t$, for every $n=1,2, \cdots$. For $t\geqslant 0$, define $T(t)$ to be the Hilbert direct sum of the operators $T_n(t)$. It is easy to see that $\{T(t)\}_{t\geqslant 0}$ is a C_0-semigroup on H (actually $\{T(t)\}_{t\geqslant 0}$ can be extended to a C_0-group on H). It is evident that $\|T(t)\|\leqslant \sup_n\|T_n(t)\|\leqslant e^t$, $t\geqslant 0$, but we claim that even $\|T(t)\|=e^t$. For this purpose, we define $v_n\in l_n^2$ to be the element $(n^{-1/2},n^{-1/2}, \ldots ,n^{-1/2})$. Then

$$\|T(t)(0, \ldots ,0,v_n,0, \cdots)\| = \|T_n(t)v_n\|\rightarrow e^t,$$

as $n\rightarrow\infty$. This implies that $\|T(t)\|=e^t$.

Now we will determine the generator A and its spectrum $\sigma(A)$. Let $\lambda\in\mathbf{C}$, $\lambda\neq in$ $(n=1,2,...)$, and let $R_n(\lambda):l_n^2\rightarrow l_n^2$ be defined by

$$R_n(\lambda) = (\lambda I_n - A_n)^{-1} = ((\lambda - in)I_n - Q_n)^{-1} = \sum_{k=0}^{n-1} \frac{Q_n^k}{(\lambda-in)^{k+1}},$$

where we have used that $Q_n^n=0$. This implies that for every n satisfying $|\lambda-in|\geqslant r>1$ we have $\|R_n(\lambda)\|\leqslant r/(r-1)$, and consequently $\sup_n\|R_n(\lambda)\|<\infty$. Now let $R(\lambda)$ be the Hilbert space direct sum of $R_n(\lambda)$. It is not difficult to show that

$$\mathfrak{R}(R(\lambda)) = \{\mathbf{x} = (x_1,x_2,...)\in H : \sum_{n=1}^{\infty} n^2\|x_n\|^2<\infty\}.$$

If $\mathbf{x}\in\oplus_{n=1}^{\infty} l_n^2$ (the algebraic direct sum of all l_n^2), then it follows easily that $\mathbf{x}\in\mathfrak{D}(A)$ (i.e. $\lim_{t\downarrow 0}t^{-1}(T(t)\mathbf{x}-\mathbf{x})$ exists) and $A\mathbf{x}=(A_1x_1,A_2x_2,...)$. This shows that $(\lambda I-A)R(\lambda)\mathbf{x}=\mathbf{x}$ for all $\mathbf{x}\in\oplus_{n=1}^{\infty} l_n^2$, $\lambda\neq in$. Since A is closed, $(\lambda I-A)R(\lambda)=I$ for all $\lambda\neq in$, which implies that $R(\lambda)=R(\lambda,A)$ for every $\lambda\in\rho(A)$. Now

$$\mathfrak{D}(A) = \mathfrak{R}(R(\lambda)) = \{\mathbf{x} : \Sigma_{n=1}^{\infty} n^2\|x_n\|^2<\infty\}.$$

We find that for all $\mathbf{x} \in \mathfrak{D}(A)$, $A\mathbf{x} = (A_1 x_1, A_2 x_2 \cdots)$, and it follows that $R(\lambda)(\lambda - A)\mathbf{x} = \mathbf{x}$ for every $\lambda \neq in$. Hence, $\sigma(A) = \{in : n = 1,2,...\}$, and thus

$$\sigma(T(t)) \neq e^{t\sigma(A)}. \quad \square$$

In this section we shall describe the various connections between the spectrum of a strongly continuous semigroup $\{T(t)\}_{t \geq 0}$ on the Banach space X and that of its infinitesimal generator A. In this respect the following identities turn out to be very useful.

$$(e^{\lambda t} - T(t))x = (\lambda - A) \int_0^t e^{\lambda(t-s)} T(s) x ds, \lambda \in \mathbb{C}, x \in X. \tag{8.1a}$$

$$(e^{\lambda t} - T(t))x = \int_0^t e^{\lambda(t-s)} T(s)(\lambda - A) x ds, \lambda \in \mathbb{C}, x \in \mathfrak{D}(A). \tag{8.1b}$$

These formulas can be established in the following way. Let $\lambda \in \mathbb{C}$ and $t \geq 0$ be fixed. The mapping $x \to \int_0^t e^{\lambda(t-s)} T(s) x ds$ defines a bounded linear operator on X, and for every $x \in X$ and $h > 0$ we have:

$$\frac{1}{h}(T(h) - I) \int_0^t e^{\lambda(t-s)} T(s) x ds = \frac{e^{\lambda h} - 1}{h} \int_h^t e^{\lambda(t-s)} T(s) x ds$$

$$+ \frac{e^{\lambda h}}{h} \int_t^{t+h} e^{\lambda(t-s)} T(s) x ds - \frac{1}{h} \int_o^h e^{\lambda(t-s)} T(s) x ds,$$

and as $h \downarrow 0$ this expression converges to

$$\lambda \int_0^t e^{\lambda(t-s)} T(s) x ds + T(t)x - e^{\lambda t} x,$$

from which we conclude that

$$\int_0^t e^{\lambda(t-s)} T(s) x ds \in \mathfrak{D}(A),$$

and

$$A \int_0^t e^{\lambda(t-s)} T(s) x ds = \lambda \int_0^t e^{\lambda(t-s)} T(s) x ds + T(t)x - e^{\lambda t} x,$$

which implies (8.1a). Now (8.1b) follows from the observation that, for $x \in \mathfrak{D}(A)$:

$$A \int_0^t e^{\lambda(t-s)} T(s) x ds = \int_0^t e^{\lambda(t-s)} T(s) A x ds.$$

For a subset Y of X we denote by $\operatorname{span} Y$ the linear span of Y, and by $\overline{\operatorname{span}}\, Y$ its closure.

LEMMA 8.2.
a) $\mathfrak{N}(\lambda - A) = \bigcap_{t \geq 0} \mathfrak{N}(e^{\lambda t} - T(t))$.

b) $\mathfrak{N}(e^{\lambda t} - T(t)) = \overline{\text{span}} \bigcup_{k\in\mathbf{Z}} \mathfrak{N}(\lambda + \frac{2\pi i k}{t} - A).$

PROOF. a) From (8.1b) we see that $\mathfrak{N}(\lambda - A) \subseteq \bigcap_{t\geq 0} \mathfrak{N}(e^{\lambda t} - T(t))$. On the other hand, if $T(t)x = e^{\lambda t}x$, $t \geq 0$, then $t^{-1}(T(t)x - x) \to \lambda x$ as $t\downarrow 0$, and hence $x \in \mathfrak{N}(\lambda - A)$.
b) Let $\lambda \in \mathbf{C}$. The inclusion

$$\overline{\text{span}} \bigcup_{k\in\mathbf{Z}} \mathfrak{N}(\lambda + 2\pi i k t^{-1} - A) \subseteq \mathfrak{N}(e^{\lambda t} - T(t))$$

is implied by a). It requires some more effort to establish the converse inclusion. Let $\tau > 0$ be fixed. For every nonnegative integer k, every $t \geq 0$, and every $x \in \mathfrak{N}(e^{\lambda\tau} - T(\tau))$ we have

$$T(t + k\tau)x = T(t)T(k\tau)x = e^{k\lambda\tau}T(t)x.$$

We define $S(t)$ to be the restriction of $e^{-\lambda t}T(t)$ to $Z := \mathfrak{N}(e^{\lambda\tau} - T(\tau))$. Since $\{T(t)\}_{t\geq 0}$ leaves Z invariant, $\{S(t)\}_{t\geq 0}$ defines a τ-periodic semigroup on Z. We can therefore extend $\{S(t)\}_{t\geq 0}$ to a group by putting

$$S(-t) = S(t)^{-1} = S(n\tau - t) \text{ if } 0 \leq t \leq n\tau.$$

If B is the generator of $\{S(t)\}_{t\in\mathbf{R}}$, then we get from (8.1a):

$$R(\mu, B) = (1 - e^{-\mu\tau})^{-1} \int_0^\tau e^{-\mu t} S(t)x\,dt, \quad \mu \in \rho(B).$$

Thus $R(\mu, B)$ is a meromorphic function on $\mathbf{C}$ with poles of order at most one at the points $\lambda_k = 2\pi k i \tau^{-1}$, $k \in \mathbf{Z}$. For the residue P_k in λ_k we find,

$$P_k x = \frac{1}{\tau} \int_0^\tau e^{-\lambda_k t} S(t)x\,dt,$$

i.e. P_k is the k'th Fourier coefficient of the periodic function $t \to S(t)$. The following properties hold

(i) $P_k P_l = P_k \delta_{kl}$, for $k, l \in \mathbf{Z}$

(ii) $x = \sum_{k=-\infty}^{\infty} P_k x$, if $x \in \mathfrak{D}(B)$

(iii) $Z = \overline{\text{span}}(\bigcup_{k\in\mathbf{Z}} P_k Z)$

(iv) $S(t)x = \sum_{k=-\infty}^{\infty} e^{\lambda_k t} P_k x$, if $x \in \mathfrak{D}(B)$

(v) $Bx = \sum_{k=-\infty}^{\infty} \lambda_k P_k x$, if $x \in \mathfrak{D}(B^2)$.

(i) is easy to prove. We confine ourselves to (ii). The other properties follow immediately. Let $x \in \mathfrak{D}(B)$. We show that the series $\Sigma_{k=-\infty}^{\infty} P_k x$ is summable. Let $y = Bx$, then $P_k y = P_k Bx = \lambda_k P_k x$. For all positive integers N, M and every $x^* \in Z^*$ we have

$$|\sum_{k=-M}^{N} <P_k x, x^*>| = |\sum_{k=-M}^{N} \frac{1}{\lambda_k} <P_k y, x^*>|$$
$$\leqslant (\sum_{k=-M}^{N} \frac{1}{|\lambda_k|^2})^{\frac{1}{2}} (\sum_{k=-M}^{N} |<P_k y, x^*>|^2)^{\frac{1}{2}}.$$

From Bessel's inequality we get

$$\sum_{K=-M}^{N} |<P_k y, x^*>|^2 \leqslant \frac{1}{\tau} \int_0^{\tau} |<S(t)y, x^*>|^2 dt \leqslant C^2 \|x^*\|^2,$$

where

$$C = (\frac{1}{\tau} \int_0^{\tau} \|S(t)y\|^2 dt)^{\frac{1}{2}}.$$

Thus

$$|\sum_{k=-M}^{N} <P_k x, x^*>| \leqslant C \|x^*\| \cdot (\sum_{K=-M}^{N} \frac{1}{|\lambda_k|^2})^{\frac{1}{2}},$$

from which we conclude that

$$\|\sum_{K=-M}^{N} P_k x\| \leqslant C(\sum_{k=-M}^{N} \frac{1}{|\lambda_k|^2})^{\frac{1}{2}}.$$

This proves the summability of the series. Set

$$z := \sum_{k=-\infty}^{\infty} P_k x.$$

Evidently, the Fourier coefficients of the continuous functions $t \to <S(t)z, x^*>$ and $t \to <S(t)x, x^*>$ are the same for each $x^* \in Z^*$. Therefore the two functions are identical, so in particular

$$<S(0)z, x^*> = <S(0)x, x^*>$$

for every $x^* \in Z^*$ and we get that

$$x = z = \sum_{k=-\infty}^{\infty} P_k x,$$

which takes care of (ii).
From (v) it follows immediately that $\mathcal{R}(P_k) = \mathcal{N}(\lambda_k - B)$ and from (ii) we deduce that

$$\overline{\text{span}}(\bigcup_{k \in \mathbf{Z}} \mathcal{N}(\lambda_k - B)) = Z = \mathcal{N}(e^{\lambda\tau} - T(\tau)).$$

Clearly

$$\overline{\text{span}}(\bigcup_{k \in \mathbf{Z}} \mathcal{N}(\lambda_k - B)) \subseteq \overline{\text{span}}(\bigcup_{k \in \mathbf{Z}} \mathcal{N}(\lambda + \lambda_k - A)),$$

and the result follows. □

Now we are in a position to prove the following spectral mapping theorems.

PROPOSITION 8.3. *For every* $t \geqslant 0$,

a) $e^{t\sigma(A)} \subseteq \sigma(T(t))$

b) $e^{t\sigma_p(A)} = \sigma_p(T(t)) \setminus \{0\}$

c) $e^{t(\sigma_p(A) \cup \sigma_r(A))} = (\sigma_p(T(t)) \cup \sigma_r(T(t))) \setminus \{0\}$

d) $e^{t\sigma_{ap}(A)} \subseteq \sigma_{ap}(T(t))$

e) *If* $\lambda \in \sigma_c(A)$ *and none of the* $\lambda + 2\pi i k t^{-1}$ $(k \in \mathbb{Z})$ *is contained in* $\sigma_p(A) \cup \sigma_r(A)$, *then* $e^{\lambda t} \in \sigma_c(T(t))$.

PROOF.

a) Suppose $e^{\lambda t} \in \rho(T(t))$, so $(e^{\lambda t} - T(t))^{-1}$ exists. From (8.1a) and (8.1b) we infer that

$$(\lambda - A)\int_0^t e^{\lambda(t-s)} T(s)(e^{\lambda t} - T(t))^{-1} x ds = x, \; x \in X,$$

$$(e^{\lambda t} - T(t))^{-1} \int_0^t e^{\lambda(t-s)} T(s)(\lambda - A) x ds = x, \; x \in \mathfrak{D}(A).$$

This latter equality can also be written as

$$\int_0^t e^{\lambda(t-s)} T(s)(e^{\lambda t} - T(t))^{-1}(\lambda - A) x ds = x, \; x \in \mathfrak{D}(A).$$

We may conclude that $\lambda \in \rho(A)$, and that

$$(\lambda - A)^{-1} x = \int_0^t e^{\lambda(t-s)} T(s)(e^{\lambda t} - T(t))^{-1} x ds, \; x \in X,$$

from which a) follows.

b) follows from Lemma 8.2.

c) We use duality to prove c). In general, the dual semigroup $\{T^*(t)\}_{t \geqslant 0}$ on X^* is not strongly continuous. Let $X^{\odot} = \{x^* \in X^* : \lim_{t \downarrow 0} T^*(t) x^* = x^*\}$. Notice that $X^{\odot}$ is invariant under $T^*(t)$. If $T^{\odot}(t)$ is the restriction, then $\{T^{\odot}(t)\}_{t \geqslant 0}$ defines a C_0-semigroup on $X^{\odot}$, and its generator $A^{\odot}$ is the part of A^* in $X^{\odot}$. See also Section 3.4. The following relations between the spectra hold:

$$\sigma_r(A) \subseteq \sigma_p(A^{\odot}) \subseteq \sigma_r(A) \cup \sigma_p(A),$$

$$\sigma_r(T(t)) \subseteq \sigma_p(T^{\odot}(t)) \subseteq \sigma_r(T(t)) \cup \sigma_p(T(t)), \quad t \geqslant 0.$$

These results can be found in HILLE & PHILLIPS [1957]. From a combination of these inclusions and the fact that $\sigma_p(T^{\odot}(t)) \setminus \{0\} = e^{t\sigma_p(A^{\odot})}$ (which was proved in b) the result follows.

d) follows from the identities (8.1a,b), and

e) follows from a), b) and c). □

A similar result holds for the *(Browder) essential spectrum,* which is discussed in Appendix A.3.1. For the sake of convenience, we recall its definition here. For a closed linear operator L, the complex value λ belongs to the (Browder) essential spectrum $\sigma_{ess}(L)$ if at least one of the following conditions is satisfied:

(C_1) λ is a limit point of $\sigma(L)$

(C_2) $M_\lambda(L)=\bigcup_{k=1}^{\infty}\mathfrak{N}((\lambda-L)^k)$ is infinite-dimensional

(C_3) $\mathfrak{R}(\lambda-L)$ is not closed.

PROPOSITION 8.4. *For every $t\geq 0$, $e^{t\sigma_{ess}(A)}\subseteq\sigma_{ess}(T(t))$.*

PROOF. Let $t>0$ be fixed and suppose that $e^{\lambda t}\in\sigma(T(t))\setminus\sigma_{ess}(T(t))$ for some $\lambda\in\mathbb{C}$. We show that $\lambda\notin\sigma_{ess}(A)$. This is equivalent to showing that neither of the three conditions $(C_1)-(C_3)$ is satisfied.

1. Suppose that λ is a limit point of $\sigma(A)$. Let $\lambda_k\in\sigma(A)$, $\lambda_k\neq\lambda$, $k=1,2,...$, and, $\lambda_k\to\lambda$ as $k\to\infty$. Then $e^{\lambda_k t}\to e^{\lambda t}$, $k\to\infty$ and $e^{\lambda_k t}\neq e^{\lambda t}$ whenever k is large enough, as $e^{\lambda_k t}=e^{\lambda t}$ implies that $\mathrm{Re}\,\lambda_k=\mathrm{Re}\,\lambda$ and $\mathrm{Im}\,\lambda_k=\mathrm{Im}\,\lambda+2m\pi t^{-1}$ for some $m\in\mathbb{Z}$. Since $e^{\lambda t}$ is an isolated point of $\sigma(T(t))$, this is impossible. Therefore (C_1) is not satisfied.

2. We show that $M_\lambda(A)\subseteq M_{e^{\lambda t}}(T(t))$. From Lemma 8.2 we know that $\mathfrak{N}(\lambda-A)\subseteq\mathfrak{N}(e^{\lambda t}-T(t))$. Suppose that $\mathfrak{N}((\lambda-A)^k)\subseteq\mathfrak{N}((e^{\lambda t}-T(t))^k)$ for some $k\geq 1$. Let $x\in\mathfrak{N}((\lambda-A)^{k+1})$, then $(\lambda-A)^kx\in\mathfrak{N}(\lambda-A)$, and hence $(\lambda-A)^kx\in\mathfrak{N}(e^{\lambda t}-T(t))$. Therefore

$$0=(e^{\lambda t}-T(t))(\lambda-A)^kx=(\lambda-A)^k(e^{\lambda t}-T(t))x,$$

and consequently

$$(e^{\lambda t}-T(t))x\in\mathfrak{N}((\lambda-A)^k)\subseteq\mathfrak{N}((e^{\lambda t}-T(t))^k),$$

which yields that

$$x\in\mathfrak{N}((e^{\lambda t}-T(t))^{k+1}).$$

Since dim $M_{e^{\lambda t}}(T(t))$ is finite, the same holds for $M_\lambda(A)$ and therefore (C_2) does not hold either.

3. Finally we show that $\mathfrak{R}(\lambda-A)$ is closed. Since $\mathfrak{N}(e^{\lambda t}-T(t))$ is finite-dimensional, there exists a closed subspace Y of X such that

$$\mathfrak{N}(e^{\lambda t}-T(t))\oplus Y=X.$$

It suffices to show that $(\lambda-A)Y$ is closed, because $(\lambda-A)\mathfrak{N}(e^{\lambda t}-T(t))$ is finite-dimensional and therefore closed. From the closed-graph-theorem and the closedness of $\mathfrak{R}(e^{\lambda t}-T(t))$ it follows that there is a constant $c>0$ such that $\|e^{\lambda t}x-T(t)x\|\geq c\|x\|$, for every $x\in Y$. From (8.1b) we obtain that for every $x\in\mathfrak{D}(A)$:

$$\|e^{\lambda t}x-T(t)x\|\leq m\|\lambda x-Ax\|,$$

for some positive constant m, which may depend on λ. Combination of both inequalities gives us:

$$\|\lambda x - Ax\| \geqslant \frac{c}{m}\|x\|, \; x \in Y \cap \mathfrak{D}(A),$$

and from the fact that $\lambda - A$ is closed, we conclude that $\mathfrak{R}(\lambda - A)$ is closed. Therefore (C_3) is not true, and the result follows. □

In some situations, the relation

$$e^{t\sigma(A)} = \sigma(T(t)) \setminus \{0\}, \; t \geqslant 0,$$

always holds. First of all, if $\{T(t)\}_{t\geqslant 0}$ is eventually compact (i.e. compact after finite time), then for every $t \geqslant 0$, $\sigma(T(t)) \setminus \{0\} = \sigma_p(T(t)) \setminus \{0\}$, and the relation follows from Proposition 8.3. It is rather easy to show that $\{T(t)\}_{t\geqslant 0}$ is eventually uniformly continuous whenever $\{T(t)\}_{t\geqslant 0}$ is eventually compact. The following result is stated without proof. See DAVIES [1980].

PROPOSITION 8.5. *If* $\{T(t)\}_{t\geqslant 0}$ *is eventually uniformly continuous then* $e^{t\sigma(A)} = \sigma(T(t)) \setminus \{0\}$, $t \geqslant 0$.

We recall the following two characterizations of the *type* (or *growth bound*) $\omega_0(A)$ of the semigroup $\{T(t)\}_{t\geqslant 0}$:

$$\omega_0(A) = \inf_{t>0} \frac{1}{t} \log \|T(t)\| = \lim_{t\to\infty} \frac{1}{t} \log \|T(t)\|. \tag{8.2}$$

For every $\omega > \omega_0(A)$ there exists a constant $M(\omega) \geqslant 1$ such that for every $t \geqslant 0$:

$$\|T(t)\| \leqslant M(\omega) e^{\omega t}, \; t \geqslant 0.$$

For $\lambda \in \mathbb{C}$ with $\operatorname{Re}\lambda > \omega_0(A)$ the resolvent operator $R(\lambda, A)$ can be computed from

$$R(\lambda, A)x = \int_0^\infty e^{-\lambda t} T(t) x \, dt, \tag{8.3}$$

the integral converging in norm. The spectral radius $r(T(t))$ of $T(t)$ is

$$r(T(t)) = e^{\omega_0 t}, \; t \geqslant 0, \tag{8.4}$$

where $\omega_0 = \omega_0(A)$. A similar result holds for the essential spectral radius $r_{ess}(T(t))$ (see (A.3.6) in Appendix A.3):

$$r_{ess}(T(t)) = e^{\omega_{ess} t}, \; t \geqslant 0, \tag{8.5}$$

where $\omega_{ess} = \omega_{ess}(A)$ is the *essential type* or *essential growth bound,* given by

$$\omega_{ess}(A) = \inf_{t>0} \frac{1}{t} \log |T(t)|_\alpha = \lim_{t\to\infty} \frac{1}{t} \log |T(t)|_\alpha. \tag{8.6}$$

Here $|\cdot|_\alpha$ stands for the measure-of-noncompactness (Appendix A.3.1). From $|T(t)|_\alpha \leqslant \|T(t)\|$ (see Lemma A.3.5a) it follows immediately that

$$\omega_{ess}(A) \leqslant \omega_0(A).$$

If the semigroup $\{T(t)\}_{t\geqslant 0}$ is eventually compact, then $\omega_{ess}(A) = -\infty$. We recall the definition of the spectral bound $s(A)$ of the generator A.

$$s(A) = \sup\{\mathrm{Re}\lambda : \lambda \in \sigma(A)\}, \tag{8.7}$$

if $\sigma(A) \neq \varnothing$, and $-\infty$ if $\sigma(A) = \varnothing$. We define the number $s_1(A)$ by

$$s_1(A) = \sup\{\mathrm{Re}\lambda : \lambda \in \sigma(A) \setminus \sigma_{ess}(A)\}. \tag{8.8}$$

The following inequalities always hold:

$$s_1(A) \leqslant s(A) \leqslant \omega_0(A).$$

In Example 8.1 we had $s(A)=0$ and $\omega_0(A)=1$. The following result holds.

PROPOSITION 8.6.
a) $\sup\{\mathrm{Re}\lambda : \lambda \in \sigma_{ess}(A)\} \leqslant \omega_{ess}(A)$.
b) $\omega_0(A) = \max\{s(A), \omega_{ess}(A)\} = \max\{s_1(A), \omega_{ess}(A)\}$.

PROOF. a) Assume $\lambda \in \sigma_{ess}(A)$, then, by Proposition 8.4, $e^{\lambda t} \in \sigma_{ess}(T(t))$, hence $e^{\omega_{ess} t} = r_{ess}(T(t)) \geqslant e^{\lambda t}$, $t \geqslant 0$. This yields a).
b) Let $\gamma := \max\{s_1(A), \omega_{ess}(A)\}$. Then $\gamma \leqslant \omega_0(A)$. To prove $\gamma \geqslant \omega_0(A)$ it suffices to show that $r(T(t)) \leqslant e^{\gamma t}$ because of (8.4). Let $t>0$ and $z \in \sigma(T(t))$, $z \neq 0$. If $z \in \sigma_{ess}(T(t))$, then $|z| \leqslant r_{ess}(T(t)) = e^{\omega_{ess}(A)t} \leqslant e^{\gamma t}$. If $z \notin \sigma_{ess}(T(t))$, then $z \in \sigma_p(T(t))$ by virtue of Theorem A.3.3. Proposition 8.3 now gives that $z = e^{\lambda t}$ for some $\lambda \in \sigma_p(A)$. If λ would be an element of $\sigma_{ess}(A)$, then z would be an element of $\sigma_{ess}(T(t))$ because of Proposition 8.4, which is not true by assumption. Therefore $\lambda \in \sigma(A) \setminus \sigma_{ess}(A)$, and $|z| = e^{t\mathrm{Re}\lambda} \leqslant e^{ts_1(A)} \leqslant e^{\gamma t}$ and we have proved that $\gamma = \omega_0(A)$. Since $s_1(A) \leqslant s(A) \leqslant \omega_0(A)$ the result follows. □

2. SPECTRAL PROPERTIES OF POSITIVE SEMIGROUPS

Throughout this section we assume that X is a Banach lattice (real or complex) with X^+ the cone of positive elements. We call a subset V of $\mathbb{C}$ *multiplicatively cyclic,* if $\lambda \in V$ implies that $(\lambda/|\lambda|)^k . |\lambda| \in V$, for every $k \in \mathbb{Z}$, and we call V *additively cyclic* if $\lambda \in V$ implies that $\mathrm{Re}\lambda + ik\mathrm{Im}\lambda \in V$, for every $k \in \mathbb{Z}$.

We first repeat some results which are stated in Chapter 7. Let $\{T(t)\}_{t\geqslant 0}$ be a C_0-semigroup of positive operators on X (or positive C_0-semigroup), with infinitesimal generator A. Then for every $\lambda > s(A)$, $R(\lambda, A)$ is a positive operator. This follows from the identity

$$R(\lambda, A)x = \int_0^\infty e^{-\lambda t} T(t)x\,dt\,, \; x \in X\,, \; \mathrm{Re}\lambda > s(A), \tag{8.9}$$

where the integral is norm convergent as an improper Riemann integral (see Theorem 7.4). The *peripheral spectrum* $\sigma_+(A)$ of A is defined as

$$\sigma_+(A) = \{\lambda \in \sigma(A) : \mathrm{Re}\lambda = s(A)\}.$$

Obviously $\sigma_+(A) \subseteq \partial\sigma(A)$, where $\partial\sigma(A)$ is the boundary of $\sigma(A)$.

Theorem 8.7. *Let $\{T(t)\}_{t\geq 0}$ be a positive C_0-semigroup with generator A, then*
a) $s(A)\in\sigma(A)$ *if* $\sigma(A)\neq\varnothing$.
b) $R(\lambda,A)\geq 0$ *for every* $\lambda>s(A)$.
c) *If* $s(A)$ *is a pole of* $R(\lambda,A)$ *of order p, and if* $\mu\in\sigma_+(A)$ *is another pole, then its order* $\leq p$.

Proof. For a) and b) we refer to Theorem 7.4 and Corollary 7.5. Here we prove c). Let $s(A)=0$ (this can be assumed without loss of generality) be a pole of order p of $R(\lambda,A)$ and let $i\nu\in\sigma_+(A)$ be another pole. For $x\in X$ and $\epsilon>0$ we have

$$\epsilon^{p+1}\|R(\epsilon+i\nu,A)x\|\leq\epsilon^{p+1}\|R(\epsilon,A)|x|\,\|,$$

and the right-hand side goes to zero as $\epsilon\downarrow 0$. This yields the desired result. □

Before we proceed let us recall some notions from the theory of Banach lattices (see Appendix 2). For $u\in X$ we denote by X_u the principal ideal generated by u. Let π_u be the unique element in $Z(X_u)$ (see Appendix 2.4) such that $\pi_u|u|=u$. We define the sequence $\{u^{(n)}\}_{n\in\mathbb{Z}}$ in X by

$$u^{(n)}=\pi_u^n|u|\,,\; n\in\mathbb{Z}. \tag{8.10}$$

Here $\pi_u^{-1}=\pi_{\bar{u}}$, where $\bar{u}$ is the conjugate element of u. Let J be a closed ideal in X and $L\in\mathcal{L}(X)$. Then J is called L-invariant if $LJ\subseteq J$. If $\{T(t)\}_{t\geq 0}$ is a C_0-semigroup, then J is called $\{T(t)\}_{t\geq 0}$-invariant if $T(t)J\subseteq J$ for every $t\geq 0$. Finally, if $R(\lambda)$ defines a pseudo resolvent on D, then J is called $R(\lambda)$-invariant if $R(\lambda)J\subseteq J$ for every $\lambda\in D$. The following lemma relates positivity to contractivity.

Lemma 8.8. *Let u be a quasi-interior point of* X^+, *and let L be a bounded linear operator on X satisfying* $Lu=u$. *Then* $L\geq 0$ *if and only if* $LX_u\subseteq X_u$ *and* $\|L\|_u\leq 1$.

Proof. If $L\geq 0$ then $LX_u\subseteq X_u$ (this follows from $Lu=u$) and $\|L\|_u=\|Lu\|_u=\|u\|_u=1$. Before proving the converse we note that for every element x in the complete Banach lattice X, the following two assertions are equivalent
(i) $-u\leq x\leq u$
(ii) $\|x-iru\|_u^2\leq 1+r^2$, for every $r\in\mathbb{R}$.
Now suppose that $\|L\|_u\leq 1$ and let $x\in X_u\cap X^+$. Without loss of generality we may assume that $0\leq x\leq 2u$, or equivalently, $-u\leq x-u\leq u$. Hence

$$\|x-u-iru\|_u\leq\sqrt{1+r^2}$$

for every $r\in\mathbb{R}$, and therefore $\|Lx-u-iru\|_u\leq\sqrt{1+r^2}$ which implies that $-u\leq Lx-u\leq u$, and thus $Lx\geq 0$. Since $\overline{X}_u=X$, this holds for every $x\in X^+$ and the result follows. □

LEMMA 8.9. *Let $\Lambda, L \in \mathfrak{L}(X)$ satisfy $|\Lambda x| \leqslant L|x|$ for every $x \in X$. Let $u \in X$ be such that $|u|$ is a quasi-interior point of X^+, $L|u| = |u|$ and $\Lambda u = u$. Then $L = \pi_u^{-1} \Lambda \pi_u$.*

PROOF. If we put $\tilde{L} := \pi_u^{-1} \Lambda \pi_u$, then for $x \in X$

$$|\tilde{L}x| = |\pi_u^{-1} \Lambda \pi_u x| = |\Lambda \pi_u x| \leqslant L|\pi_u x| = L|x|. \qquad (*)$$

Since $L|u| = |u|$ and $L \geqslant 0$ we obtain from the former lemma that $\|L\|_{|u|} \leqslant 1$. Then also $\|\tilde{L}\|_{|u|} \leqslant 1$ by inequality (*). Moreover

$$\tilde{L}|u| = \pi_u^{-1} \Lambda \pi_u |u| = \pi_u^{-1} \Lambda u = \pi_u^{-1} u = |u|,$$

and applying Lemma 8.8 once more, we find $\tilde{L} \geqslant 0$. From (*) we infer that

$$0 \leqslant \tilde{L} \leqslant L,$$

and thus

$$\|L - \tilde{L}\|_{|u|} = \|(L - \tilde{L})|u|\|_{|u|} = 0.$$

The proof is complete. □

PROPOSITION 8.10. *Let L be a positive linear operator on X (and therefore bounded) and suppose that $Lu = \alpha u$ and $L|u| = |u|$, for some $u \in X$ and $\alpha \in \mathbb{C}$, $|\alpha| = 1$. Then, for every $k \in \mathbb{Z}$ we have $Lu^{(k)} = \alpha^k u^{(k)}$.*

PROOF. Obviously, L leaves $\overline{X}_u$ invariant. Let L_0 be the restriction of L to $\overline{X}_u$ and $\Lambda_0 := \alpha^{-1} L_0$. Then $\overline{X}_u, u, \Lambda_0$ and L_0 obey the conditions of Lemma 8.9 and from that result we get that $L_0 = \pi_u^{-1} \Lambda_0 \pi_u = \alpha^{-1} \pi_u^{-1} L_0 \pi_u$. Iteration yields $L_0^k = \alpha^{-k} \pi_u^{-k} L_0 \pi_u^k$. Now, application to $|u|$ gives the desired result. □

A consequence of this result is the following

COROLLARY 8.11. *Let L be a lattice homomorphism on X. Then $\sigma_p(L)$ is multiplicatively cyclic.*

PROOF. Let $\lambda \in \sigma_p(L)$, $\lambda \neq 0$, and let $u \in X$, $u \neq 0$ satisfy $Lu = \lambda u$. Now obviously, $\tilde{L} = |\lambda|^{-1} L$ obeys the assumptions of Proposition 8.10 with $\alpha = \lambda |\lambda|^{-1}$, and the result follows. □

PROPOSITION 8.12. *Let $u \in X$, $\nu \in \mathbb{R}$, D a subset of $\mathbb{C}$ containing all λ with* $\mathrm{Re}\,\lambda > 0$, *and $R(\lambda)$ a pseudo resolvent on D satisfying*
(i) $|R(\lambda)x| \leqslant R(\mathrm{Re}\,\lambda)|x|$, $x \in X$, $\mathrm{Re}\,\lambda > 0$,
(ii) $\lambda R(\lambda + i\nu)u = u$, *for some* $\lambda \in \mathbb{C}$ *with* $\mathrm{Re}\,\lambda > 0$,
(iii) $\lambda R(\lambda)|u| = |u|$, *for some* $\lambda \in \mathbb{C}$, *with* $\mathrm{Re}\,\lambda > 0$.
Then the following holds
a) $\lambda R(\lambda + ik\nu)u^{(k)} = u^{(k)}$, $k \in \mathbb{Z}$, $\mathrm{Re}\,\lambda > 0$.
b) $R(\lambda) = \pi_u^{-1} R(\lambda + i\nu) \pi_u$, $\mathrm{Re}\,\lambda > 0$, *if $|u|$ is a quasi-interior point of X^+.*

PROOF. First we note that in (ii) and (iii) we may replace the word "some" by "all".

a) The ideal $J=\overline{X}_u$ is invariant under $R(\lambda)$ for every λ with $\mathrm{Re}\lambda>0$. Let $R_J(\lambda)$ be the restriction of $R(\lambda)$ to J. Then $\lambda R_J(\lambda)$ and $\lambda R_J(\lambda+i\nu)$ satisfy the conditions of Proposition 8.9 for $\lambda>0$, and hence

$$\lambda R_J(\lambda)=\pi_u^{-1}\lambda R_J(\lambda+i\nu)\pi_u\,,\ \lambda>0.$$

Therefore the analytic functions $\lambda\mapsto\lambda R_J(\lambda)$ and $\lambda\mapsto\pi_u^{-1}\lambda R_J(\lambda+i\nu)\pi_u$ coincide on $\mathbb{R}_+$, hence

$$\lambda R_J(\lambda)=\pi_u^{-1}\lambda R_J(\lambda+i\nu)\pi_u\,,\ \mathrm{Re}\lambda>0. \tag{*}$$

Iteration yields

$$\lambda R_J(\lambda)=\pi_u^{-k}\lambda R_J(\lambda+ik\nu)\pi_u^k,\ \mathrm{Re}\lambda>0,\ k\in\mathbb{Z}.$$

Now, application to $|u|$ gives

$$\pi_u^{-k}\lambda R(\lambda+ik\nu)u^{(k)}=\lambda R(\lambda)|u|=|u|,\ \mathrm{Re}\lambda>0,\ k\in\mathbb{Z},$$

and a) follows. If $|u|$ is a quasi-interior point of X^+, then $R(\lambda)=R_J(\lambda)$, since $J=\overline{X}_u=X$, and in this case b) follows from (*). □

Now if $\{T(t)\}_{t\geq 0}$ is a positive C_0-semigroup, then the resolvent $R(\lambda,A)$ satisfies condition (i) of Proposition 8.12:

$$|R(\lambda,A)x|\leq R(\mathrm{Re}\lambda,A)|x|,\ x\in X,\ \mathrm{Re}\lambda>s(A). \tag{8.11}$$

This follows trivially from relation (8.9). Suppose that $s(A)>-\infty$. Replacing, if necessary, A by $A-s(A).I$, we may always assume that $s(A)=0$. If there exist $u\in X$ and $\nu\in\mathbb{R}$ such that $A|u|=0$ and $Au=i\nu u$, then $R(\lambda)=R(\lambda,A)$ satisfies conditions (ii) and (iii) of Proposition 8.12 as well. Therefore, for every $k\in\mathbb{Z}$ and $\lambda\in\mathbb{C}$ with $\mathrm{Re}\lambda>0$, we have

$$\lambda R(\lambda+ik\nu,A)u^{(k)}=u^{(k)}$$

which is equivalent to

$$Au^{(k)}=ik\nu u^{(k)}.$$

If, in addition, $|u|$ is a quasi-interior point of X^+, then

$$R(\lambda,A)=\pi_u^{-1}R(\lambda+i\nu,A)\pi_u,$$

for every λ with $\mathrm{Re}\lambda>0$. This implies $\pi_u\mathfrak{D}(A)=\mathfrak{D}(A)$ and $A=\pi_u^{-1}(A-i\nu)\pi_u$ on $\mathfrak{D}(A)$. We summarize these results in the following proposition.

PROPOSITION 8.13. *Let A be the generator of a positive C_0-semigroup and $s(A)=0$. Assume that there exists a $u\in X$ and $\nu\in\mathbb{R}$ such that $A|u|=0$ and $Au=i\nu u$. Then*

a) $Au^{(k)}=ik\nu u^{(k)},\ k\in\mathbb{Z}$.

b) *If, in addition, $|u|$ is a quasi-interior point of X^+, then $\pi_u\mathfrak{D}(A)=\mathfrak{D}(A)$ and $A=\pi_u^{-1}(A-i\nu)\pi_u$ on $\mathfrak{D}(A)$.*

Note that, if b) is satisfied, the corresponding semigroup $\{T(t)\}_{t\geq 0}$ obeys

$$T(t) = \pi_u^{-1} e^{-i\nu t} T(t)\pi_u \, , t\geq 0,$$

hence $\sigma(T(t))=e^{-i\nu t}\sigma(T(t))$, for $t\geq 0$.
Now we are ready to prove the main result of this section.

THEOREM 8.14. *Let $\{T(t)\}_{t\geq 0}$ be a positive C_0-semigroup on X and suppose that $s(A)>-\infty$ is a pole of the resolvent $R(\lambda,A)$. Then $\sigma_+(A)$ is additively cyclic.*

PROOF. Without loss of generality we may assume that $s(A)=0$. To start with, let us assume that 0 is a pole of $R(\lambda,A)$ of order one. Suppose that $i\nu\in\sigma_+(A)$ for some $\nu\in\mathbb{R}$. Then for every $\lambda\in\rho(A)$ we have $(\lambda-i\nu)^{-1}\in\partial\sigma(R(\lambda,A))\subseteq\sigma_{ap}(R(\lambda,A))$. Note that $\sigma(R(\lambda,A))\setminus\{0\}=\{(\lambda-\mu)^{-1}: \mu\in\sigma(A)\}$. Let $\hat{X}$ be the extension of X as described in Appendix 3.3, and let $\hat{R}(\lambda)$ be the pseudoresolvent on $\hat{X}$ induced by $R(\lambda,A)$. It follows immediately that

$$|\hat{R}(\lambda)\hat{x}|\leq\hat{R}(\mathrm{Re}\lambda)|\hat{x}|\, , \hat{x}\in\hat{X}, \mathrm{Re}\lambda>0. \tag{i}$$

We have $(\lambda-i\nu)^{-1}\in\sigma_p(\hat{R}(\lambda))$ if $\mathrm{Re}\lambda>0$. Now fix $\lambda\in\mathbb{C}$ with $\mathrm{Re}\lambda>0$. There exists an element $\hat{u}\in\hat{X}$ such that

$$\hat{R}(\lambda)\hat{u} = (\lambda - i\nu)^{-1}\hat{u}.$$

We obtain from Proposition A.3.9 that this relation is satisfied for every λ with $\mathrm{Re}\lambda>0$, as well. Hence

$$\lambda\hat{R}(\lambda + i\nu)\hat{u} = \hat{u}, \quad \mathrm{Re}\lambda>0$$

$$\lambda\hat{R}(\lambda)|\hat{u}| \geq |\hat{u}|, \quad \lambda>0. \tag{ii}$$

The second relation follows from (i). There exists an element $\psi\in(\hat{X}^+)^*$ such that $<|\hat{u}|,\psi>>0$. By assumption $\lambda=0$ is a first-order pole of $R(\lambda,A)$, and the same is true for $\hat{R}(\lambda)$. Hence the subset $\{\lambda\hat{R}(\lambda)^*\psi:0<\lambda<1\}$ of $\hat{X}^*$ is norm-bounded. Let $\{\lambda_n\}_{n=1}^{\infty}$ be a sequence converging to zero as $n\to\infty$. Then the closure of $\{\lambda_n R(\lambda_n)\psi\}_{n=1}^{\infty}$ is weakly * compact, and therefore has a weakly * accumulation point, ϕ say. We show that

$$<x,\lambda\hat{R}(\lambda)\phi> = <x,\phi>, \text{ for all } x\in X.$$

Let $x\in X$. Since $\{\lambda\lambda_n\hat{R}(\lambda)^*\hat{R}(\lambda_n)^*\psi\}_{n=1}^{\infty}$ is norm bounded, there exists a subsequence $\{\lambda\lambda_{n_k}\hat{R}(\lambda)^*\hat{R}(\lambda_{n_k})^*\psi\}_{k=1}^{\infty}$ (which may depend on x) such that

$$<x,\lambda\lambda_{n_k}\hat{R}(\lambda)^*\hat{R}(\lambda_{n_k})^*\psi> = <x,\frac{\lambda\lambda_{n_k}}{\lambda-\lambda_{n_k}}\hat{R}(\lambda_{n_k})^*\psi> - <x,\frac{\lambda\lambda_{n_k}}{\lambda-\lambda_{n_k}}\hat{R}(\lambda)^*\psi>$$

$$\to <x,\phi> \text{ as } k\to\infty,$$

but also,

$$<x,\lambda\lambda_{n_k}\hat{R}(\lambda)^*\hat{R}(\lambda_{n_k})^*\psi>\to<x,\lambda\hat{R}(\lambda)^*\phi> \text{ as } k\to\infty,$$

yielding that

$$\phi = \lambda\hat{R}(\lambda)^*\phi.$$

Moreover, we get from (ii) that

$$0 < <|\hat{u}|,\psi> \leqslant <\lambda\hat{R}(\lambda)|\hat{u}|,\psi> = <|\hat{u}|,\lambda\hat{R}(\lambda)^*\psi>$$

which results in

$$<|\hat{u}|,\phi> > 0.$$

We have thus proved that

$$\lambda\hat{R}(\lambda)^*\phi = \phi,\ \mathrm{Re}\lambda > 0,\text{ and } <|\hat{u}|,\phi> > 0. \qquad \text{(iii)}$$

The ideal $\hat{J} = \{\hat{x}\in\hat{X}: <|\hat{x}|,\phi> = 0\}$ is $\hat{R}(\lambda)$ - invariant because of (i) and (ii). Let $\tilde{R}(\lambda)$ be the pseudo resolvent on $\tilde{X} = \hat{X}/\hat{J}$ induced by $\hat{R}(\lambda)$. Then

$$|\tilde{R}(\lambda)\tilde{x}| \leqslant \tilde{R}(\mathrm{Re}\lambda)|\tilde{x}|,\ \tilde{x}\in\tilde{X},\ \mathrm{Re}\lambda > 0. \qquad \text{(i')}$$

Let $\tilde{u}$ be the canonical image of $\hat{u}$. Obviously

$$\lambda\tilde{R}(\lambda + i\nu)\tilde{u} = \tilde{u},\ \mathrm{Re}\lambda > 0.$$

From $<|\hat{u}|,\phi> > 0$ we gather that $\hat{u}\notin\hat{J}$, or, equivalently, $\tilde{u}\neq 0$. If $\lambda > 0$, then because of (iii):

$$<\lambda\hat{R}(\lambda)|\hat{u}| - |\hat{u}|,\phi> = 0,$$

hence $\lambda\hat{R}(\lambda)|\hat{u}| - |\hat{u}|\in\hat{J}$, and therefore

$$\lambda\tilde{R}(\lambda)|\tilde{u}| = |\tilde{u}|.$$

Thus we have

$$\lambda\tilde{R}(\lambda + i\nu)\tilde{u} = \tilde{u},\quad \mathrm{Re}\lambda > 0,$$
$$\lambda\tilde{R}(\lambda)|\tilde{u}| = |\tilde{u}|,\quad \lambda > 0, \qquad \text{(ii')}$$

and we obtain from Proposition 8.12 that

$$\lambda\tilde{R}(\lambda + ik\nu)\tilde{u}^{(k)} = \tilde{u}^{(k)},\ k\in\mathbb{Z},\ \mathrm{Re}\lambda > 0,$$

and hence $(\lambda - ik\nu)^{-1}\in\sigma(\tilde{R}(\lambda))$ for every $k\in\mathbb{Z}$. This yields (e.g. Proposition A.3.10) that $(\lambda - ik\nu)^{-1}\in\sigma(\hat{R}(\lambda)) = \sigma(R(\lambda,A))$, i.e. $ik\nu\in\sigma(A)$.

Now suppose that we can prove the present result for poles of order less than p ($p\geqslant 2$) and suppose that $s(A) = 0$ is a pole of order p of $R(\lambda,A)$. Let

$$Q := \lim_{\lambda\downarrow 0}\lambda^p R(\lambda,A),$$

i.e. $Q = B_{-p}$, where B_{-p} is given by (A.3.4). Then $Q\in\mathcal{L}(X)$ and $Q > 0$. Moreover

$$QA = AQ = \lim_{\lambda\downarrow o}\lambda^p AR(\lambda,A) = \lim_{\lambda\downarrow 0}(-\lambda^p I + \lambda^{p+1}R(\lambda,A)) = 0.$$

Let the ideal J in X be defined by

$$J = \{x \in X : Q|x| = 0\}.$$

Since $T(t)Qx = QT(t)x = x$, for every $x \in X$, we obtain that J is $\{T(t)\}$-invariant. Let $\tilde{X} = X/J$, then $\tilde{Q} = 0$ and 0 is a pole of $R(\lambda, \tilde{A})$ of order less than p. From this observation and Proposition A.3.10 it is not difficult to see that the result is valid for poles of arbitrary order. □

REMARK 8.15. It is possible to prove Theorem 8.14 under weaker assumptions. A C_0-semigroup $\{T(t)\}_{t \geq 0}$ is said to satisfy the (w)-condition if the family of operators $\{(\lambda - s(A))R(\lambda, A) : \lambda > s(A)\}$ is uniformly bounded. The semigroup is (w)-solvable if there exists a chain

$$\{0\} = J_0 \subseteq J_1 \subseteq \cdots \subseteq J_{n-1} \subseteq J_n = X$$

of $\{T(t)\}$-invariant ideals such that for every $k = 1, \ldots, n$ the semigroup $\{T_k(t)\}_{t \geq 0}$ on $X_k = J_k / J_{k-1}$ induced by $\{T(t)\}_{t \geq 0}$ satisfies the (w)-condition. If $s(A)$ is a pole of the resolvent, then the semigroup is (w)-solvable. It can be shown that the conclusions of the theorem remain valid if we merely assume that $\{T(t)\}_{t \geq 0}$ is a positive semigroup, which is (w)-solvable.

3. Spectral properties of irreducible semigroups

We recall that a positive C_0-semigroup $\{T(t)\}_{t \geq 0}$ is called irreducible if $\{0\}$ and X are the only closed $\{T(t)\}_{t \geq 0}$-invariant ideals in X. The irreducibility of the semigroup is equivalent to the irreducibility of the resolvent of its generator; see Proposition 7.6.

For irreducible semigroups a much stronger result than Theorem 8.14 holds. Before we state this result we work out a rather special but illuminating example.

EXAMPLE 8.16. *Markov semigroups*

Let $X = C(K)$ be the Banach lattice of continuous functions on the compact Hausdorff space K. A Markov semigroup on $C(K)$ is by definition a positive C_0-semigroup $\{T(t)\}_{t \geq 0}$ satisfying

$$T(t)\mathbf{1} = \mathbf{1},\ t \geq 0,$$

where $\mathbf{1}$ is the function on K which is identically one. Obviously, the generator A of $\{T(t)\}_{t \geq 0}$ satisfies $A\mathbf{1} = 0$. If $f \in C(K)$, $\|f\| \leq 1$, then $|f| \leq \mathbf{1}$, hence $|T(t)f| \leq T(t)|f| \leq T(t)\mathbf{1} = \mathbf{1}$, from which it follows that $\|T(t)\| = 1$, $t \geq 0$. Thus

$$s(A) = \omega_0(A) = 0.$$

Before we proceed, we remark that Markov semigroups play a very important role in the theory of Markov processes.

For the rest of this example we assume that $\{T(t)\}_{t \geq 0}$ defines an *irreducible* Markov semigroup. We shall concentrate on the so-called peripheral point spectrum $\sigma_{p+}(A)$ which is defined by

$$\sigma_{p+}(A) = \sigma_p(A) \cap \sigma_+(A).$$

Obviously $0 \in \sigma_{p+}(A)$ since $A\mathbf{1}=0$. Note that $\sigma_{p+}(A)=\sigma_+(A)$ if $\omega_{ess}(A)<0$. Suppose that

$$Af = i\nu f,$$

where $\nu \in \mathbb{R}$ and $f \in C(K)$ with $\|f\|=1$. Then

$$T(t)f = e^{i\nu t} f,\ t \geq 0,$$

hence

$$0 < |f| = |T(t)f| \leq T(t)|f| \leq T(t)\mathbf{1} = \mathbf{1},\ t \geq 0,$$

or, equivalently

$$0 \leq T(t)(\mathbf{1} - |f|) \leq \mathbf{1} - |f|,\ t \geq 0.$$

Let $J = \{h \in C(K): h(k)=0 \text{ if } |f(k)|=1\}$. Then $J=\overline{X}_u$, where $u = \mathbf{1} - |f|$. Obviously X_u is $\{T(t)\}_{t\geq 0}$-invariant because $0 \leq T(t)u \leq u$. This implies that J is a closed $\{T(t)\}_{t\geq 0}$-invariant ideal of $C(K)$. Since $\{T(t)\}_{t\geq 0}$ is irreducible and $J \neq X$ (in particular $\mathbf{1} \notin J$) we find that $J=\{0\}$, which is possible only if

$$|f| = \mathbf{1}.$$

So we have

$$Af = i\nu f \text{ and } A|f| = 0.$$

Now we infer from Proposition 8.13 that
(i) $Af^{(k)} = ik\nu f^{(k)},\ k \in \mathbb{Z}$
(ii) $A = \pi_f^{-1}(A - i\nu)\pi_f,\ \mathfrak{D}(A) = \pi_f \mathfrak{D}(A)$.
Suppose there exists an $\tilde{f} \in C(K)$ such that $A\tilde{f} = i\nu\tilde{f}$. Defining, for each $k \in K$, the function $u_k \in C(K)$ by

$$u_k = \tilde{f}(k)f - f(k)\tilde{f},$$

we get $Au_k = i\nu u_k$ and similar arguments as those above give us that $|u_k| = c\mathbf{1}$ for some constant c. Since $u_k(k)=0$, we find $c=0$, hence $u_k=0$. This yields the geometric simplicity of $i\nu$.

Now suppose that

$$Ag = i\mu g,$$

where $\mu \in \mathbb{R}$ and $g \in C(K)$ with $\|g\|=1$. Using (ii) we find that

$$i\mu g = Ag = \pi_f^{-1}(A - i\nu)\pi_f g,$$

which implies that

$$A(\bar{f}g) = i(\mu - \nu)\bar{f}g.$$

Here we have used that $\pi_f^{-1} g = \bar{f}g$, which follows from $|f| = \mathbf{1}$. Therefore $\sigma_{p+}(A)$ is an additive subgroup of $i\mathbb{R}$.

From (ii) we find that for every $\lambda \in \rho(A)$

$$R(\lambda, A) = \pi_f^{-1} R(\lambda + i\nu, A)\pi_f,$$

which means that every $i\nu\in\sigma_{p+}(A)$ is a pole of $R(\lambda,A)$ if 0 is a pole, and, moreover, they are of the same order.

Now suppose that 0 is indeed a pole of $R(\lambda,A)$ of order $p\geqslant 1$. If $p>1$ would hold, then

$$0\neq Q=\lim_{\lambda\downarrow 0}\lambda^p R(\lambda,A)$$

is a positive operator. Moreover

$$Q\mathbf{1}=\lim_{\lambda\downarrow 0}\lambda^p R(\lambda,A)\mathbf{1}=\lim_{\lambda\downarrow 0}\lambda^{p-1}\mathbf{1}=0.$$

Now, if $h\in C(K)$, then $|h|\leqslant\|h\|\mathbf{1}$, hence $|Qh|\leqslant Q|h|\leqslant\|h\|Q\mathbf{1}=0$, which yields that $Q=0$. But this is a contradiction. We may conclude that $p=1$.

We end this example with a brief summary of our findings.

Let $\{T(t)\}_{t\geqslant 0}$ be an irreducible Markov semigroup on $C(K)$. Then $\sigma_{p+}(A)$ is a subgroup of $i\mathbb{R}$, or equivalently, there exists a $\nu\in\mathbb{R}$ such that $\sigma_{p+}(A)=i\nu\mathbb{Z}$. The geometric multiplicity of every element in $\sigma_{p+}(A)$ is one. If, in addition, $s(A)=0$ is a pole of the resolvent, then $\sigma_{p+}(A)$ consists only of poles of $R(\lambda,A)$ of order one. □

We now state the main result of this section.

THEOREM 8.17. *Let $\{T(t)\}_{t\geqslant 0}$ be a positive irreducible C_0-semigroup and $s(A)>-\infty$ a pole of the resolvent $R(\lambda,A)$. Then*

a) *$s(A)$ is a first-order pole with geometric multiplicity one; moreover there exists a quasi-interior point x_0 of X^+ such that*

$$Ax_0=s(A)x_0,$$

and a strictly positive point x_0^* of $(X^+)^*$ such that

$$A^*x_0^*=s(A)x_0^*.$$

b) *$\sigma_+(A)=s(A)+i\nu\mathbb{Z}$ for some $\nu\geqslant 0$ and these elements are all first-order poles of $R(\lambda,A)$ with algebraic multiplicity one.*

PROOF a) Assume for convenience that $s(A)=0$, and let p be the order of the pole 0 of $R(\lambda,A)$. Define

$$Q:=\lim_{\lambda\downarrow 0}\lambda^p R(\lambda,A),$$

then $Q>0$ and $QT(t)=T(t)Q=Q$, $t\geqslant 0$. Therefore

$$J=\{x\in X: Q|x|=0\}$$

is a $\{T(t)\}_{t\geqslant 0}$-invariant ideal, hence $J=\{0\}$ because $\{T(t)\}_{t\geqslant 0}$ is irreducible and $J\neq X$ (as $Q>0$). If $p>1$ then $Q^2=0$ ($Q^2=B_{-p}^2=B_{-2p+1}$; see Appendix A.3.a) which is a contradiction. Therefore $p=1$.

Let $x\in X^+$ be such that $x_0=Qx>0$, then $Ax_0=0$, hence $T(t)x_0=x_0$, and $\overline{X}_{x_0}$

is a $\{T(t)\}_{t\geq 0}$-invariant ideal. Therefore $\overline{X}_{x_0} = X$, yielding that x_0 is a quasi-interior point of X^+. Let $x_0^* > 0$ obey $A^* x_o^* = 0$ (x_0^* is of the form $Q^* x^*$), then $T(t)^* x_0^* = x_0^*$. Now

$$J = \{x \in X : <|x|, x_0^*> = 0\}$$

is a $\{T(t)\}_{t\geq 0}$-invariant ideal, from which we conclude that $J = \{0\}$. Hence x_0^* is strictly positive. By renormalization we get $<x_0, x_0^*> = 1$.

Now let $x > 0$ satisfy $Ax = 0$ and $<x, x_0^*> = 1$. We prove that $x = x_0$. Clearly, we have the inequality $T(t)|x - x_0| \geq |T(t)(x - x_0)| = |x - x_0|$. If we would have $T(t)|x - x_0| > |x - x_0|$, then $<T(t)|x - x_0|, x_0^*> > <|x - x_0|, x_0^*>$ which is a contradiction on account of $<T(t)|x - x_0|, x_0^*> = <|x - x_0|, T(t)^* x_0^*> = <|x - x_0|, x_0^*>$. Thus we get $T(t)|x - x_0| = |x - x_0|$ and $A|x - x_0| = 0$. Now let $u = |x - x_0| + (x - x_0)$, $v = |x - x_0| - (x - x_0)$. Then $u, v \in X^+$, $Au = Av = 0$ and u and v are disjoint. Since $Au = 0$, u is a quasi-interior point of X^+ or $u = 0$, and the same observation holds for v. Since any quasi-interior point is a weak order unit, we get $u = 0$ or $v = 0$ (see Appendix 2.4). Suppose $v = 0$ (the case $u = 0$ is treated similarly). Then $|x - x_0| = x - x_0$ yielding that $<|x - x_0|, x_0^*> = <x - x_0, x_0^*> = 1 - 1 = 0$ and it follows that $x = x_0$.

Next suppose that $y \in X$ obeys $Ay = 0$. Let $y = y^+ - y^-$ where $y^-, y^+ \in X^+$. Obviously $T(t)|y| \geq |y|$ and as above we can exclude the possibility $T(t)|y| > |y|$. Therefore $T(t)|y| = |y|$, i.e. $T(t)(y^- + y^+) = y^- + y^+$. From $T(t)y = y$ we find $T(t)(y^+ - y^-) = y^+ - y^-$ and therefore $T(t)y^+ = y^+$ and $T(t)y^- = y^-$. The arguments above learn that $y^+ = <y^+, x_0^*> x_0$, $y^- = <y^-, x_0^*> x_0$, hence $y = <y, x_0^*> x_0$, which proves the geometric simplicity of $s(A) = 0$.

b) Assume that $\sigma_+(A) \subseteq \sigma_p(A)$. This can alway be obtained by means of a transition to the extended Banach space $\hat{X}$; compare the proof of Theorem 8.14. Let $i\nu \in \sigma_+(A)$ and $Au = i\nu u$ for some $u \in X$, $u \neq 0$. Then $T(t)u = e^{i\nu t}u$, hence $T(t)|u| \geq |T(t)u| = |u|$, and by the same arguments as used in the proof of a) we get equality. Thus $A|u| = 0$ and we may apply Proposition 8.13 which gives us the following:

$$Au^{(k)} = ik\nu u^{(k)}, \quad k \in \mathbb{Z},$$

$$A = \pi_u^{-1}(A - i\nu)\pi_u, \quad \pi_u \mathfrak{D}(A) = \mathfrak{D}(A),$$

from which it follows that

$$R(\lambda, A) = \pi_u^{-k} R(\lambda + ik\nu, A)\pi_u^k,$$

and we find that $ik\nu$ ($k \in \mathbb{Z}$) are first-order poles of $R(\lambda, A)$ with geometric multiplicity one. In order to show that $\sigma_+(A)$ is a subgroup of $i\mathbb{R}$ we take $i\mu \in \sigma_+(A)$ and $v \in X$ such that $Av = i\mu v$. Then

$$R(\lambda, A) = \pi_v^{-1} R(\lambda + i\mu, A)\pi_v.$$

Combining this with

$$R(\lambda, A) = \pi_u^{-1} R(\lambda + i\nu, A)\pi_u,$$

we find

$$R(\lambda,A) = \pi_u \pi_v^{-1} R(\lambda + i\mu - i\nu, A) \pi_v \pi_u^{-1},$$

and the result follows. □

Chapter 9

Asymptotic Behaviour

1. Stability

Suppose that A is a complex $n \times n$-matrix and consider the semigroup $\{T(t)\}_{t\geqslant 0}$ in $\mathbb{C}^n$ defined by $T(t)=e^{tA}$. We consider in $\mathbb{C}^n$ some norm and the corresponding operator norm on the matrices. As well-known, the following statements are equivalent:

(i) $\|e^{tA}\|\to 0$ as $t\to\infty$
(ii) $\|e^{tA}x\|\to$ as $t\to\infty$ for all $x\in\mathbb{C}^n$
(iii) $\mathrm{Re}\,\lambda_i<0$ for all eigenvalues $\lambda_1, \ldots, \lambda_k$ of A.

Note that (ii) is equivalent to saying that every solution $y=y(t)$ of the Cauchy problem $y'=Ay$, $y(0)=x$ converges to 0 as $t\to\infty$ (i.e. 0 is an asymptotically stable solution of the system $y'=Ay$). Now consider an arbitrary (complex) Banach space X and let $\{T(t)\}_{t\geqslant 0}$ be a C_0-semigroup with generator A. Clearly, in this situation condition (i) above implies (ii), but in general these two statements are not equivalent, as is shown in the following examples.

Example 9.1.

(i) Let $X=C_0(\mathbb{R}^+)$, the Banach space of all continuous functions on $\mathbb{R}^+$ vanishing at ∞, with the sup-norm. Define $T(t)f(x)=f(x+t)$ for all $f\in C_0(\mathbb{R}^+)$ and all $t\geqslant 0$. Then $\|T(t)\|=1$ for all $t\geqslant 0$, and $\|T(t)f\|_\infty\to 0$ as $t\to\infty$ for all $f\in C_0(\mathbb{R}^+)$.

(ii) Let $X=l_1$ and let A be the multiplication by the sequence $(-\alpha_1,-\alpha_2,\ldots)$ where $\alpha_j>0$ and $\alpha_j\to 0$ as $j\to\infty$. Then $T(t)=e^{tA}$ is multiplication by the sequence $(e^{-\alpha_1 t},e^{-\alpha_2 t},\ldots)$ and so $\|T(t)\|=\|(e^{-\alpha_1 t},e^{-\alpha_2 t},\ldots)\|_\infty=1$ for all $t\geqslant 0$. However, if $x\in l_1$, then $\|T(t)x\|_1\to 0$ as $t\to\infty$.

Clearly, a semigroup $\{T(t)\}_{t\geqslant 0}$ with $T(t)x \to 0$ as $t\to\infty$ for all $x\in X$, is bounded. A C_0-semigroup $\{T(t)\}_{t\geqslant 0}$ for which $\|T(t)\|\to 0$ as $t\to\infty$ is called *uniformly stable.* Recall that the type (growth bound) of $\{T(t)\}_{t\geqslant 0}$ is defined by

$$\omega_0(A)=\omega_0=\lim_{t\to\infty}\frac{\log\|T(t)\|}{t}=\inf_{t>0}\frac{\log\|T(t)\|}{t}.$$

For $\omega>\omega_0$ there exists $M=M_\omega\geqslant 0$ such that $\|T(t)\|\leqslant Me^{\omega t}$ for all $t\geqslant 0$. Now it is clear that $\{T(t)\}_{t\geqslant 0}$ is uniformly stable if and only if $\omega_0<0$.
Furthermore we recall that the spectral bound $s(A)$ of the generator A is defined by

$$s(A)=\sup\{\operatorname{Re}\lambda:\lambda\in\sigma(A)\},$$

and that $-\infty\leqslant s(A)\leqslant\omega_0<\infty$. Condition (iii) above corresponds to $s(A)<0$. By the above observations it follows that uniform stability of $\{T(t)\}_{t\geqslant 0}$ implies that $s(A)<0$. Now we consider the problem to what extent $s(A)<0$ implies the uniform stability of the semigroup $\{T(t)\}_{t\geqslant 0}$, i.e. implies that $\omega_0<0$. This leads to the problem whether $\omega_0=s(A)$ holds for C_0-semigroups. This relation is sometimes called the *spectrum determined growth condition.*

As observed in Section 8.1 (formula (8.4)), the spectral radius of $T(t)$ is given by $r(T(t))=e^{\omega_0 t}$. This implies that $\omega_0=s(A)$ holds for semigroups with bounded generator (i.e., for uniformly continuous semigroups). Indeed, if A is bounded, then the spectral mapping theorem yields that $\sigma(T(t))=e^{t\sigma(A)}$, which implies that $r(T(t))=e^{ts(A)}$, and hence $s(A)=\omega_0$. As illustrated in Example 8.1, such a spectral mapping theorem does not hold for unbounded generators. In fact, the semigroup in this example satisfies $s(A)=0$ and $\omega_0=1$, so $s(A)<\omega_0$ is possible, even in Hilbert space.

There are, however, some classes of semigroups for which $s(A)=\omega_0$ holds. Recall that $e^{t\sigma(A)}=\sigma(T(t))\setminus\{0\}$, $t\geqslant 0$, holds for any C_0-semigroup which is eventually uniformly continuous (as already observed in Proposition 8.5). By the same argument as above we obtain, therefore, the following result.

PROPOSITION 9.2. *If A is the generator of an eventually uniformly continuous C_0-semigroup, then $s(A)=\omega_0$. In particular, $s(A)=\omega_0$ holds for analytic semigroups and for eventually compact semigroups.*

Now we shall consider the problem whether $s(A)=\omega_0$ holds for positive C_0-semigroups in Banach lattices. It is shown in Theorem 7.4 that for a positive C_0-semigroup $\{T(t)\}_{t\geqslant 0}$ with generator A, the equality

$$s(A)=\sup\{\lambda\in\mathbb{R}:\lambda\in\sigma(A)\}$$

holds, and that the resolvent is given by

$$R(\lambda,A)x=\int_0^\infty e^{-\lambda t}T(t)x\,dt,$$

for all $x \in X$ and all $\lambda \in \mathbb{C}$ with $\mathrm{Re}\lambda > s(A)$ (where the integral exists as a norm convergent improper integral). The next example shows that $s(A) < \omega_0$ may occur for positive C_0-semigroups in Banach lattices.

EXAMPLE 9.3. Consider $\mathbb{R}^+$ with Lebesque measure m and define $L_\rho = L_p(\mathbb{R}^+, dm) \cap L_1(\mathbb{R}^+, e^x dm)$, $1 < p < \infty$, which is a Banach function space with respect to the function norm

$$\rho(f) = \Big(\int_0^\infty |f(x)|^p \, dm(x)\Big)^{1/p} + \int_0^\infty |f(x)| e^x \, dm(x).$$

Note that ρ is order continuous and has the Fatou property, but that L_ρ is not reflexive. For $t \geqslant 0$ define the operators $T(t)$ in L_ρ by $T(t)f(x) = f(x+t)$. It is easy to see that $\{T(t)\}_{t \geqslant 0}$ is a C_0-semigroup in L_ρ and $\|T(t)\| \leqslant 1$ for all $t \geqslant 0$. Furthermore, using that $p > 1$, it is not difficult to show that $\|T(t)\| = 1$ for all $t \geqslant 0$, and so $\omega_0 = 0$ (note that the translation semigroup in $L_1(\mathbb{R}^+, e^x dm)$ has type equal to -1). Denote the generator of $\{T(t)\}_{t \geqslant 0}$ by A. Then $\mathfrak{D}(A)$ consists of all (locally) absolutely continuous functions f in L_ρ with $f' \in L_\rho$ and $Af = f'$ for all $f \in \mathfrak{D}(A)$. Evidently, $\lambda \in \sigma_p(A)$ whenever $\mathrm{Re}\lambda < -1$, and so $s(A) \leqslant -1$. We assert that $s(A) = -1$. To this end it is sufficient to show that for $\mathrm{Re}\lambda > -1$ the equation $\lambda f - f' = g$ has a unique solution $f \in L_\rho$ for every $g \in L_\rho$. A solution of this equation is given by

$$f(x) = \int_x^\infty e^{\lambda(x-s)} g(s) ds, \quad x \geqslant 0$$

(since $g \in L_1(\mathbb{R}^+, e^x dx)$, the integral converges for all $\mathrm{Re}\lambda > -1$), and a simple calculation shows that $f \in L_\rho$. The solution is unique since the function $e^{\lambda x}$ does not belong to L_ρ for $\mathrm{Re}\lambda > -1$. We may conclude, therefore, that $s(A) = -1$, whereas $\omega_0 = 0$.

It is possible to modify the above example in such a way that we get a positive C_0-semigroup $\{T(t)\}_{t \geqslant 0}$ in a reflexive Banach lattice with $s(A) = -\infty$ and $\omega_0 = 0$. In fact, let $\{T(t)\}_{t \geqslant 0}$ be the translation semigroup in the Banach function space

$$L_\rho = L_q(\mathbb{R}^+, dm) \cap L_p(\mathbb{R}^+, e^{px^2} dm)$$

($1 < p < q < \infty$), where the function norm ρ is defined in the same way as above.

It is a consequence of Theorem 7.29 that, for a positive C_0-semigroup in an ordered Banach space X with normal cone X^+ and $\mathrm{int}\, X^+ \neq \varnothing$, the equality $\omega_0 = s(A)$ holds. This result applies of course, in particular, to the situation that X is the Banach lattice $C(\Omega)$ for some compact Hausdorff space Ω. In order to study some other special situations in which $s(A) = \omega_0$ holds, the following result will be useful.

PROPOSITION 9.4. *Let $\{T(t)\}_{t\geq 0}$ be a C_0-semigroup in the Banach space X with generator A. If there exists $1\leq p<\infty$ such that*

$$\int_0^\infty \|T(t)x\|^p dt<\infty \text{ for all } x\in X,$$

then $\{T(t)\}_{t\geq 0}$ is uniformly stable (i.e., $\omega_0<0$).

PROOF. First we make the following observation. Suppose that F is a positive continuous real function on $\mathbb{R}^+$ for which there exists a constant $M\geq 0$ such that $F(t+s)\leq MF(t)$ for all $0\leq s\leq 1$. Then $\int_0^\infty F(t)dt<\infty$ implies that $F(t)\to 0$ as $t\to\infty$. Indeed, if $\lim_{t\to\infty} F(t)=0$ does not hold, then there exist $\delta>0$ and $t_1<t_2<...\uparrow\infty$ such that $F(t_j)\geq\delta$ for all j. We may assume that $t_{j+1}-t_j>1$ for $j=1,2,....$ It follows from the hypothesis on F that $F(t_j)\leq MF(t)$ for all $t\in[t_j-1,t_j]$, and hence $F(t)\geq M^{-1}\delta>0$ for all $t\in[t_j-1,t_j]$, and for all $j=1,2,....$ This clearly contradicts the assumption that $\int_0^\infty F(t)dt<\infty$.
For the proof of the proposition, fix $x\in X$, and consider the function $F(t)=\|T(t)x\|^p$, $t\geq 0$. There exists a constant $C>0$ such that $\|T(t)\|\leq C$ for all $0\leq t\leq 1$, and hence $F(t+s)\leq C^p F(t)$ for all $0\leq s\leq 1$. Now it follows from the above observation that $F(t)\to 0$ as $t\to\infty$. We thus have shown that $\|T(t)x\|\to 0$ as $t\to\infty$ for all $x\in X$, which implies in particular that $\|T(t)\|\leq K$ for all $t\geq 0$ and some constant K. Furthermore, if $t>0$ and $x\in X$, then

$$t\|T(t)x\|^p = \int_0^t \|T(t)x\|^p ds\leq\int_0^t \|T(t-s)\|^p\|T(s)x\|^p ds$$

$$\leq K^p\int_0^\infty \|T(s)x\|^p ds.$$

This shows that for any $x\in X$ there exists a constant M_x such that $\|t^{1/p}T(t)x\|\leq M_x$ for all $t\geq 0$, and hence $t^{1/p}\|T(t)\|\leq M$ for all $t\geq 0$ and some constant M. Therefore, $\|T(t)\|\to 0$ as $t\to\infty$, and we may conclude that $\omega_0<0$.
□

As a first application of the above proposition we prove the following result.

THEOREM 9.5. *If A is the generator of a positive C_0-semigroup $\{T(t)\}_{t\geq 0}$ in the Banach lattice $L_1(\Omega,\mu)$, then $s(A)=\omega_0$.*

PROOF. A moment's reflection shows that it is sufficient to prove that $s(A)<0$ implies that $\omega_0<0$. By Theorem 7.4, if $s(A)<0$, then

$$\int_0^\infty T(t)f dt$$

exists for all $f\in L_1(\Omega,\mu)$ as a norm convergent improper integral. Take any $0\leq f\in L_1(\Omega,\mu)$. Since the L_1-norm is additive on positive elements, it follows that

$$\int_0^R \|T(t)f\|_1 dt = \|\int_0^R T(t)f dt\|_1$$

for all $R \geq 0$, and hence

$$\int_0^\infty \|T(t)f\|_1 dt = \|\int_0^\infty T(t)f dt\|_1 < \infty.$$

Therefore $\int_0^\infty \|T(t)f\|_1 dt < \infty$ for all $f \in L_1(\Omega,\mu)$, and in view of Proposition 9.4 we conclude that $\omega_0 < 0$. □

The next case we will discuss is $X = L_2(\Omega,\mu)$. We start with some remarks and introduce some notation. First suppose that $\{T(t)\}_{t\geq 0}$ is a C_0-semigroup in the Banach space X, with generator A. For any $z \in \mathbb{C}$ with $\operatorname{Re} z > \omega_0$ we have

$$R(z,A)x = \int_0^\infty e^{-zt} T(t)x dt, \ x \in X$$

(absolutely convergent integral). Since for any $\omega > \omega_0$ there exists a constant $M = M_\omega$ such that $\|T(t)\| \leq Me^{\omega t}$ for all $t \geq 0$, it follows that for every $\alpha > \omega_0$ there exists a constant C_α such that

$$\|R(\alpha + i\beta, A)\| \leq C_\alpha \text{ for all } \beta \in \mathbb{R},$$

i.e., the family $\{R(\alpha + i\beta, A) : \beta \in \mathbb{R}\}$ is uniformly bounded for each $\alpha > \omega_0$.

Now suppose that X is a Banach lattice and that $\{T(t)\}_{t\geq 0}$ is a positive semigroup. For $z \in \mathbb{C}$ with $\operatorname{Re} z > s(A)$ we have

$$R(z,A)x = \int_0^\infty e^{-zt} T(t)x dt,$$

(improper integral) for all $x \in X$, and so, writing $z = \alpha + i\beta$ with $\alpha, \beta \in \mathbb{R}$, we find that $|R(\alpha + i\beta, A)x| \leq R(\alpha, A)|x|$ for all $x \in X$ (see Corollary 7.5). Therefore,

$$\|R(\alpha + i\beta, A)\| \leq \|R(\alpha, A)\| \text{ for all } \beta \in \mathbb{R}$$

so the family $\{R(\alpha + i\beta, A) : \beta \in \mathbb{R}\}$ is uniformly bounded for every $\alpha > s(A)$. We note already that, as a consequence of the next theorem, combined with Example 8.1, the latter assertion is in general false for C_0-semigroups which are not positive.

In preparation of the next theorem we recall some simple facts concerning the vector valued Fourier transform. For any Banach space X the space of X-valued Bochner-integrable functions on $\mathbb{R}$ (with Lebesque measure), is denoted by $L_1(\mathbb{R}, X)$. For any $F \in L_1(\mathbb{R}, X)$ the Fourier transform, is, as usual defined by

$$\hat{F}(\tau) = \frac{1}{\sqrt{2\pi}} \int_{-\infty}^{+\infty} e^{-i\tau t} F(t) dt, \ \tau \in \mathbb{R},$$

which is a continuous function from $\mathbb{R}$ into X. If $\mathcal{H}$ is a Hilbert space and $F \in L_1(\mathbb{R},\mathcal{H}) \cap L_2(\mathbb{R},\mathcal{H})$, then $\hat{F} \in L_2(\mathbb{R},\mathcal{H})$ and $\|\hat{F}\|_2 = \|F\|_2$ (where $\|\cdot\|_2$ denotes the norm in the space $L_2(\mathbb{R},\mathcal{H})$). This fact can be obtained easily from the scalar-valued case by using orthonormal systems. Therefore the Fourier transform extends to an isometry from $L_2(\mathbb{R},\mathcal{H})$ onto itself. This assertion is false if $\mathcal{H}$ is replaced by an arbitrary Banach space.

THEOREM 9.6. *If $\{T(t)\}_{t\geqslant 0}$ is a C_0-semigroup in the Hilbert space $\mathcal{H}$, with generator A, then*

$$\omega_0 = \inf\{\alpha\in\mathbb{R} : \alpha + i\beta\in\rho(A) \text{ for all } \beta\in\mathbb{R}, \text{ and } \{R(\alpha + i\beta, A)\}_{\beta\in\mathbb{R}} \text{ is uniformly bounded}\}.$$

PROOF. Denoting the right-hand side infimum by s_0, it is clear from the above remarks that $s_0 \leqslant \omega_0$. Without loss of generality we may assume that $\omega_0 = 0$ (replace, if necessary, A by $A - \omega_0 I$). For the sake of convenience we denote $R(z) = R(z,A)$ for all $z \in \rho(A)$ and write $z = \alpha + i\beta$ with $\alpha, \beta \in \mathbb{R}$. By definition, for every $\alpha > s_0$ there exists a constant C_α such that $\|R(\alpha + i\beta)\| \leqslant C_\alpha$ for all β. For $z = \alpha + i\beta$ with $\alpha > s_0$ and $x \in \mathcal{H}$ we define the functions

$$F_x^z(t) = e^{-z|t|}T(|t|)x,\ t\in\mathbb{R}$$

$$G_x^z(\tau) = \frac{1}{\sqrt{2\pi}}\{R(z + i\tau)x + R(z - i\tau)x\},\ \tau\in\mathbb{R}.$$

If $\alpha > 0 = \omega_0$, then $F_x^z \in L_1(\mathbb{R},\mathcal{H})$, and $\hat{F}_x^z = G_x^z$. Note that $F_x^z \in L_2(\mathbb{R},\mathcal{H})$, so $G_x^z \in L_2(\mathbb{R},\mathcal{H})$ and $\|G_x^z\|_2 = \|F_x^z\|_2$ whenever $\alpha > 0$. In fact

$$\|F_x^z\|_2^2 = \int_{-\infty}^{\infty} e^{-2\alpha|t|}\|T(|t|)\|^2 dt \leqslant K_\alpha^2 \|x\|^2$$

for some constant $K_\alpha \geqslant 0$, depending on $\alpha > 0$ only. Now assume that $s_0 < \omega_0 = 0$, and let $x \in \mathcal{D}(A^2)$ be fixed for the moment. For $z \in \mathbb{C}$ with $\alpha > s_0$ and all $\tau \neq 0$ we get from the resolvent equation that

$$\begin{aligned}
&\sqrt{2\pi}\,G_x^z(\tau) = R(z + i\tau)x + R(z - i\tau)x = \\
&R(z + i\tau)R(z)(zI - A)x + R(z - i\tau)R(z)(zI - A)x = \\
&-\frac{1}{i\tau}[R(z+i\tau) - R(z)](zI - A)x + \frac{1}{i\tau}[R(z-i\tau) - R(z)](zI - A)x = \\
&-\frac{1}{i\tau}[R(z + i\tau) - R(z - i\tau)](zI - A)x,
\end{aligned}$$

and repeating this argument we find that

$$\sqrt{2\pi}\,G_x^z(\tau) = -\frac{1}{\tau^2}[R(z + i\tau) + R(z - i\tau) - 2R(z)](zI - A)^2 x.$$

Since $z = \alpha + i\beta$, $\alpha > s_0$ and $\tau \in \mathbb{R}$, it follows from the definition of s_0 that

$$\sqrt{2\pi}\,\|G_x^z(\tau)\| \leqslant \frac{2}{\tau^2}(C_\alpha + \|R(z)\|)\|(zI - A)^2 x\|.$$

This estimate, together with the continuity of G_x^z, shows that $G_x^z \in L_1(\mathbb{R},\mathcal{H}) \cap L_2(\mathbb{R},\mathcal{H})$. Therefore we can consider the inverse Fourier transform $\tilde{G}_x^z$ of G_x^z, which is given by

$$\tilde{G}_x^z(t) = \frac{1}{2\pi}\int_{-\infty}^{+\infty} e^{i\tau t}\{R(z+i\tau)x + R(z-i\tau)x\}d\tau.$$

As observed before, if $\alpha > 0$, then $G_x^z = \hat{F}_x^z$, and hence $\tilde{G}_x^z = F_x^z$ whenever $\alpha > 0$. We claim that $\tilde{G}_x^z = F_x^z$ holds for all $z = \alpha + i\beta$ with $\alpha > s_0$. Indeed, fix $t \in \mathbb{R}$ momentarily. Evidently, $F_x^z(t)$ is an analytic function of z. We will show now that the function $z \to \tilde{G}_x^z(t)$ is analytic on the half-plane $\mathrm{Re}\, z = \alpha > s_0$. To this end, let Γ be a closed contour in $\{z \in \mathbb{C} : \mathrm{Re}\, z > s_0\}$. Since the function $z \to G_x^z(\tau)$ is analytic, it follows by Fubini's theorem that

$$\oint_\Gamma \tilde{G}_x^z(t)dz = \frac{1}{\sqrt{2\pi}}\oint_\Gamma\{\int_{-\infty}^{+\infty} e^{i\tau t}G_x^z(\tau)d\tau\}dz =$$

$$\frac{1}{\sqrt{2\pi}}\int_{-\infty}^{+\infty} e^{i\tau t}\{\oint_\Gamma G_x^z(\tau)dz\}d\tau = 0.$$

Therefore, by Morera's theorem, $\tilde{G}_x^z(t)$ is an analytic function of z in $\{z \in \mathbb{C} : \mathrm{Re}\, z > s_0\}$. By the uniqueness of analytic extensions we may conclude that $\tilde{G}_x^z(s) = F_x^z(s)$ for all $\mathrm{Re}\, z = \alpha > s_0$.

We thus have shown that $\tilde{G}_x^z = F_x^z$ for all $\alpha > s_0$ and all $x \in \mathcal{D}(A^2)$. This implies in particular that $\|F_x^z\|_2 = \|G_x^z\|_2$ for all $x \in \mathcal{D}(A^2)$ and all $\alpha > s_0$. Now take $\alpha \in \mathbb{R}$ such that $s_0 < \alpha < 0$ and $2|\alpha|\|R(\alpha)\| < 1$. Using that

$$R(\alpha + i\tau) = [I - 2\alpha R(\alpha + i\tau)]R(-\alpha + i\tau),$$

$$R(\alpha - i\tau) = [I - 2\alpha R(\alpha - i\tau)]R(-\alpha - i\tau),$$

a simple calculation shows that

$$[I - 2\alpha R(\alpha - i\tau)]\cdot[I - 2\alpha R(\alpha + i\tau)]G_x^{-\alpha}(\tau) =$$

$$[I - 2\alpha R(\alpha)]G_x^\alpha(\tau).$$

By the choice of α, $[I - 2\alpha R(\alpha)]^{-1}$ exists in $\mathcal{L}(\mathcal{H})$, so

$$G_x^\alpha(\tau) = [I - 2\alpha R(\alpha)]^{-1}[I - 2\alpha R(\alpha - i\tau)]\cdot[I - 2\alpha R(\alpha + i\tau)]G_x^{-\alpha}(\tau).$$

Since $\|R(\alpha \pm i\tau)\| \leqslant C_\alpha$ for all $\tau \in \mathbb{R}$, there exists a constant M_α (independent of x and τ) such that $\|G_x^\alpha(\tau)\| \leqslant M_\alpha \|G_x^{-\alpha}(\tau)\|$ for all $\tau \in \mathbb{R}$, and hence

$$\|G_x^\alpha\|_2 \leqslant M_\alpha \|G_x^{-\alpha}\|_2$$

for all $x \in \mathcal{D}(A^2)$. Combining the above results we obtain

$$\|F_x^\alpha\|_2 = \|G_x^\alpha\|_2 \leqslant M_\alpha \|G_x^{-\alpha}\|_2 = M_\alpha \|F_x^{-\alpha}\|_2 \leqslant M_\alpha K_{-\alpha}\|x\|,$$

for all $x \in \mathcal{D}(A^2)$, i.e.,

$$\int_{-\infty}^{+\infty} e^{-2\alpha|t|}\|T(|t|)x\|^2 dt \leqslant M_\alpha^2 K_{-\alpha}^2 \|x\|^2, \; x \in \mathfrak{D}(A^2).$$

Since $\mathfrak{D}(A^2)$ is dense in $\mathcal{H}$, a simple argument shows that

$$\int_{-\infty}^{\infty} e^{-2\alpha|t|}\|T(|t|)x\|^2 dt < \infty,$$

for all $x \in \mathcal{H}$, and hence

$$\int_0^{\infty} \|e^{-\alpha t} T(t)x\|^2 dt < \infty, \; x \in \mathcal{H}.$$

An application of Proposition 9.4 now shows that the semigroup $\{e^{-\alpha t}T(t)\}_{t\geqslant 0}$ is strongly stable, i.e., $\omega_0 - \alpha < 0$. This implies that $-\alpha < 0$, which contradicts the fact that $s_0 < \alpha < 0$. We may conclude, therefore that $s_0 = 0 = \omega_0$, and the theorem is completely proved. □

In combination with some of the remarks preceding Theorem 9.6, the above theorem yields the following result.

THEOREM 9.7. *If $\{T(t)\}_{t\geqslant 0}$ is a positive C_0-semigroup with generator A in the Banach lattice $\mathcal{H} = L_2(\Omega, \mu)$, then $\omega_0 = s(A)$.*

In the last part of the present section we will characterize the spectral bound $s(A)$ of the generator A of a positive semigroup $\{T(t)\}_{t\geqslant 0}$ in terms of the exponential growth bounds of the solutions of the Cauchy problem

$$u'(t) = Au(t), \; u(0) = x \in \mathfrak{D}(A),$$

i.e., in terms of the exponential growth bounds of the orbits $\{T(t)x\}_{t\geqslant 0}$ with $x \in \mathfrak{D}(A)$.

We recall some of the relevant facts. Let X be a (complex) Banach space and suppose that $\{T(t)\}_{t\geqslant 0}$ is a C_0-semigroup in X with generator A. For every $x \in X$ the *exponential growth bound* $\omega(x)$ of $\{T(t)x\}_{t\geqslant 0}$ is defined by

$$\omega(x) = \limsup_{t\to\infty} \frac{\log\|T(t)x\|}{t}.$$

It is not difficult to show that

$$\omega(x) = \inf\{\omega \in \mathbb{R} : \exists \text{ constant } M_\omega \text{ such that } \|T(t)x\| \leqslant M_\omega e^{\omega t}, \; t \geqslant 0\},$$

and that $\omega(x)$ is the abscissa of absolute convergence of the Laplace transform of the function $T(t)x$, i.e.,

$$\omega(x) = \inf\{\lambda \in \mathbb{R} : \int_0^{\infty} e^{-\lambda t}\|T(t)x\| dt < \infty\}$$

(cf. the first part of the proof of Proposition 9.4). It should be noted that the type (exponential growth bound) ω_0 of the semigroup $\{T(t)\}_{t\geqslant 0}$ is given by

$$\omega_0 = \sup\{\omega(x) : x \in X\}.$$

Besides the type ω_0 we consider the *exponential growth bound of the solutions of the Cauchy problem,* which is defined by

$$\omega_1 = \sup\{\omega(x) : x \in \mathfrak{D}(A)\},$$

and it is easily seen that

$$\omega_1 = \inf\{\lambda \in \mathbb{R} : \int_0^\infty e^{-\lambda t}\|T(t)x\|dt < \infty \text{ for all } x \in \mathfrak{D}(A)\}.$$

In preparation of the following proposition, we recall that for any $\lambda \in \mathbb{C}$ with $\mathrm{Re}\lambda > \omega_0$ the resolvent $R(\lambda, A)$ of A is given by the absolutely convergent Laplace transform

$$R(\lambda, A)x = \int_0^\infty e^{-\lambda t}T(t)x dt, \quad x \in X.$$

However, as we have noted before, it is possible that for $\lambda \in \mathbb{C}$ with $\mathrm{Re}\lambda \leqslant \omega_0$ the integral

$$\int_0^\infty e^{-\lambda t}T(t)x dt$$

exists for all $x \in X$ as an improper norm convergent integral. Such λ belong to $\rho(A)$ and $R(\lambda, A)$ is given by the above integral. Note that if $\int_0^\infty e^{-\lambda t}T(t)x dt$ converges for some $\lambda \in \mathbb{C}$, then $\int_0^\infty e^{-\mu t}T(t)x dt$ converges for all $\mu \in \mathbb{C}$ with $\mathrm{Re}\mu > \mathrm{Re}\lambda$.

PROPOSITION 9.8. *If $\{T(t)\}_{t\geqslant 0}$ is a C_0-semigroup in the Banach space X, then*

$$\omega_1 = \inf\{\lambda \in \mathbb{R} : \int_0^\infty e^{-\lambda t}T(t)x dt \text{ exists for all } x \in X\}.$$

PROOF. Denote the infimum on the right hand side by s_1, and take any $\lambda > s_1$. By formula (8.1b),

$$(e^{-\lambda t}I - T(t))x = \int_0^t e^{\lambda(t-s)}T(s)(\lambda I - A)x ds,$$

for all $x \in \mathfrak{D}(A)$, and so

$$T(t)x = e^{\lambda t}\{x - \int_0^t e^{-\lambda s}T(s)(\lambda I - A)x ds\},$$

for $x \in \mathfrak{D}(A)$. Since $\lambda > s_1$, for any $x \in \mathfrak{D}(A)$ there exists a constant M_x such that

$$\|\int_0^t e^{-\lambda s}T(s)(\lambda I - A)x ds\| \leqslant M_x,$$

for all $t \geq 0$, and hence $\|T(t)x\| \leq e^{\lambda t}(\|x\| + M_x)$ for all $t \geq 0$. This shows that $\lambda > \omega_1$, and consequently $s_1 \geq \omega_1$.

For the converse inequality, take $\lambda > \omega_1$, and put

$$S(t)x = \int_0^t e^{-\lambda s} T(s)x ds,$$

for all $x \in X$ and $t \geq 0$. In view of $\lambda > \omega_1$, it is clear that $\lim_{t\to\infty} S(t)x = \int_0^\infty e^{-\lambda s} T(s)xds$ exists for all $x \in \mathfrak{D}(A)$. Therefore, in order to show that $\lim_{t\to\infty} S(t)x$ exists for all $x \in X$, it is sufficient to prove that $\{S(t)\}_{t\geq 0}$ is uniformly bounded, as $\mathfrak{D}(A)$ is dense in X. For this purpose let $x \in X$ be fixed momentarily and set $z = S(1)x$. It follows easily that $z \in \mathfrak{D}(A)$ and

$$S(n)x = \sum_{k=0}^{n-1} e^{-\lambda k} T(k)z.$$

Using once again that $\lambda > \omega_1$ it is easy to verify that the series $\Sigma_{k=0}^{\infty} e^{-\lambda k} T(k)z$ is convergent. We thus have shown that the sequence $\{S(n)x : n = 1,2,...\}$ is convergent for all $x \in X$, and hence the linear operators $\{S(n)\}_{n=1}^{\infty}$ are uniformly bounded. Finally, since

$$S(t) = S(t-n) + e^{-\lambda(t-n)} T(t-n) S(n)$$

whenever $n \leq t < n+1$, we may conclude that $\{S(t)\}_{t \geq 0}$ are uniformly bounded, which shows that

$$\lim_{t\to\infty} S(t)x = \int_0^\infty e^{-\lambda s} T(s)xds$$

exists for all $x \in X$. Therefore $\lambda > s_1$ and consequently $s_1 \leq \omega_1$. By this the proof of the proposition is complete. □

The following result is now an immediate consequence of Theorem 7.4 and the above proposition.

COROLLARY 9.9. *If A is the generator of the positive C_0-semigroup $\{T(t)\}_{t\geq 0}$ in a Banach lattice, then*

$$s(A) = \omega_1 = \sup\{\omega(x) : x \in \mathfrak{D}(A)\}.$$

2. Asymptotic behaviour of positive semigroups

Throughout this section we assume that X is a Banach lattice and that $\{T(t)\}_{t\geq 0}$ is a strongly continuous semigroup on X with infinitesimal generator A. In Chapter 8 we presented a rather detailed analysis of the peripheral spectrum $\sigma_+(A)$ of A under the additional assumption that $\{T(t)\}_{t\geq 0}$ is a positive or even irreducible semigroup. If, moreover, the inequality $\omega_{ess}(A) < \omega_0(A)$ holds, then we can say a lot more. Note already that, by Proposition 8.6, $\omega_{ess}(A) < \omega_0(A)$ implies that $\omega_0(A) = s(A)$.

THEOREM 9.10. *Let* $\{T(t)\}_{t\geqslant 0}$ *be a positive* C_0*-semigroup for which the inequality* $\omega_{ess}(A)<\omega_0(A)$ *holds. Then*

a) $\sigma_+(A)=\{s(A)\}$, *and there exists an* $\epsilon>0$ *such that*

 i) $\omega_{ess}(A)\leqslant s(A)-\epsilon$

 ii) $\mathrm{Re}\lambda\leqslant s(A)-\epsilon$, *for every* $\lambda\in\sigma(A),\lambda\neq s(A)$.

b) *If, in addition,* $\{T(t)\}_{t\geqslant 0}$ *is irreducible, then* $s(A)$ *is a first-order pole of* $R(\lambda,A)$ *with geometric multiplicity one.*

PROOF. a) From Proposition 8.6a we conclude that $s(A)$ is a pole of the resolvent $R(\lambda,A)$. We may assume without loss of generality that $s(A)=0$. Suppose that $i\nu\in\sigma_+(A)$, for some $\nu\in\mathbb{R}\setminus\{0\}$. Then, by Theorem 8.14, $ik\nu\in\sigma_+(A)$ for every $k\in\mathbb{Z}$. Let $t>0$ be such that $\nu t/2\pi$ is irrational. From Proposition 8.3 we easily deduce that $e^{ik\nu t}\in\sigma(T(t))$ for every $k\in\mathbb{Z}$, and therefore we have that $\{z\in\mathbb{C}:|z|=1\}\subseteq\sigma(T(t))$, because the spectrum of $T(t)$ is a closed set. This, however, implies that $1\leqslant r_{ess}(T(t))=e^{\omega_{ess}(A)t}$, hence $0\leqslant\omega_{ess}(A)=s(A)=\omega_0(A)$. But this is a contradiction. Therefore $\sigma_+(A)=\{s(A)\}$.

Now suppose that $\lambda_k\in\sigma(A)$, with $\mathrm{Re}\lambda_k$ strictly increasing and $\lim_{k\to\infty}\mathrm{Re}\lambda_k=s(A)=0$. Let $t>0$ be fixed. Then $e^{\lambda_k t}\in\sigma(T(t))$, $|e^{\lambda_k t}|<|e^{\lambda_{k+1}t}|$ and $|e^{\lambda_k t}|\to 1$. Therefore $\sigma(T(t))$ has a limit point on $\{z\in\mathbb{C}:|z|=1\}$, yielding that $\omega_{ess}(A)\geqslant 0$. As we have seen, this is a contradiction and the result follows. b) follows from Theorem 8.17. □

Suppose that $\{T(t)\}_{t\geqslant 0}$ is a positive semigroup with $\omega_{ess}(A)<\omega_0(A)$. Then $s(A)=\omega_0(A)$ by Proposition 8.6b. Let $\lambda_0:=s(A)$ and let $\epsilon\geqslant 0$ be as in Theorem 9.10a. Since $\lambda_0\in\sigma(A)\setminus\sigma_{ess}(A)$, we have that λ_0 is a pole of $R(\lambda,A)$ and we can decompose the space X in the following way (Theorem A.3.1)

$$X=\mathcal{N}((\lambda_0-A)^p)\oplus\mathcal{R}((\lambda_0-A)^p), \tag{9.1}$$

where p is the order of the pole. Both subspaces in (9.1) are invariant under the action of $T(t)$. Let P be the (finite-rank) projection on $\mathcal{N}((\lambda_0-A)^p)$ corresponding to this decomposition, i.e. $P=B_{-1}$ where B_{-1} is given by (A.3.4). Let

$$\tilde{T}(t)=T(t)(I-P),\ t\geqslant 0.$$

Clearly, $\tilde{T}(t)$ is just the restriction of $T(t)$ to the invariant subspace $\mathcal{R}((\lambda_0-A)^p)$, and $\{\tilde{T}(t)\}_{t\geqslant 0}$ defines a strongly continuous semigroup on this space. Its infinitesimal generator $\tilde{A}$ is the restriction of A to $\mathcal{R}((\lambda_0-A)^p)$. The spectrum of $\tilde{A}$ is given by

$$\sigma(\tilde{A})=\sigma(A)\setminus\{\lambda_0\}.$$

As a consequence

$$s(\tilde{A})\leqslant\lambda_0-\epsilon.$$

Obviously

$$\omega_{ess}(\tilde{A})\leqslant\omega_{ess}(A)\leqslant\lambda_0-\epsilon,$$

and now Proposition 8.6b yields

$$\omega_0(\tilde{A}) \leqslant \lambda_0 - \epsilon.$$

For every $\eta \in (0,\epsilon)$ there exists therefore a constant $\tilde{M}(\eta) \geqslant 1$ such that

$$\|\tilde{T}(t)\tilde{x}\| \leqslant \tilde{M}(\eta) e^{(\lambda_0 - \eta)t} \|\tilde{x}\|,$$

for every $\tilde{x} \in \mathcal{R}((\lambda_0 - A)^p)$. This implies that for every $x \in X$,

$$\|T(t)(I-P)x\| = \|\tilde{T}(t)(I-P)x\|$$
$$\leqslant \tilde{M}(\eta) e^{(\lambda_0 - \eta)t} \|(I-P)x\| \leqslant M(\eta) e^{(\lambda_0 - \eta)t} \|x\|, \; t \geqslant 0.$$

Here $M(\eta) = \tilde{M}(\eta)\|I-P\|$.

If we know in addition that λ_0 is a simple pole, then $p=1$ and

$$AP = PA = \lambda_0 P,$$

hence

$$T(t)P = PT(t) = e^{\lambda_0 t} P, \; t \geqslant 0.$$

It follows that for every $x \in X$ and $t \geqslant 0$,

$$T(t)x = T(t)(Px + (I-P)x) = e^{\lambda_0 t} Px + T(t)(I-P)x,$$

and therefore

$$\|e^{-\lambda_0 t} T(t)x - Px\| \leqslant M(\eta) e^{-\eta t} \|x\|, \; t \geqslant 0, \tag{9.2}$$

for every $\eta \in (0,\epsilon)$. In particular, if the semigroup $\{T(t)\}_{t \geqslant 0}$ is irreducible, then λ_0 is a simple pole of $R(\lambda,A)$. Moreover, its geometric multiplicity is one. From Theorem 8.17 we know that there exists a quasi-interior point $x_0 \in X^+$ and a strictly positive functional $x_0^* \in (X^*)^+$ such that

$$A_0 x_0 = \lambda_0 x_0, \; A^* x_0^* = \lambda_0 x_0^*, \; \langle x_0, x_0^* \rangle = 1. \tag{9.3}$$

The projection P is now an operator of rank one, given by

$$Px = \langle x, x_0^* \rangle x_0.$$

We summarize our findings in the following theorem.

THEOREM 9.11. *Let $\{T(t)\}_{t \geqslant 0}$ be a positive C_0-semigroup satisfying the inequality $\omega_{ess}(A) < \omega_0(A)$. Let p be the order of the pole $\lambda_0 = s(A)$ of $R(\lambda,A)$, and let P be the projection on $\mathcal{N}((\lambda_0 - A)^p)$ along $\mathcal{R}((\lambda_0 - A)^p)$. Finally, let $\epsilon > 0$ be as in Theorem 9.10.*

a) *For every $\eta \in (0,\epsilon)$ there is a constant $M(\eta) \geqslant 1$ such that*

$$\|T(t)(I-P)\| \leqslant M(\eta) e^{(\lambda_0 - \eta)t}, t \geqslant 0.$$

b) *If $p=1$, then for every $\eta \in (0,\epsilon)$,*

$$\|e^{-\lambda_0 t} T(t) - P\| \leqslant M(\eta) e^{-\eta t}, \; t \geqslant 0.$$

c) *If $\{T(t)\}_{t \geqslant 0}$ is irreducible and x_0, x_0^* are given by (9.3) then for every*

$\eta \in (0,\epsilon)$ *and* $x \in X$,

$$\|e^{-\lambda_0 t}T(t)x - <x,x_0^*>x_0\| \leqslant M(\eta)e^{-\eta t}\|x\|, \; t \geqslant 0.$$

3. COMPACTNESS AND IRREDUCIBILITY OF PERTURBED SEMIGROUPS

In this section we investigate compactness and irreducibility properties of perturbed semigroups (with "perturbed" in the sense of Section 3.5). The results that we get can sometimes be used to prove correctness of the hypotheses of Theorem 9.11, in particular the inequality $\omega_{ess}(A) < \omega_0(A)$, and the irreducibility of the semigroup. Our situation is the situation described in Section 3.5. However, we point out immediately that all results trivially carry over to the case that the perturbation is a bounded linear operator from X into X.

We briefly recall the situation of Section 3.5. Let $(X, \|\cdot\|)$ be a Banach space, $\{T_0(t)\}_{t \geqslant 0}$ a C_0-semigroup with generator A_0, and assume that X is $\odot$-reflexive with respect to A_0. Let $B: X \to X^{\odot *}$ be a bounded linear operator, and let $\{T(t)\}_{t \geqslant 0}$ be the semigroup which is obtained from the variation-of-constants formula:

$$T(t)x = T_0(t)x + \int_0^t T_0^{\odot *}(t-s)BT(s)x ds \, , \; t \geqslant 0 \, , \; x \in X. \tag{9.4}$$

We refer to Section 3.5 for an interpretation of this formula. $\{T_0(t)\}_{t \geqslant 0}$ is called the unperturbed semigroup, B the perturbation, and $\{T(t)\}_{t \geqslant 0}$ the perturbed semigroup. By the method of successive approximations, one finds the following norm convergent series expansion:

$$T(t) = \sum_{k=0}^{\infty} T_k(t), \quad t \geqslant 0, \tag{9.5}$$

where

$$T_k(t)x = \int_0^t T_0^{\odot *}(t-s)BT_{k-1}(s)x ds, \tag{9.6}$$

for every $t \geqslant 0$, $k \geqslant 1$, and $x \in X$. We let

$$U(t) = T(t) - T_0(t), \quad t \geqslant 0,$$

or equivalently,

$$U(t)x = \int_0^t T_0^{\odot *}(t-s)BT(s)x ds \, , \; t \geqslant 0 \, , \; x \in X. \tag{9.7}$$

Laplace transformation of (9.4) yields

$$R(\lambda,A) = R(\lambda,A_0) + R(\lambda,A_0^{\odot *})BR(\lambda,A), \tag{9.8}$$

for every $\lambda \in \rho(A_0) \cap \rho(A)$. For $\mathrm{Re}\lambda$ large enough, to be precise, if $\|R(\lambda,A_0)\| < \|B\|^{-1}$, then

$$R(\lambda,A) = [I - R(\lambda,A_0^{\odot *})B]^{-1}R(\lambda,A_0),$$

or equivalently

$$R(\lambda,A) = \sum_{k=0}^{\infty}[R(\lambda,A_0^{\odot *})B]^k R(\lambda,A_0). \tag{9.9}$$

As a matter of fact, this explicit expression for $R(\lambda,A)$ can be used to show that A obeys the Hille-Yosida conditions, yielding that A is again the generator of a strongly continuous semigroup. The mapping

$$\lambda \to R(\lambda,A_0^{\odot *})B$$

defines an analytic function in the half-plane $\mathrm{Re}\lambda > \omega_0(A_0)$, with values in $\mathcal{L}(X)$.

The following result is an immediate consequence of the series expansion (9.8) and the fact that the (weakly) compact operators on X form a closed, two-sided ideal in $\mathcal{L}(X)$.

PROPOSITION 9.12. *The resolvent operator $R(\lambda,A_0)$ is (weakly) compact for every $\lambda \in \rho(A_0)$ if and only if $R(\lambda,A)$ is (weakly) compact for every $\lambda \in \rho(A)$.*

PROPOSITION 9.13. *If for every λ with $\mathrm{Re}\lambda > s(A_0)$, $R(\lambda,A_0^{\odot *})B$ is a compact operator, then $\{\lambda \in \sigma(A) : \mathrm{Re}\lambda > s(A_0)\}$ contains at most countably many isolated points λ_k and each λ_k lies outside the essential spectrum of A, or equivalently, $\sigma_{ess}(A) \subseteq \{\lambda : \mathrm{Re}\lambda \leqslant s(A_0)\}$.*

PROOF. It suffices to show that $\sigma_{ess}(A) \cap \{\lambda : \mathrm{Re}\lambda > s(A_0)\} = \varnothing$. Suppose there exists an element $\lambda \in \sigma_{ess}(A)$ with $\mathrm{Re}\lambda > s(A_0)$. Choose $\gamma \in \rho(A)$ such that $\mathrm{Im}\gamma = \mathrm{Im}\lambda$ and $\mathrm{Re}\gamma > s(A)$. From the spectral mapping theorem for the essential spectrum (Theorem A.3.6) we get that $(\gamma - \lambda)^{-1} \in \sigma_{ess}(R(\gamma,A))$. Therefore

$$r_{ess}(R(\gamma,A)) \geqslant \left|\frac{1}{\gamma-\lambda}\right| \geqslant \frac{1}{\mathrm{Re}\gamma - \mathrm{Re}\lambda} > \frac{1}{\mathrm{Re}\gamma - s(A_0)}.$$

On the other hand,

$$r_{ess}(R(\gamma,A)) = r_{ess}(R(\gamma,A_0)),$$

since $R(\gamma,A) - R(\gamma,A_0) = R(\gamma,A_0^{\odot *})BR(\gamma,A)$ is a compact operator. But

$$r_{ess}(R(\gamma,A_0)) \leqslant r(R(\gamma,A_0)) \leqslant \frac{1}{\mathrm{Re}\gamma - s(A_0)},$$

which is a contradiction. The desired result is immediate. □

For $\lambda \in \rho(A_0)$ we define the bounded linear operator $K(\lambda) : X \to X$ by

$$K(\lambda) = R(\lambda,A_0^{\odot *})B. \tag{9.10}$$

Below it will be made clear that the question under what conditions the operators $U(t)$ are eventually compact is an important one: see Proposition 9.20. In order to be able to give an answer to this question we need the following two lemma's.

LEMMA 9.14. *For every $\lambda \in \rho(A_0)$ and $x \in X$,*

$$R(\lambda, A_0)U(t)x = \int_0^t T_0(t-s)K(\lambda)T(s)x\,ds, \quad t \geq 0.$$

PROOF. Let $x \in X$, $x^{\odot} \in X^{\odot}$, $t \geq 0$ and $\lambda \in \rho(A_0)$. Then

$$< \int_0^t T_0(t-s)K(\lambda)T(s)x\,ds\,,\, x^{\odot} > =$$

$$\int_0^t < T_0(t-s)R(\lambda, A_0^{\odot *})BT(s)x\,,\, x^{\odot} > ds =$$

$$\int_0^t < R(\lambda, A_0^{\odot})T_0^{\odot}(t-s)x^{\odot}\,,\, BT(s)x > ds =$$

$$\int_0^t < T_0^{\odot}(t-s)R(\lambda, A_0^{\odot})x^{\odot}\,,\, BT(s)x > ds =$$

$$\int_0^t < R(\lambda, A_0^{\odot})x^{\odot}\,,\, T_0^{\odot *}(t-s)BT(s)x > ds =$$

$$< \int_0^t T_0^{\odot *}(t-s)BT(s)x\,ds\,,\, R(\lambda, A_0^{\odot})x^{\odot} > =$$

$$< R(\lambda, A_0^{\odot *})\int_0^t T_0^{\odot *}(t-s)BT(s)x\,ds\,,\, x^{\odot} >,$$

from which the result follows. □

Let $I \subseteq [0, \infty)$ and let $V(t)$ be a mapping from I into $\mathcal{L}(X)$. We say that V is *u-continuous* on I, if V is continuous on I relative to the uniform operator topology on $\mathcal{L}(X)$.

LEMMA 9.15. *If $U(t)$ is u-continuous on $[t_0, \infty)$, then*

$$U(t) = \lim_{\lambda \to \infty} \lambda[R(\lambda, A)T(t) - R(\lambda, A_0)T_0(t)],$$

for every $t \geq t_0$, where the limit holds with respect to the uniform operator topology.

PROOF. If $\lambda \in \mathbb{R}$ is large enough, then

$$R(\lambda,A)T(t)x - R(\lambda,A_0)T_0(t)x =$$

$$\int_0^\infty e^{-\lambda s}\{T(s)T(t)x - T_0(s)T_0(t)x\}ds =$$

$$\int_0^\infty e^{-\lambda s}U(t+s)xds\,,\ t \geqslant 0\,,\ x \in X.$$

Thus we get that

$$\lambda R(\lambda,A)T(t)x - \lambda R(\lambda,A_0)T_0(t)x - U(t)x =$$

$$\lambda \int_0^\infty e^{-\lambda s}\{U(t+s) - U(t)\}xds\,,\ t \geqslant 0\,,\ x \in X.$$

For every $t \geqslant 0$, $x \in X$, and $h > 0$:

$$\|\lambda R(\lambda,A)T(t)x - \lambda R(\lambda,A_0)T_0(t)x - U(t)x\| \leqslant$$

$$\int_0^h \lambda e^{-\lambda s}\|U(t+s) - U(t)\|\,\|x\|ds +$$

$$\int_h^\infty \lambda e^{-\lambda s}\|U(t+s)x - U(t)x\|ds \leqslant$$

$$\sup_{0 \leqslant s \leqslant h}\|U(t+s) - U(t)\|\,\|x\| + \frac{4M\lambda}{\lambda - \omega}e^{\bar{\omega}t}e^{-(\lambda-\bar{\omega})h}\|x\|.$$

Here we have used that

$$\|U(t+s)x - U(t)x\| \leqslant (\|T_0(t+s)\| + \|T(t+s)\| +$$
$$\|T_0(t)\| + \|T(t)\|)\|x\|,$$

and $\|T_0(t)\|$, $\|T(t)\| \leqslant Me^{\bar{\omega}t}$, $t \geqslant 0$, if $\|T_0(t)\| \leqslant Me^{\omega t}$, and $\bar{\omega} = \omega + M\|B\|$ (see Theorem 3.20). Thus

$$\|\lambda R(\lambda,A)T(t) - \lambda R(\lambda,A_0)T_0(t) - U(t)\| \leqslant$$

$$\sup_{0 \leqslant s \leqslant h}\|U(t+s) - U(t)\| + C \cdot e^{-\lambda h}\,,\ t \geqslant 0,$$

for every $h > 0$, and the result follows from the *u*-continuity of $U(t)$. □

Now we can prove the following theorem.

THEOREM 9.16. *Let $U(t)$ be u-continuous on $[t_0, \infty)$, and $K(\lambda)$ a compact operator, for $\lambda \in \mathbb{R}$ sufficiently large. Then $U(t)$ is compact for every $t \geqslant t_0$.*

PROOF. Because of the previous lemma, and the fact that the compact operators form a closed ideal of $\mathfrak{L}(X)$, is suffices to show that the operators

$$\lambda[R(\lambda,A)T(t) - R(\lambda,A_0)T_0(t)] =$$
$$\lambda[R(\lambda,A) - R(\lambda,A_0)]T(t) + \lambda R(\lambda,A_0)U(t),\ t \geqslant t_0,$$

are compact for $\lambda \in \mathbb{R}$ sufficiently large. Since

$$R(\lambda,A) - R(\lambda,A_0) = \sum_{n=1}^{\infty} K(\lambda)^n R(\lambda,A_0),$$

is a compact operator, we "only" have to show that

$$R(\lambda,A_0)U(t)$$

is compact if λ is sufficiently large. We use Lemma 9.14.

$$R(\lambda,A_0)U(t)x = \int_0^t T_0(t-s)K(\lambda)T(s)x\,ds.$$

We set $V_\lambda(t) = R(\lambda,A_0)U(t)$, $t \geqslant 0$. Let $\mathbb{B} = \{x \in X : \|x\| \leqslant 1\}$. We have to show that $V_\lambda(t)\mathbb{B}$ is a relatively compact subset of X, for every $t \geqslant 0$. Now let $t \geqslant 0$ be fixed. Suppose we can show that

$$\bigcup_{0 \leqslant s \leqslant t} T_0(t-s)K(\lambda)T(s)\mathbb{B}$$

is a relatively compact subset of X. For $x \in \mathbb{B}$

$$V_\lambda(t)x \in t \cdot \overline{\text{co}}\,\Big(\bigcup_{0 \leqslant s \leqslant t} T_0(t-s)K(\lambda)T(s)\mathbb{B}\Big),$$

where for a subset Y of X, $\overline{\text{co}}\,(Y)$ denotes its closed convex hull (see Appendix A.3). By Mazur's theorem,

$$\overline{\text{co}}\,\Big(\bigcup_{0 \leqslant s \leqslant t} T_0(t-s)K(\lambda)T(s)\mathbb{B}\Big)$$

is compact and the proof will be complete.
Now we prove that

$$\bigcup_{0 \leqslant s \leqslant t} T_0(t-s)K(\lambda)T(s)\mathbb{B}$$

is indeed a relatively compact subset of X. Let

$$V := K(\lambda)\Big(\bigcup_{0 \leqslant s \leqslant t} T(s)\mathbb{B}\Big),$$

then by the compactness of $K(\lambda)$ and the boundedness of $T(s)$ $(0 \leqslant s \leqslant t)$ we have that V is relatively compact. Obviously

$$\bigcup_{0 \leqslant s \leqslant t} T_0(t-s)K(\lambda)T(s)\mathbb{B} \subseteq \bigcup_{0 \leqslant s \leqslant t} T_0(s)V.$$

Below we shall prove that the restriction of $T_0(s)$ to V is u-continuous on $[0,t]$. From this fact, the result follows immediately.

Let $L > 0$ be such that $\|T_0(s)\| \leqslant L$, $s \in [0,t]$. Let $\epsilon > 0$. Since V is relatively compact, there exist $v_1, \ldots, v_m \in V$ such that $V \subseteq \bigcup_{i=1}^m B(v_i, \epsilon/3L)$, where

$B(v_i,\epsilon/3L)$ is the closed ball with center v_i and radius $\epsilon/3L$. For any $i=1,\ldots,m$, the function $s\to T_0(s)v_i$ is uniformly continuous on $[0,t]$. Thus there exists a $\delta_i>0$ such that $\|T_0(s)v_i-T_0(\sigma)v_i\|\leqslant\epsilon/3$ if $|s-\sigma|<\delta_i$. Define $\delta=\min\{\delta_1,\ldots,\delta_m\}$. Now let $v\in V$, and choose $i\in\{1,\ldots,m\}$ such that $v\in B(v_i,\epsilon/3L)$. Then

$$\|T_0(s)v - T_0(\sigma)v\|\leqslant\|T_0(s)v - T_0(s)v_i\| + \|T_0(s)v_i - T_0(\sigma)v_i\|$$
$$+\|T_0(\sigma)v_i - T_0(\sigma)v\|<\frac{\epsilon}{3}+\frac{\epsilon}{3}+\frac{\epsilon}{3}=\epsilon.$$

This proves the u-continuity on $[0,t]$ of $T_0(s):V\to X$. □

REMARKS 9.17.

i) An important consequence of this theorem is the following corollary. If $T_0(t)$ is u-continuous on $[t_0,\infty)$, and has a compact resolvent $R(\lambda,A_0)$, then $T_0(t)$ is compact if $t\geqslant t_0$. To see this take $B=I$. Then $U(t)=(e^t-1)T_0(t)$, and the result follows from Theorem 9.16.
ii) Note that the proof becomes trivial if we know in addition that $R(\lambda,A_0)$ is compact.
iii) The assumption of the previous theorem that $U(t)$ is u-continuous cannot be missed. Even more, if $U(t)$ is compact on $\mathbb{R}_+$ then $U(t)$ is u-continuous on $\mathbb{R}_+$. This follows by similar arguments as those used in the proof above.

We now present a criterion to check the u-continuity of $U(t)$.

PROPOSITION 9.18. *Let $k\geqslant 1$ and $t_0\geqslant 0$. If $T_1(t)=\cdots=T_{k-1}(t)=0$ for $t\geqslant t_0$ and $T_k(t)$ is u-continuous on $[0,\infty)$, then $U(t)$ is u-continuous on $[t_0,\infty)$.*

PROOF. Below we shall prove that for every integer $i\geqslant 1$, $T_i(t)$ is u-continuous on $[0,\infty)$ if $T_{i-1}(t)$ is u-continuous on $[0,\infty)$. Now, if $T_k(t)$ is u-continuous on $[0,\infty)$, then we obtain by induction that $T_{k+1}(t)$, $T_{k+2}(t)$ etc., are likewise u-continuous on $[0,\infty)$. We have $U(t)=\Sigma_{i=1}^{\infty}T_i(t)$, $t\geqslant 0$, and if $t\geqslant t_0$, then $U(t)=\Sigma_{i=k}^{\infty}T_i(t)$. Therefore $U(t)$ is u-continuous or $[t_0,\infty)$.

Let $T_i(t)$ be u-continuous on $[0,\infty)$. Now

$$\|T_{i+1}(t+h)-T_{i+1}(t)\|\leqslant\|\int_t^{t+h} T_0^{\odot *}(s)BT_i(t+h-s)ds\| +$$
$$\|\int_0^t T_0^{\odot *}(s)B\{T_i(t+h-s)-T_i(t-s)\}ds\|.$$

It is easy to see that the first expression at the right-hand-side goes to zero as $h\downarrow 0$. As to the second expression we get that

$$\|\int_0^t T_0^{\odot *}(s)B\{T_i(t+h-s)-T_i(t-s)\}ds\|\leqslant$$

$$\frac{M}{\omega}(e^{\omega t}-1)\|B\|\cdot \sup_{0\leqslant s\leqslant t}\|T_i(s+h)-T_i(s)\|,$$

which goes to zero as $h\downarrow 0$, because of the u-continuity of $T_i(t)$. This concludes the proof. □

REMARKS 9.19.

i) As an important special case of this result we mention: if $T_1(t)$ is u-continuous on $\mathbb{R}_+$, then $U(t)$ is u-continuous on $\mathbb{R}_+$.

ii) If $U(t)$ is u-continuous on $\mathbb{R}_+$, then $U^{\odot *}(t)$ maps $X^{\odot *}$ into X. To see this, let $x^{\odot *}\in X^{\odot *}$ and define $y_h\in X$ as: $y_h=h^{-1}\int_0^h U^{\odot *}(t+s)x^{\odot *}ds$, where the integral is a weak * Riemann integral (see Section 3.5). It is easily understood that $y_h\to U^{\odot *}(t)x^{\odot *}$, as $h\downarrow 0$, relative to the norm topology of $X^{\odot *}$. But X is a closed subspace of $X^{\odot *}$, and the assertion follows.

If $U(t)$ is eventually compact, say for $t\geqslant t_0$, then

$$|T(t)|_\alpha=|T_0(t)+U(t)|_\alpha=|T_0(t)|_\alpha\leqslant\|T_0(t)\|,$$

for $t\geqslant t_0$ (see Lemma A.3.8), and we obtain from (8.2) and (8.6) that

$$\omega_{ess}(A)\leqslant\omega_0(A_0).$$

PROPOSITION 9.20. *If* $U(t)=T(t)-T_0(t)$ *is eventually compact, then* $\omega_{ess}(A)\leqslant\omega_0(A_0)$.

Finally we shall concentrate ourselves in this section on positivity and irreducibility properties of the semigroups. Our main result relates the irreducibility of the perturbed semigroup $\{T(t)\}_{t\geqslant 0}$ to a sort of joint irreducibility of the unperturbed semigroup $\{T_0(t)\}_{t\geqslant 0}$ and the perturbation B, or better, the operators $K(\lambda)$. Throughout the rest of this section we assume that X is a Banach lattice. So X^* is a Banach lattice as well. It is unknown, in general, whether $X^\odot$ is a Banach lattice again. However, as a closed subspace of the Banach lattice X^*, $X^\odot$ is a Banach space with partial ordering. If $\{T_0(t)\}_{t\geqslant 0}$ is a positive semigroup on X, and $K(\lambda)$ given by (9.10) is a positive linear operator (or equivalently: $B:X\to X^{\odot *}$ is positive), then $U(t)$ defines a positive operator for every $t\geqslant 0$. This follows from the series expansion

$$U(t)=\sum_{k=1}^{\infty}T_k(t),\quad t\geqslant 0,$$

where $T_k(t)$ is given by (9.6). Note that the embedding operator $i:X\to X^{\odot *}$ (see Section 3.4) maps the cone X^+ into $(X^{\odot *})^+$. As a consequence $\{T(t)\}_{t\geqslant 0}$ defines likewise a positive semigroup, and moreover, for every $t\geqslant 0$

$$0\leqslant T_0(t)\leqslant T(t).$$

This yields that

$$\omega_0(A_0)\leqslant\omega_0(A).$$

For $\lambda \in \mathbb{R}$ sufficiently large we have

$$0 \leqslant R(\lambda, A_0) \leqslant R(\lambda, A),$$

which implies that

$$s(A_0) \leqslant s(A).$$

We shall now prove an important criterion for the irreducibility of the perturbed semigroup $\{T(t)\}_{t\geqslant 0}$. The demonstration of this result is based on the fact that a closed ideal J in X is invariant under $\{T(t)\}_{t\geqslant 0}$ if and only if J is invariant under $R(\lambda, A)$ for $\lambda > s(A)$ (c.f. Proposition 7.6).

Theorem 9.21. *Let X be a Banach lattice, $\{T_0(t)\}_{t\geqslant 0}$ a positive semigroup, and B a positive operator. The perturbed semigroup $\{T(t)\}_{t\geqslant 0}$ is irreducible if and only if $J=\{0\}$ and $J=X$ are the only closed ideals satisfying*
(i) $T_0(t)J \subseteq J$, $t \geqslant 0$
(ii) $K(\lambda)J \subseteq J$.

Proof. First suppose that $\{T(t)\}_{t\geqslant 0}$ is irreducible. Suppose that J is a closed ideal in X satisfying (i) and (ii). Then $R(\lambda, A_0)J \subseteq J$ and $K(\lambda)J \subseteq J$ and from (9.9) we get that $R(\lambda, A)J \subseteq J$, hence $J=\{0\}$ or $J=X$.

As for the proof of the converse result, assume that $\{0\}$ and X are the only closed ideals satisfying (i) and (ii). We prove that $\{T(t)\}_{t\geqslant 0}$, or equivalently, $R(\lambda, A)$ (where $\lambda > s(A)$) is irreducible. Let J be a closed ideal in X satisfying $R(\lambda, A)J \subseteq J$, $\lambda \geqslant s(A)$. We intend to show that this implies that $J=\{0\}$ or X. For every $x \in J$ we have

$$|R(\lambda, A_0)x| \leqslant R(\lambda, A_0)|x| \leqslant R(\lambda, A)|x| \in J,$$

from which we get that $R(\lambda, A_0)J \subseteq J$, or equivalently, $T_0(t)J \subseteq J$, $t \geqslant 0$. Once more, let $x \in J$. Then, for $\lambda > s(A)$

$$\begin{aligned} |R(\lambda, A_0^{\odot *})BR(\lambda, A)x| &\leqslant R(\lambda, A_0^{\odot *})BR(\lambda, A)|x| \\ &= (R(\lambda, A) - R(\lambda, A_0))|x| \\ &\leqslant R(\lambda, A)|x| \in J, \end{aligned}$$

hence $R(\lambda, A_0^{\odot *})BR(\lambda, A)x \in J$. If $\mu > s(A)$, then also

$$R(\mu, A_0)R(\lambda, A_0^{\odot *})BR(\lambda, A)x \in J,$$

and from the resolvent equation we get,

$$(R(\mu, A_0^{\odot *}) - R(\lambda, A_0^{\odot *}))BR(\lambda, A)x \in J.$$

Now we multiply by λ, let $\lambda \to \infty$ and use that J is closed and that $\lambda R(\lambda, A)x \to x$ as $\lambda \to \infty$. This gives us $R(\mu, A_0^{\odot *})Bx \in J$, i.e. $K(\mu)x \in J$. So (i) and (ii) are satisfied, which implies that $J=\{0\}$ or X, whence it follows that $\{T(t)\}_{t\geqslant 0}$ is irreducible. □

COROLLARY 9.22. *Let* $\{T_0(t)\}_{t\geq 0}$ *be a positive* C_0*-semigroup on the Banach lattice X, with generator* A_0. *Let* $B:X\to X$ *be a positive bounded linear operator, and let* $\{T(t)\}_{t\geq 0}$ *be the semigroup generated by* $A=A_0+B$. *Then* $\{T(t)\}_{t\geq 0}$ *is irreducible if and only if* $J=\{0\}$ *and* $J=X$ *are the only closed ideals satisfying*
(i) $T_0(t)J\subseteq J$, $t\geq 0$,
(ii) $BJ\subseteq J$.

PROOF. If $B\in\mathfrak{L}(X)$ then (ii) of Theorem 9.21 says that $R(\lambda,A_0)Bx\in J$ if $x\in J$. Multiplying with λ, letting $\lambda\to\infty$ and using the closedness of J, we find $Bx\in J$. This yields the result. □

Chapter 10

Two Examples from Structured Population Dynamics

1. Size-dependent cell growth and division

Consider a cell population whose individuals are characterized by their size s. Here size can represent length, volume, DNA-content or any other quantity which is conserved at division. Every cell grows deterministically according to the ordinary differential equation

$$\frac{ds}{dt} = g(s).$$

We assume that g is such that cells cannot grow beyond $s=1$ (see Assumptions 10.1). Furthermore we conceive of death and division as random processes, and let $\mu(s)$ and $b(s)$ stand for the probability per unit of time that a mother cell with size s dies or divides into two daughter cells each having size $s/2$. We assume that cells cannot divide at a size less than a $(0<a<1)$, and consequently cells with a size less than $a/2$ do not exist. Concerning the rates g, μ and b we make the following assumptions.

Assumptions 10.1.

$[A_g]$ $g \in C^1[\frac{1}{2}a, 1]$, $g(s)>0, s \in [\frac{1}{2}a, 1)$; $g(1)=1$ and $g'(1) \neq 0$.

$[A_\mu]$ μ is a Lipschitz-continuous, nonnegative function on $[\frac{1}{2}a, 1]$.

$[A_b]$ b is a Lipschitz-continuous function on $[\frac{1}{2}a, 1]$;
$b(s)=0$, $s \in [\frac{1}{2}a, a]$; $b(s)>0$, $s \in (a, 1]$.

These assumptions are certainly not the weakest under which the theory works. However, in this chapter we do not aim for more generality.

Let $u(t,s)$ be the size density at time t, i.e. $\int_{s_1}^{s_2} u(t,s)ds$ is the number of individuals with size between s_1 and s_2. Then $u(t,s)$ satisfies the first-order PDE:

$$\frac{\partial u}{\partial t}(t,s) + \frac{\partial}{\partial s}(g(s)u(t,s)) = -\mu(s)u(t,s) - b(s)u(t,s) + \qquad (10.1)$$

$$+ 4b(2s)u(t, 2s),$$

and the boundary condition

$$u(t,a/2) = 0, \; t>0, \qquad (10.2)$$

which guarantees that there is no influx of cells with size $s=a/2$. The last term at the right-hand-side of (10.1) should be interpreted as zero for $s \geqslant 1/2$. We impose an initial condition

$$u(0,s) = \phi(s). \qquad (10.3)$$

We assume that $\phi \in X := L_1[a/2,1]$ and look for solutions $u(t,\cdot) \in X$ such that $t \to u(t,\cdot)$ is continuous with respect to the norm on X, in this particular case the L_1-norm. We can write (10.1)-(10.3) as an abstract Cauchy problem on X:

$$\frac{du}{dt}(t) = Au(t), \; u(0) = \phi, \qquad (10.4)$$

where the closed operator A with domain

$$\mathcal{D}(A) = \{\psi \in X : g\psi \text{ is absolutely continuous and } \psi(a/2) = 0\} \quad (10.5)$$

is given by

$$(A\psi)(s) = -\frac{d}{ds}(g(s)\psi(s)) - \mu(s)\psi(s) - b(s)\psi(s) + 4b(2s)\psi(2s). \; (10.6)$$

We write

$$A = A_0 + B, \qquad (10.7)$$

where the closed operator A_0 with domain $\mathcal{D}(A_0) = \mathcal{D}(A)$ is given by

$$(A_0\psi)(s) = -\frac{d}{ds}(g(s)\psi(s)) - \mu(s)\psi(s) - b(s)\psi(s),$$

and $B: X \to X$ is the bounded linear operator

$$(B\psi)(s) = 4b(2s)\psi(2s).$$

Let $S(t,s)$ be the solution of the ODE

$$\frac{dS}{dt}(t,s) = g(S(t,s)), \; S(0,s) = s.$$

Then $S(t,s) = G^{inv}(t + G(s))$, where

$$G(s) = \int_{a/2}^{s} \frac{d\xi}{g(\xi)},$$

and where G^{inv} denotes the inverse function of G. We can interpret $S(t,s)$ as the size of an individual at time t given its size at time zero were s. Note that for every $a/2 \leqslant s < 1$ we have

$$\lim_{t\to\infty} S(t,s) = 1.$$

Let

$$E(s) = \exp\left(-\int_{a/2}^{s} \frac{\mu(\xi)+b(\xi)}{g(\xi)} d\xi\right), \quad a/2 \leqslant s \leqslant 1. \tag{10.8}$$

Note that $E(1)=0$. For $\sigma \leqslant s$ we can interpret $E(s)/E(\sigma)$ as the probability that a cell with size σ will reach s without dying or dividing. Obviously A_0 generates a C_0-semigroup $\{T_0(t)\}_{t\geqslant 0}$ for which we can write down an explicit expression:

$$\begin{aligned}(T_0(t)\phi)(s) &= \frac{E(s)}{g(s)} \cdot \frac{g(S(-t,s))}{E(S(-t,s))} \cdot \phi(S(-t,s)), \quad \frac{a}{2} < S(-t,s), \\ &= 0, \text{ elsewhere.}\end{aligned} \tag{10.9}$$

The variation-of-constants formula (e.g. (3.37)) now takes the form

$$\begin{aligned}u(t,s) = \frac{E(s)}{g(s)} \cdot \{ &(\frac{g}{E}\phi)(S(-t,s)) + \\ &4\int_0^t \frac{g(S(-\tau,s))}{E(S(-\tau,s))} - b(2S(-\tau,s)).n(t-\tau,2S(-\tau,s))d\tau\}.\end{aligned} \tag{10.10}$$

The series

$$u(t) = \sum_{k=0}^{\infty} T_k(t)\phi,$$

where $T_k(t)$ is given by (3.38), can be interpreted as a *generation expansion:* $u_0(t) = T_0(t)\phi$ represents the cells of the zero'th generation, i.e. all cells which were present at time zero and have not yet died or divided. Inductively $u_k(t) = T_k(t)\phi$ represents the cells of the k'th generation consisting of all daughters of cells in the $(k-1)$'th generation.

From (10.10) we can compute the solution $u(t,\cdot) = T(t)\phi$ of (10.4). Here $\{T(t)\}_{t\geqslant 0}$ is the semigroup generated by $A = A_0 + B$. We wish to apply Theorem 9.11 which gives a characterization of the large-time-behaviour of solutions. Before doing so we have to verify two conditions

(i) $\omega_{ess}(A) < \omega_0(A)$

(ii) $\{T(t)\}_{t\geqslant 0}$ is irreducible.

First we will show that (i) is satisfied if we make the following assumption on the growth rate g.

ASSUMPTION 10.2. $g(2s) < 2g(s)$, $s \in [a/2, 1/2]$.

Note that this assumption has already been satisfied in a neighbourhood of $s = 1/2$ because of Assumption 10.1-[A_g].

LEMMA 10.3. *Let $U(t)$ be given by (3.42), i.e.*

$$U(t)=T(t)-T_0(t), \quad t\geqslant 0.$$

Then $U(t)$ is compact for all $t\geqslant 0$.

PROOF. Obviously $K(\lambda)=R(\lambda,A_0)B$ is compact, and Theorem 9.16 with Proposition 9.18 yields that it suffices to show that $T_1(t)$ is continuous on $\mathbb{R}_+$, with respect to the uniform operator topology. We have the following explicit expression:

$$(T_1(t)\phi)(s)=4\frac{E(s)}{g(s)}\int_0^t(\frac{g}{E})(S(-\tau,s))\cdot(\frac{b}{g}\cdot E)(2S(-\tau,s))\times$$

$$\times(\frac{g}{E}\cdot\phi)(S(-t+\tau,2S(-\tau,s)))d\tau.$$

We substitute $\xi=S(-t+\tau,2S(-\tau,s))$ for τ and find

$$\frac{d\xi}{d\tau}=\{g(2z)-2g(z)\}\frac{g(\xi)}{g(2z)},$$

where $z=S(-\tau,s)$. Because of Assumption 10.2,

$$|\frac{d\xi}{d\tau}|\neq 0,$$

and proving u-continuity of $T_1(t)$ on $[0,\infty)$ has now become an easy exercise which is left to the reader. □

Because of Proposition 9.20

$$\omega_{ess}(A)\leqslant\omega_0(A_0). \tag{10.11}$$

LEMMA 10.4. $\omega_0(A_0)=-\mu(1)-b(1)=:-\beta.$

PROOF. Let $c:=-g(1)>0$. From Assumption 10.1 it follows that there exist positive constants m_1,m_2,m_3,m_4 such that

$$m_1(1-s)e^{-ct}\leqslant 1-S(t,s)\leqslant m_2(1-s)e^{-ct},$$

$$m_3(1-s)^{\beta/c}\leqslant E(s)\leqslant m_4(1-s)^{\beta/c}.$$

Now (10.9) yields that for $\phi\in X$

$$\|T_0(t)\phi\|=\int_{S(0,t)}^1\frac{E(s)}{g(s)}\cdot(\frac{g}{E}|\phi|)(S(-t,s))ds=$$

$$\int_0^1\frac{E(S(t,\xi))}{E(\xi)}|\phi(\xi)|d\xi\leqslant\int_0^1\frac{m_4(m_2(1-\xi)e^{-ct})^{\beta/c}}{m_3(1-\xi)^{\beta/c}}|\phi(\xi)|d\xi=Me^{-\beta t}\|\phi\|,$$

where $M=\frac{m_4}{m_3}\cdot m_2^{\beta/c}$. Similarly we can show that

$$\|T_0(t)\phi\| \geqslant me^{-\beta t}\|\phi\|,$$

if $\phi \geqslant 0$, where $m = \dfrac{m_3}{m_4} m_1^{\beta/c}$. This yields that $\omega_0(A_0) = -\beta$. □

PROPOSITION 10.5. $\omega_{ess}(A) < \omega_0(A)$.

PROOF. We already noted that $K(\lambda) = R(\lambda, A_0)B$ is compact for $\lambda \in \rho(A_0)$. Now, if $1 \in \sigma(K(\lambda))$ then $1 \in \sigma_p(K(\lambda))$. Let $K(\lambda)\psi = \psi$ for some $\psi \in X$, then $A\psi = \lambda\psi$. Suppose we can show that there exists a $\lambda \in \mathbb{R}, \lambda > -\beta$ such that $r(K(\lambda)) = 1$, then $1 \in \sigma_p(K(\lambda))$ (see Appendix A.3.3). (Note that $K(\lambda)$ is a positive operator if $\lambda > -\beta$.) Then $s(A) \geqslant \lambda > -\beta = \omega_0(A_0) \geqslant \omega_{ess}(A)$ and the result would follow. Since $\lambda \to r(K(\lambda))$ is a continuous function on $(-\beta, \infty)$ and $\lim_{\lambda \to \infty} r(K(\lambda)) = 0$ it suffices to show that $\lim_{\lambda \downarrow -\beta} r(K(\lambda)) > 1$. We prove this by exploiting the following well-known result due to Krein and Rutman: let $L \in \mathfrak{L}^+(X)$ and let $\psi \in X^+$, $\psi \neq 0$ be such that

$$L\psi \geqslant c\psi, \text{ for some } c > 0, \text{ then } r(L) \geqslant c.$$

As in the proof of the previous lemma we have the following estimates for E_λ:

$$l_1(1-s)^{\frac{\beta + \mathrm{Re}\lambda}{c}} \leqslant |E_\lambda(s)| \leqslant l_2(1-s)^{\frac{\beta + \mathrm{Re}\lambda}{c}},$$

for some positive constants l_1, l_2. Thus there exists a positive constant C such that for every $\lambda > -\beta$ and $\psi \in X^+$:

$$[K(\lambda)\psi](s) \geqslant C(1-s)^{\frac{\beta+\lambda}{c} - 1} \int_{a/2}^{s} b(2\xi)\psi(2\xi)d\xi.$$

Now let $p = \lambda + \beta > 0$ and define

$$\psi_p(s) = \begin{cases} 0 & , s \in [\frac{a}{2}, \frac{a+1}{2}) \\ (1-s)^{-1+p} & , s \in [\frac{a+1}{2}, 1]. \end{cases}$$

There is a positive value $\tilde{b} > 0$ such that $b(s) > \tilde{b}$ for every $s \in [\frac{a+1}{2}, 1]$. Now, for $s \geqslant \dfrac{a+1}{2}$ we have

$$[K(\lambda)\psi_p](s) \geqslant C(1-s)^{-1+p} \int_{a/2}^{1/2} b(2\xi)\psi_p(2\xi)d\xi =$$

$$\frac{C}{2}(1-s)^{-1+p} \int_{a}^{1} b(\xi)\psi_p(\xi)d\xi \geqslant$$

$$\frac{C}{2}(1-s)^{-1+p} \int_{\frac{a+1}{2}}^{1} \tilde{b}(1-\xi)^{-1+p}d\xi = \frac{\tilde{b}C}{2p}\left(\frac{1-a}{2}\right)^p \psi_p(s).$$

This implies that for every $\lambda > -\beta$,

$$r(K(\lambda)) \geqslant \frac{\tilde{b}C}{2(\lambda+\beta)}\left(\frac{1-a}{2}\right)^{\lambda+\beta},$$

and $r(K(\lambda)) \to \infty$, as $\lambda \downarrow -\beta$, and we are done. □

REMARK. Note that in this particular case

$$s(A) = \omega_0(A),$$

because X is an L_1 -space (see Theorem 9.5).

We continue our exposition with a proof of the irreducibility of the perturbed semigroup. We use Corollary 9.22 for this purpose. First, however, we give a characterization of the closed ideals in $X = L_1[a/2,1]$:

Every closed ideal J in $L_1[a/2,1]$ has the form

$$J = J_\Omega = \{\phi \in L_1[a/2,1] : \phi \text{ vanishes on } \Omega\},$$

where Ω is a measurable subset of $[a/2,1]$.

PROPOSITION 10.6. *$\{T(t)\}_{t\geqslant 0}$ is irreducible.*

PROOF. According to Corollary 9.22 we are done if we can show that $\{T_0(t):t\geqslant 0\} \cup \{B\}$ is irreducible. Let J be a closed ideal in $X = L_1[a/2,1]$ which is invariant under $\{T_0(t)\}_{t\geqslant 0}$. Hence $J = J_\Omega = \{\phi \in X : \phi$ vanishes on $\Omega\}$ for some measurable subset Ω of $[a/2,1]$. Since $T_0(t)J \subseteq J$ for every $t \geqslant 0$ it follows that

$$s \in \Omega \wedge S(-t,s) \geqslant a/2 \Rightarrow S(-t,s) \in \Omega. \qquad (*)$$

From $BJ \subseteq J$ we conclude that

$$s \in \Omega \cap (a/2, 1/2) \Rightarrow 2s \in \Omega. \qquad (**)$$

Combination of (*) and (**) yields that,

$$\Omega = \varnothing \text{ or } \Omega = [a/2,1],$$

corresponding to the cases

$$J = X \text{ or } J = \{0\},$$

and this proves the irreducibility of the semigroup $\{T(t)\}_{t\geqslant 0}$. □

Now application of Theorem 9.11 gives us the following result.

THEOREM 10.7. *There exist* $\lambda_0 \in \mathbb{R}$, $\lambda_0 > -\beta$, $\epsilon_0 > 0$, $\phi_0 \in L_1^+[a/2,1]$, *and* $\phi_0^* \in L_\infty^+[a/2,1]$ *which are positive a.e. such that for every* $\epsilon \in (0,\epsilon_0)$, *there is a constant* $M(\epsilon) \geq 1$ *such that*

$$\|e^{-\lambda_0 t} T(t)\phi - <\phi,\phi_0^*>\phi_0\| \leq M(\epsilon) e^{-\epsilon t} \|\phi\|, \; t \geq 0,$$

for every $\phi \in L_1[a/2,1]$.

We call λ_0 *the intrinsic growth rate* (or *Malthusian parameter)* and ϕ_0 the *stable size distribution.*

2. DYNAMICS OF AGE-STRUCTURED POPULATIONS

In the previous section, the perturbation B was a bounded linear operator from X into X, and we could use the standard perturbation theory for linear C_0-semigroups. In the present section we consider an example from structured population dynamics for which the theory of Section 3.5 is quite useful.

Consider the following system

$$\frac{\partial u}{\partial t}(t,a) + \frac{\partial u}{\partial a}(t,a) = -\mu(a)u(t,a), \, t>0, 0<a<1, \tag{10.12}$$

$$u(t,0) = \int_0^1 b(\alpha)u(t,\alpha)d\alpha, \, t>0, \tag{10.13}$$

$$u(0,a) = \phi(a), \quad 0 \leq \alpha \leq 1. \tag{10.14}$$

This initial value problem describes the evolution in time of an age-structured population. In this system, t is time, a is age, and $u(t,a)$ is the age-density at time t, that is $\int_{a_1}^{a_2} u(t,a)da$ is the number of individuals at time t which have age between a_1 and a_2. Furthermore, $\mu(a)$ and $b(a)$ are the per capita death rate and birth rate, respectively. The boundary condition (10.13) expresses the fact that all newborns have age zero. Finally, it is implicitly assumed that individuals older than one do not exist (which is e.g. true if μ has a non-integrable singularity in $a=1$) or do not reproduce ($b(a)=0$ if $a \geq 1$), in which case one may neglect them. For the moment we assume that $\phi \in X := L_1[0,1]$. Later on, we make clear that we may also allow for initial conditions $\phi \in M[0,1]$, the space of all regular Borel measures on $[0,1]$. Note that $L_1[0,1]$ can be continuously imbedded in $M[0,1]$, since

$$L_1[0,1] \cong M_a[0,1],$$

where $M_a[0,1]$ is the closed subspace of $M[0,1]$ containing all *absolutely continuous* measures. Throughout this section we make the following assumptions

ASSUMPTIONS 10.8. $\mu, b \in L_\infty^+[0,1]$.

Below we show that we can regard system (10.12)-(10.14) as a "bounded" (in the sense of Section 3.5) perturbation of the system that we get if we replace the boundary condition (10.13) by

$$u(t,0)=0, \quad t>0. \tag{10.15}$$

In order to attain this, we reformulate (10.12), (10.15), (10.14) as the following abstract Cauchy problem on X:

$$\frac{du}{dt}(t)=A_0u(t), \quad u(0)=\phi,$$

where A_0 is the closed linear operator on X given by

$$\mathcal{D}(A_0)=\{\phi\in W_{1,1}[0,1]:\phi(0)=0\},$$
$$A_0\phi=-\phi'-\mu\phi.$$

It is easily checked that A_0 is the generator of a C_0-semigroup $\{T_0(t)\}_{t\geqslant 0}$ given by

$$(T_0(t)\phi)(a)=\begin{cases}\dfrac{E(a)}{E(a-t)}\phi(a-t) & , \ t\leqslant a,\\ 0 & , \ t>a,\end{cases}$$

where E is given by

$$E(a)=\exp(-\int_0^a\mu(\alpha)d\alpha).$$

Clearly, $X^*=L_\infty[0,1]$ and the weak * continuous semigroup $\{T_0^*(t)\}_{t\geqslant 0}$ on X^* is given by

$$(T_0^*(t)\psi)(a)=\begin{cases}\dfrac{E(a+t)}{E(a)}\psi(a+t) & , \ a+t\leqslant 1,\\ 0 & , \ a+t>1.\end{cases}$$

Its weak * generator A_0^* is

$$\mathcal{D}(A_0^*)=\{\psi\in W_{\infty,1}[0,1]:\psi(1)=0\},$$
$$A_0^*\psi=\psi'-\mu\psi.$$

Clearly

$$X^\odot=C_0[0,1]=\{\psi\in C[0,1]:\psi(1)=0\},$$

and the action of $T_0^\odot(t)$ is the same as the action of $T_0^*(t)$. Using that $A_0^\odot$ is the part of A_0^* in $X^\odot$, we find that

$$\mathcal{D}(A_0^\odot)=\{\psi\in X^\odot\cap W_{\infty,1}[0,1]:\psi'-\mu\psi\in X^\odot\},$$
$$A_0^\odot\psi=\psi'-\mu\psi.$$

Taking duals again we find that

$$X^{\odot *}=M_0[0,1]=\{\Phi\in M[0,1]:\Phi(\{1\})=0\}.$$

As to our notation, we use the following convention. Elements of $M[0,1]$ are denoted by capitals. However, if $\Phi \in M[0,1]$ is absolutely continuous, then we denote the corresponding element of $L_1[0,1]$ by ϕ:

$$\Phi(\Omega) = \int_{\Omega} \phi(a)da,$$

for every Borel set $\Omega \subset [0,1]$.
The weak * continuous semigroup $T_0^{\odot *}(t)$ is given by

$$(T_0^{\odot *}(t)\Phi)(\Omega) = \int_{\Omega_{-t}} \frac{E(a+t)}{E(a)}\Phi(da), \quad t \geq 0,$$

where $\Omega_{-t} = \{a - t : a \in \Omega, a \geq t\}$, for every Borel set $\Omega \subset [0,1]$. It is easily checked that

$$\mathfrak{D}(A_0^{\odot *}) = \{\Phi \in M_a[0,1] : \phi(a) = F[0,a] \text{ for some } F \in M[0,1]\},$$

$$(A_0^{\odot *}\Phi)(\Omega) = -F(\Omega) - \int_{\Omega} \mu(a)\Phi(da), \quad \Phi \in \mathfrak{D}(A_0^{\odot *}).$$

Clearly

$$X^{\odot\odot} = \{\Phi \in X^{\odot *} : \Phi \text{ is absolutely continuous}\}.$$

Thus $X^{\odot\odot} = M_a[0,1]$, and can therefore be identified with $X = L_1[0,1]$. Thus we get:

PROPOSITION 10.9. *X is $\odot$-reflexive with respect to A_0.*

Let $B : X \to X^{\odot *}$ be the bounded linear operator given by

$$B\phi = \Delta \cdot \int_0^1 b(\alpha)\phi(\alpha)d\alpha,$$

where $\Delta \in X^{\odot *} = M[0,1]$ is the Dirac measure concentrated at zero. Let A be the perturbation of A_0 by B as given by Theorem 3.22-(a), i.e.

$$\mathfrak{D}(A) = \{\Phi \in \mathfrak{D}(A_0^{\odot *}) : A_0^{\odot *}\Phi + B\Phi \in X\},$$

$$A\Phi = A_0^{\odot *}\Phi + B\Phi.$$

Identifying $M_a[0,1]$ and $L_1[0,1]$ it follows that

$$\mathfrak{D}(A) = \{\phi \in W_{1,1}[0,1] : \phi(0) = \int_0^1 b(a)\phi(a)da\},$$

$$A\phi = -\phi' - \mu\phi.$$

From Section 3.5 we know that A is the generator of a strongly continuous semigroup $\{T(t)\}_{t \geq 0}$ on X, and this semigroup corresponds to solutions of the original system (10.12)-(10.14). Note that in this example all information about B is contained in the domain of A, whereas the action of A is the same as the action of A_0.

By duality, we also obtain a solution of the system (10.12)-(10.14) if ϕ is not L_1 but a Borel measure, just by putting

$$u(t,\cdot;\Phi) = T^{\odot *}(t)\Phi, \quad t \geq 0,$$

for $\Phi \in X^{\odot *} = M_0[0,1]$.

A different way to obtain the results of this section would be to start with the dual (or *backward*) system

$$\frac{dv}{dt}(t) = A^{\odot} v(t),\, v(0) = \psi \in X^{\odot},$$

or written out:

$$\frac{\partial v}{\partial t}(t,a) - \frac{\partial v}{\partial a}(t,a) = -\mu(a)v(t,a) + b(a)v(t,0),$$

$$v(t,1) = 0,$$

$$v(0,a) = \psi(a).$$

Note that the last term at the right-hand-side of the PDE yields the perturbation $B^*: X^{\odot} \to X^*$ given by:

$$(B^*\psi)(a) = b(a)\psi(0).$$

Like in the previous section, one can determine the asymptotic behaviour of solutions $u(t,\cdot;\phi)$ by applying the results of Sections 9.2 and 9.3. It can be shown that the semigroup $\{T(t)\}_{t\geq 0}$ is eventually compact, and, under the extra assumption that $b(a)>0$ a.e. on $(1-\epsilon,1)$, that $\{T(t)\}_{t\geq 0}$ is irreducible.

We conclude this section by pointing out that, in this particular example, there is yet another method to prove existence and uniqueness of solutions, and find their asymptotic behaviour. Namely, it is easy to write down a Volterra integral equation for the birth function $y(t) = \int_0^1 b(a)u(t,a)da$. The relation with the semigroup $\{T(t)\}_{t\geq 0}$ is given by

$$(T(t)\phi)(a) = \begin{cases} \dfrac{E(a)}{E(a-t)}\phi(a-t) \;, & t \leq a, \\ E(a)y(t-a) \;, & t > a. \end{cases}$$

Appendices

Appendix 1

Convex Functions

In this appendix we collect some of the basic properties of convex functions on linear spaces. In particular we will deal with continuity, subdifferentials and differentiability. In the special case of a Hilbert space, the subdifferential mappings of convex functions provide us with examples of maximal monotone (m-accretive) mappings. All vector spaces considered here are real.

1. Definitions and elementary properties

In this section we consider functions F defined on a *real* vector space X with values in $(-\infty,\infty]=\mathbb{R}\cup\{\infty\}$. For such a function F the set

$$\operatorname{dom}(F) = \{x\in X : F(x)<\infty\}$$

is called the *effective domain* of F. If $\operatorname{dom}(F)\neq\emptyset$, then F is called *proper.* The *epigraph* of F is the subset of $X\times\mathbb{R}$ defined by

$$\operatorname{epi}(F) = \{(x,\alpha)\in X\times\mathbb{R} : F(x)\leqslant\alpha\}.$$

Clearly, $\operatorname{dom}(F)$ is precisely the projection of $\operatorname{epi}(F)$ onto X.

Definition A.1.1. The function $F:X\to(-\infty,\infty]$ is called *convex* if $F(\lambda x+(1-\lambda)y)\leqslant\lambda F(x)+(1-\lambda)F(y)$ for all $x,y\in X$ and all $0\leqslant\lambda\leqslant 1$.

Remark A.1.2. The following observations are straightforward.

(i) If $F:X\to(-\infty,\infty]$ is convex, then $F(\Sigma_{i=1}^n\lambda_i x_i)\leqslant\Sigma_{i=1}^n\lambda_i F(x_i)$ for all $x_i\in X$ $(i=1,\ldots,n)$ and all $0\leqslant\lambda_i\leqslant 1$ with $\Sigma_{i=1}^n\lambda_i=1$.

(ii) If F is convex then $\operatorname{dom}(F)$ is a convex subset of X, i.e., $\lambda x+(1-\lambda)y\in\operatorname{dom}(F)$ whenever $x,y\in\operatorname{dom}(F)$ and $0\leqslant\lambda\leqslant 1$. Furthermore, if $\mathfrak{U}\subseteq X$ and $f:\mathfrak{U}\to\mathbb{R}$, and if we define $F(x)=f(x)$ for $x\in\mathfrak{U}$ and $F(x)=\infty$ for $x\notin\mathfrak{U}$, then F is convex if and only if $\mathfrak{U}$ and f are both convex.

(iii) The function $F:X\to(-\infty,\infty]$ is convex if and only if $\operatorname{epi}(F)$ is a convex subset of $X\times\mathbb{R}$. If F is convex, then the sets $\{x\in X:F(x)\leqslant\alpha\}$ and $\{x\in X:F(x)<\alpha\}$ are convex for all $\alpha\in\mathbb{R}$.

(iv) If the functions $F,G:X\to(-\infty,\infty]$ are convex, and $\lambda\geqslant 0$, then $F+G$ and λF are convex as well. Moreover, if $\{F_\tau\}$ is a collection of convex functions, then the function F defined by $F(x)=\sup_\tau F_\tau(x)$ for $x\in X$, is convex.

A proper convex function F from X into $(-\infty,\infty]$ will be called *strictly convex* if $F(\lambda x+(1-\lambda)y)<\lambda F(x)+(1-\lambda)F(y)$ for all $x,y\in\operatorname{dom}(F)$ with $x\neq y$ and all $0<\lambda<1$.

Now assume that X is a locally convex topological vector space. As usual, the function $F:X\to(-\infty,\infty]$ is called *lower semi-continuous* (l.s.c.) if the sets $\{x\in X:F(x)\leqslant\alpha\}$ are closed for all $\alpha\in\mathbb{R}$. It is easy to see that F is l.s.c. if and only if $\operatorname{epi}(F)$ is a closed subset of $X\times\mathbb{R}$. Furthermore, the pointwise supremum of a family of l.s.c. functions is likewise l.s.c.

Denote by X^* the topological dual of the space X. The well-known fact that a convex subset in X is closed if and only if the set is $\sigma(X,X^*)$-closed, has the following immediate, but interesting consequence: a convex function $F:X\to(-\infty,\infty]$ is l.s.c. if and only if F is $\sigma(X,X^*)$-l.s.c.

Definition A.1.3. For a locally convex space X we denote the collection of all l.s.c. convex functions $F:X\to(-\infty,\infty]$ by $\Gamma_\infty(X)$, and by $\Gamma(X)$ we denote the set of all proper functions in $\Gamma_\infty(X)$.

It follows from the above remarks that $\alpha F+\beta G\in\Gamma_\infty(X)$ whenever $F,G\in\Gamma_\infty(X)$ and $\alpha,\beta\geqslant 0$, and that the pointwise supremum of a family of functions in $\Gamma_\infty(X)$ belongs to $\Gamma_\infty(X)$ as well. The simplest functions in $\Gamma(X)$ are, of course, the *continuous affine functions* on X, i.e., functions g given by $g(x)=\langle x,x^*\rangle+\alpha$, with $x^*\in X^*$ and $\alpha\in\mathbb{R}$. The collection of all these affine functions will be denoted by $A(X)$. It follows from the above observations that the pointwise supremum of any subset of $A(X)$ is a function in $\Gamma_\infty(X)$. The next proposition shows that every function in $\Gamma_\infty(X)$ is of this form.

Proposition A.1.4. *If X is a locally convex space, and if $F\in\Gamma_\infty(X)$, then* $F(x)=\sup\{g(x):g\in A(X)$ and $g\leqslant F\}$ *for all $x\in X$.*

Before indicating the proof of the above proposition, we recall the following two separation theorems, which will be used in the sequel. Suppose that X is a locally convex space and that A and B are non-empty convex subsets of X. Then

(1) if $\operatorname{int}A\neq\varnothing$ and $(\operatorname{int}A)\cap B=\varnothing$, then there exists a non-zero $x^*\in X^*$ such that $\sup\{\langle x,x^*\rangle:x\in A\}\leqslant\inf\{\langle x,x^*\rangle:x\in B\}$,

(2) if A is closed, B is compact and $A \cap B = \emptyset$, then there exists $x^* \in X^*$ such that $\sup\{<x,x^*>: x \in A\} < \inf\{<x,x^*>: x \in B\}$ (as usual, we tacitly assume that locally convex spaces are Hausdorff).

PROOF OF PROPOSITION A.1.4. Take $F \in \Gamma(X), x_0 \in X$ and $\alpha_0 \in \mathbb{R}$ such that $\alpha_0 < F(x_0)$. We have to show that there exists a $g \in A(X)$ such that $g \leqslant F$ and $\alpha_0 < g(x_0)$. Since epi(F) is a closed convex subset of the space $X \times \mathbb{R}$, and $(x_0, \alpha_0) \notin$ epi(F), there exists a linear functional $\Phi \in (X \times \mathbb{R})^*$ such that $<(x_0,\alpha_0),\Phi> \; < \gamma < <(x,\alpha),\Phi>$ for all $(x,\alpha) \in$ epi(F) and some real γ. Using the canonical identification of $(X \times \mathbb{R})^*$ with $X^* \times \mathbb{R}$, we thus find $x^* \in X^*$ and $\beta \in \mathbb{R}$ such that

$$<x_0,x^*> + \beta\alpha_0 < \gamma < <x,x^*> + \beta\alpha,$$

for all $(x,\alpha) \in$ epi(F). First assume that $\beta \neq 0$. Take any $x \in$ dom(F). Then $\gamma < <x,x^*> + \beta\alpha$ for all $\alpha \geqslant F(x)$, which implies that $\beta > 0$. Now the affine function g, defined by $g(x) = -<x,\beta^{-1}x^*> + \beta^{-1}\gamma$, fulfils the requirements. Observe that $x_0 \in$ dom(F) implies that $\beta \neq 0$, since $(x_0, F(x_0)) \in$ epi(F) in that case. Now assume that $\beta = 0$, so $<x_0,x^*> < \gamma << x,x^*>$ for all $x \in$ dom(F). Define $h \in A(X)$ by $h(x) = -<x,x^*> + \gamma$, then $h(x_0) > 0$ and $h(x) < 0$ for all $x \in$ dom(F). Since dom$(F) \neq \emptyset$, it follows from the first part of the proof that there exists $f \in A(X)$ such that $f \leqslant F$. Then $f + \lambda h \leqslant F$ for all $\lambda > 0$. Now take $\lambda > 0$ so large that $f(x_0) + \lambda h(x_0) > \alpha_0$, and let $g = f + \lambda h$. □

It should be observed that it follows in particular from the above proposition that for any $F \in \Gamma(X)$ there exists $g \in A(X)$ such that $g \leqslant F$.

REMARK A.1.5. As above, let X be a locally convex space and let F be a function from X into $(-\infty, \infty]$. The Γ-*regularization* G of F is defined by

$$G(x) = \sup\{g(x): g \in A(X) \text{ and } g \leqslant F\}$$

for $x \in X$. If F does not have a minorant $g \in A(X)$, then the function G is identically equal to $-\infty$. Therefore we will restrict ourselves to functions F which have a continuous affine minorant. Then it is clear that $G \in \Gamma_\infty(X)$, and G is in fact the largest function in $\Gamma_\infty(X)$ which is majorized by F. Furthermore, it is not difficult to see that epi(G) is precisely the closed convex hull of epi(F). In Section A.1.3 another description of the Γ-regularization of F in terms of conjugate functions will be given.

We end this section with some examples.

EXAMPLES A.1.6. We assume that X is a locally convex space.
(i) Let K be a convex subset of X containing 0. For $x \in X$ define

$$\rho_K(x) = \inf\{\lambda > 0: x \in \lambda K\},$$

where as usual $\inf \emptyset = \infty$. Then ρ_K is a *sublinear functional* on X, with values in $[0, \infty]$, i.e., $\rho_K(x+y) \leqslant \rho_K(x) + \rho_K(y)$ and $\rho_K(\alpha x) = \alpha\rho_K(x)$ for all

$x,y \in X$ and $\alpha \geqslant 0$. In particular, ρ_K is a convex function on X, and ρ_K is called the *gauge functional* or *Minkowski functional* of K. Note that

$$\operatorname{dom}(\rho_K) = \bigcup_{\lambda \geqslant 0} \lambda K,$$

the cone generated by K, and that $\operatorname{dom}(\rho_K) = X$ if and only if K is absorbing. Furthermore, it is clear that

$$\{x \in X : \rho_K(x) < 1\} \subseteq K \subseteq \{x \in X : \rho_K(x) \leqslant 1\},$$

and it is not difficult to see that ρ_K is continuous if and only if 0 is an interior point of the set K. If K is a closed convex set, then $K = \{x \in X : \rho_K(x) \leqslant 1\}$, and ρ_K is l.s.c.

(ii) For any non-empty subset K of X the *support functional* $p_K : X^* \to (-\infty, \infty]$ is defined by

$$p_K(x^*) = \sup\{<x,x^*> : x \in K\},$$

for all $x^* \in X^*$. Evidently, p_K is sublinear, and the support functional of the closed convex hull of K is equal to p_K. We assume, therefore, that K is a closed convex subset of X. Being the pointwise supremum of $\sigma(X^*, X)$-continuous functionals, p_K is $\sigma(X^*, X)$-l.s.c. The effective domain $\operatorname{dom}(p_K)$ is a cone in X^*, which is called the *barrier cone* of K. It is an immediate consequence of the above mentioned separation theorems that

$$K = \{x \in X : <x,x^*> \leqslant p_K(x^*) \text{ for all } x^* \in X^*\},$$

which shows in particular that K is determined by its support functional. As usual, the *polar* of the set K is defined by

$$K^o = \{x^* \in X^* : <x,x^*> \leqslant 1 \text{ for all } x \in K\},$$

which is a $\sigma(X^*, X)$-closed convex subset of X^* with $0 \in K^o$. Similarly, for any subset M of X^* the polar M^o is a $\sigma(X, X^*)$-closed convex subset of X. For $x^* \in X^*$ we have

$$\rho_{K^o}(x^*) = \inf\{\lambda > 0 : x^* \in \lambda K^o\} =$$

$$= \inf\{\lambda > 0 : <x,x^*> \leqslant \lambda \text{ for all } x \in K\} = p_K(x^*),$$

which shows that $\rho_{K^o} = p_K$. If we assume in addition that $0 \in K$ then we find analogously that $\rho_K = p_{K^o}$ (note that $K^{oo} = K$ in that case, by the bipolar theorem).

(iii) For a non-empty subset K of X the *indicator function* I_K is defined by $I_K(x) = 0$ if $x \in K$ and $I_K(x) = \infty$ otherwise. Then $\operatorname{dom}(I_K) = K$ and I_K is convex if and only if K is convex. Furthermore, I_K is l.s.c. if and only if K is closed. It is easy to see that the Γ-regularization of I_K is the indicator function of the closed convex hull of K.

2. Continuity

A real valued convex function on an interval in $\mathbb{R}$, is continuous on the interior of this interval. In this section we discuss some extensions of this result to locally convex spaces, and in particular to Banach spaces. Let X be a locally convex space and let $F: X \to (-\infty, \infty]$ be a convex function. We start with a simple, but useful lemma.

Lemma A.1.7. *If F is bounded above (by a finite constant) on a neighbourhood of $x_0 \in X$, then F is continuous at x_0.*

Proof. Without loss of generality we may assume that $x_0 = 0$ and $F(0) = 0$. Suppose that $F(x) \leqslant M$ for all x in some symmetric neighbourhood U of 0. First observe that this implies that $|F(x)| \leqslant M$ for all $x \in U$. Indeed, if $x \in U$ then $-x \in U$, and $2F(0) \leqslant F(x) + F(-x)$ so $-F(x) \leqslant F(-x) \leqslant M$. Furthermore, if $0 < \epsilon < 1$ then $F(\epsilon x) \leqslant \epsilon F(x)$ for all $x \in X$, which shows that $F(x) \leqslant \epsilon M$ for all $x \in \epsilon U$. Since ϵU is a symmetric neighbourhood of 0, this shows that $|F(x)| \leqslant \epsilon M$ for all $x \in \epsilon U$. □

Proposition A.1.8. *The following two statements are equivalent.*

(i) *F is bounded above on some non-empty open subset of X.*

(ii) *F is continuous on the interior of its effective domain, and this interior is non-empty.*

Proof. Evidently (ii) implies (i). Now assume that F is bounded above on some non-empty open subset of X. By the above lemma it is sufficient to show that every point of $\operatorname{int}(\operatorname{dom}(F))$ has a neighbourhood on which F is bounded above. Without loss of generality we may assume that $F(x) \leqslant M$ for all x in some open neighbourhood U of 0. Take any $y \in \operatorname{int}(\operatorname{dom}(F))$. Then there exists $\lambda > 1$ such that $\lambda y \in \operatorname{dom}(F)$. Now $y + (1-\lambda^{-1})U$ is a neighbourhood of y which is contained in $\operatorname{dom}(F)$, as $\operatorname{dom}(F)$ is convex. For $z = y + (1-\lambda^{-1})x$ with $x \in U$ we have

$$F(z) \leqslant \lambda^{-1} F(\lambda y) + (1-\lambda^{-1}) F(x) \leqslant \lambda^{-1} F(\lambda y) + (1-\lambda^{-1}) M,$$

which shows that F is bounded above on $y + (1-\lambda^{-1})U$. □

The above proposition has an immediate consequence for finite dimensional spaces. Indeed, if $\dim X < \infty$, then any point in $\operatorname{int}(\operatorname{dom}(F))$ has a neighbourhood contained in $\operatorname{dom}(F)$ which is the convex hull of finitely many points, and thus F is bounded above on this neighbourhood. This yields the following result, which extends the fact mentioned at the beginning of this section.

Corollary A.1.9. *If X is finite-dimensional, then F is continuous on the interior of its effective domain.*

It should be observed that if F is upper semi-continuous at some point p in $\operatorname{dom}(F)$, then F is bounded above on some neighbourhood of Γ, and hence F

is continuous on $\operatorname{int}(\operatorname{dom}(F))$. This is not true if we assume that F is lower semi-continuous, as is shown by the norm on the dual of Banach space X, which is $\sigma(X^*,X)$-l.s.c., but not $\sigma(X^*,X)$-continuous. However, for continuity with respect to the norm topology in a Banach space the following holds.

PROPOSITION A.1.10. *If X is a Banach space and $F:X\to(-\infty,\infty]$ is a l.s.c. convex function, then F is continuous on the interior of its effective domain.*

PROOF. By the above proposition, it is sufficient to show that F is bounded above on some non-empty open set. Without loss of generality we assume that $0\in\operatorname{int}(\operatorname{dom}(F))$ and that $F(0)=0$. Choose an arbitrary $\alpha>0$ and put $K=\{x\in X:F(x)\leqslant\alpha\}$. Then K is a closed convex subset of X, and since the restriction of F to every one-dimensional subspace of X is continuous at 0, it follows that K is absorbing and hence

$$\bigcup_{n=1}^{\infty} nK=X.$$

By the Baire category theorem there exists n_0 such that $\operatorname{int}(n_0K)\neq\varnothing$, and so $\operatorname{int}K\neq\varnothing$. Therefore, $\operatorname{int}K$ is a non-empty open set in X on which F is bounded above, and the proof is complete. □

Suppose that X is a locally convex space and that $F:X\to(-\infty,\infty]$ is convex. As observed already, F is l.s.c. if and only if $\operatorname{epi}(F)$ is a closed subset of $X\times\mathbb{R}$. Continuity of F can also be expressed in terms of $\operatorname{epi}(F)$. In fact, using Proposition A.1.8, it is easy to see that F is continuous on $\operatorname{int}(\operatorname{dom}(F))$ and $\operatorname{int}(\operatorname{dom}(F))\neq\varnothing$ if and only if $\operatorname{int}(\operatorname{epi}(F))$ is non-empty.

3. THE CONJUGATE FUNCTION

Let X be a locally convex space with topological dual space X^*. In X^* we consider the $\sigma(X^*,X)$-topology, so that the dual of X^* is X. Let F be a function from X into $(-\infty,\infty]$.

DEFINITION A.1.11. The function $F^*:X^*\to[-\infty,\infty]$, which is defined by

$$F^*(x^*)=\sup\{\langle x,x^*\rangle-F(x):x\in X\}$$

for all $x^*\in X^*$, is called the *conjugate function* of F.

Note that $F^*(x^*)=-\infty$ is possible if and only if $F\equiv+\infty$. If we assume that F is proper, then F^* takes its values in $(-\infty,\infty]$. Furthermore, it is clear that in the supremum defining F^* we may restrict x to $\operatorname{dom}(F)$. If $\alpha\in\mathbb{R}$, then $F^*(x^*)\leqslant\alpha$ is equivalent to $\langle x,x^*\rangle-\alpha\leqslant F(x)$ for all $x\in\operatorname{dom}(F)$, and so F^* is proper if and only if F has a continuous affine minorant. Evidently, if F is proper, then $\langle x,x^*\rangle\leqslant F(x)+F^*(x^*)$ for all $x\in X$ and $x^*\in X^*$.

Now assume that F is proper. Since F^* is the pointwise supremum of a collection functions in $A(X^*)$, it follows that $F^*\in\Gamma_\infty(X^*)$, and $F^*\in\Gamma(X^*)$ if and only if F has a minorant in $A(X)$. In particular, if follows from Proposition

A.1.4 that $F^* \in \Gamma(X^*)$ whenever $F \in \Gamma(X)$.

The *second conjugate* F^{**} of F is defined as the conjugate function of F^*, i.e.,

$$F^{**}(x) = \sup\{<x,x^*> - F^*(x^*) : x^* \in X^*\}.$$

If F is proper with a continuous affine minorant, then $F^* \in \Gamma(X^*)$, and hence $F^{**} \in \Gamma(X)$.

Recall from Remark A.1.5 that, if the proper function $F: X \to (-\infty, \infty]$ has a minorant in $A(X)$, then the Γ-regularization of F is the largest element in $\Gamma(X)$ which is majorized by F. The next proposition gives an alternative characterization of this Γ-regularization.

PROPOSITION A.1.12. *If F is a proper function from X into $(-\infty, \infty]$ which has a continuous affine minorant, then F^{**} is the Γ-regularization of F. In particular, if $F \in \Gamma(X)$ then $F^{**} = F$.*

PROOF. As observed above, $F^{**} \in \Gamma(X)$. If $x^* \in \mathrm{dom}(F^*)$, then it is clear from the definition of F^* that $<x,x^*> - F^*(x^*) \leqslant F(x)$ for all $x \in X$, and hence it follows from the definition of F^{**} that $F^{**}(x) \leqslant F(x)$ for all $x \in X$. Now assume that $g \in A(X)$ is such that $g \leqslant F$. Then $g(x) = <x,x^*> - \alpha$ for some $x^* \in X^*$ and $\alpha \in \mathbb{R}$, and so $<x,x^*> - \alpha \leqslant F(x)$ for all $x \in X$. Hence $F^*(x^*) \leqslant \alpha$, so that $g(x) \leqslant <x,x^*> - F^*(x^*)$ for all $x \in X$, and thus $g(x) \leqslant F^{**}(x)$ for all $x \in X$. Since every function in $\Gamma(X)$ is, by Proposition A.1.4, the pointwise supremum of functions in $A(X)$, we may conclude that F^{**} is the largest element of $\Gamma(X)$ which is majorized by F. □

It follows from the above proposition that the mapping $F \mapsto F^*$ is a bijection from $\Gamma(X)$ onto $\Gamma(X^*)$. If $F \in \Gamma(X)$ and $G \in \Gamma(X^*)$ are such that $F^* = G$ (and hence $G^* = F$), then we say that F and G are in *in duality* with each other.

EXAMPLE A.1.13. Let K be a non-empty subset of the locally convex space X. As in Example A.1.6-(iii), the indicator function of K is denoted by I_K. For $x^* \in X^*$ we then have

$$\begin{aligned} I_K^*(x^*) &= \sup\{<x,x^*> - I_K(x) : x \in X\} \\ &= \sup\{<x,x^*> : x \in K\} = p_K(x^*), \end{aligned}$$

so the conjugate function I_K^* is precisely the support functional of K (see Example A.1.6-(ii)). The second conjugate $I_K^{**} = p_K^*$ is the Γ-regularization of I_K, and hence p_K^* is the indicator function of the closed convex hull of K. Hence, if K is closed and convex, then the functions I_K and p_K are in duality. Conversely, if $p: X^* \to (-\infty, \infty]$ is a sublinear $\sigma(X^* X)$-l.s.c. functional, then p^* is the indicator function of some closed convex subset in X. Indeed, if p is sublinear then $p^*(x) = p^*(2x)$ for all $x \in X$, which shows that $p^* = I_K$ for some closed convex set K in X. since $p = p^{**} = I_K^*$, it follows that p is the support functional of K. From the definition of p^* it is also clear that

$K=\{x\in X: <x,x^*>\leq p(x^*)$ for all $x^*\in X^*\}$, as observed already in Example A.1.6-(ii). Similarly, of course, the l.s.c. sublinear functionals on X are precisely the conjugate functions of the indicator functions of $\sigma(X^*,X)$-closed convex subsets in X^*.

4. Subdifferentials

Let X be a locally convex space and suppose that F is a function from X into $(-\infty,\infty]$. The function $g\in A(X)$ is called an *exact minorant* of F at the point $x\in X$ if $g\leq F$ and $g(x)=F(x)$. Then g is given by $g(y)=<y-x,x^*>+F(x)$ for all $y\in X$, for some $x^*\in X^*$.

Definition A.1.14. The function $F:X\to(-\infty,\infty]$ is called *subdifferentiable* at $x\in X$ if there exists an exact minorant in $A(X)$ of F at the point x. The corresponding $x^*\in X^*$ is called a *subgradient* of F at x, and the collection of all subgradients of F at x is called the *subdifferential* of F at x, which is denoted by $\partial F(x)$.

The (multivalued) mapping ∂F from X into X^* is called the *subdifferential mapping* of the function F. By definition, the domain $\mathfrak{D}(\partial F)$ consists of all points at which F is subdifferentiable. In the next proposition we collect some simple properties of ∂F.

Proposition A.1.15. *Assume that* $F:X\to(-\infty,\infty]$ *is proper. Then*
(i) $x^*\in\partial F(x)$ *if and only if* $x\in\operatorname{dom}(F)$ *and* $<y-x,x^*>\leq F(y)-F(x)$ *for all* $y\in X$.
(ii) *If* $\partial F(x)\neq\varnothing$, *then* $F(x)=F^{**}(x)$; *if* $F(x)=F^{**}(x)$, *then* $\partial F(x)=\partial F^{**}(x)$.
(iii) $x^*\in\partial F(x)$ *if and only if* $<x,x^*>=F(x)+F^*(x^*)$.
(iv) $\partial F(x)$ *is a convex and* $\sigma(X^*,X)$- *closed subset of* X^*; *if* F *is continuous at* x, *then* $\partial F(x)$ *is* $\sigma(X^*,X)$-*compact.*
(v) $(\partial F)^{-1}\subseteq\partial F^*$; *if* $F\in\Gamma(X)$, *then* $(\partial F)^{-1}=\partial F^*$, *i.e.,* $x^*\in\partial F(x)$ *if and only if* $x\in\partial F^*(x^*)$.
(vi) $0\in\partial F(x)$ *if and only if* $F(x)=\min\{F(y):y\in X\}$.

Proof.
(i) This is clear from the definition, since $x^*\in\partial F(x)$ if and only if the function $g(y)=<y-x,x^*>+F(x)$ is a minorant of F.
(ii) If $\partial F(x)\neq\varnothing$, then F is proper and has an affine minorant, so by Proposition A.1.12, F^{**} is the Γ-regularization of F. For $x^*\in\partial F(x)$ we have $<y-x,x^*>+F(x)\leq F(y)$ for all $y\in X$, and hence $<y-x,x^*>+F(x)\leq F^{**}(y)$ for all $y\in X$, so $F(x)\leq F^{**}(x)$. Since $F^{**}(x)\leq F(x)$ always holds, we get $F^{**}(x)=F(x)$. The proof of the second statement goes along the same lines.
(iii) As observed already in the previous section, $<x,x^*>\leq F(x)+F^*(x^*)$ is always valid. Now assume that $<x,x^*>=F(x)+F^*(x^*)$, then

$$<y-x,x^*>\leq\{F(y)+F^*(x^*)\}-\{F(x)+F^*(x^*)\}=F(y)-F(x),$$

and so $x^*\in\partial F(x)$. Conversely, if $x^*\in\partial F(x)$, then

$$<y,x^*> - F(y) \leqslant <x,x^*> - F(x)$$

for all $y \in \mathrm{dom}(F)$, and so it follows from the definition of F^* that $F^*(x^*) = <x,x^*> - F(x)$.

(iv) Evidently, $\partial F(x)$ is convex. Furthermore, it follows from (iii) that

$$\partial F(x) = \{x^* \in X^* : F^*(x^*) - <x,x^*> \leqslant -F(x)\},$$

and since the function $x^* \mapsto F^*(x^*) - <x,x^*>$ is $\sigma(X^*,X)$-l.s.c, this implies that $\partial F(x)$ is $\sigma(X^*,X)$-closed. Moreover, if F is continuous at x, then F is bounded on a neighbourhood of x, and it follows from the Alaoglu-Bourbaki theorem that $\partial F(x)$ is $\sigma(X^*,X)$-compact.

(v) If $x^* \in \partial F(x)$, then $<x,x^*> = F(x) + F^*(x^*)$. Now it follows from $F^{**} \leqslant F$ that $F^{**}(x) + F^*(x^*) \leqslant <x,x^*>$, and since the converse inequality always holds we conclude that $x \in \partial F^*(x^*)$. This shows that $(\partial F)^{-1} \subseteq \partial F^*$. If $F \in \Gamma(X)$, then by Proposition A.1.12, $F^{**} = F$, so that $(\partial F^*)^{-1} \subseteq \partial F$ and hence $(\partial F)^{-1} = \partial F^*$.

(vi) Obvious. □

The geometrical interpretation of the subdifferential should be clear by considering the epigraph of the function. Furthermore, appropriate applications of the separation theorems for convex sets to the set epi(F) in $X \times \mathbb{R}$, yield the existence of subgradients of convex functions in some cases. This is illustrated by the next proposition.

PROPOSITION A.1.16. *If the convex function* $F: X \to (-\infty, \infty]$ *is continuous in* $\mathrm{int}(\mathrm{dom}(F))$, *then* $\partial F(x) \neq \varnothing$ *for all* $x \in \mathrm{int}(\mathrm{dom}(F))$, *or, in other words,* $\mathrm{int}(\mathrm{dom}(F)) \subseteq \mathfrak{D}(\partial F)$.

PROOF. We may assume that $\mathrm{int}(\mathrm{dom}(F)) \neq \varnothing$. As noted at the end of Section A.1.2, if F is continuous on $\mathrm{int}(\mathrm{dom}(F))$, then $\mathrm{int}(\mathrm{epi}(F))$ is non-empty. Fix an element $x \in \mathrm{int}(\mathrm{dom}(F))$. Then $(x, F(x))$ is not contained in $\mathrm{int}(\mathrm{epi}(F))$, and so by the separation theorems of convex sets (as mentioned in Section A.1.1), there exists $\Phi \neq 0$ in $(X \times \mathbb{R})^*$ such that $<(x,F(x)),\Phi> \leqslant <(y,\beta),\Phi>$ for all $(y,\beta) \in \mathrm{epi}(F)$. Hence there exist $x^* \in X^*$ and $\alpha \in \mathbb{R}$ such that $<x,x^*> + \alpha F(x) \leqslant <y,x^*> + \alpha\beta$ for all (y,β) in epi(F). We claim that $\alpha > 0$. Indeed, by taking $y = x$ in the above inequality we find that $\alpha F(x) \leqslant \alpha\beta$ for all $\beta \geqslant F(x)$, so $\alpha \geqslant 0$. If $\alpha = 0$, then $<x,x^*> \leqslant <y,x^*>$ for all $y \in \mathrm{dom}(F)$, and since $x \in \mathrm{int}(\mathrm{dom}(F))$, this implies that $x^* = 0$, which is a contradiction. Therefore $\alpha > 0$. Now $(y, F(y)) \in \mathrm{epi}(F)$ for all $y \in \mathrm{dom}(F)$, so $<x,x^*> + \alpha F(x) \leqslant <y,x^*> + \alpha F(y)$, i.e., $<y-x, -\alpha^{-1}x^*> \leqslant F(y) - F(x)$, which shows that $-\alpha^{-1}x^* \in \partial f(x)$. □

REMARK A.1.17. Let K be a closed convex subset of the space X. A point $x \in K$ is called a *support point* of K if there exists $x^* \neq 0$ in X^* such that $<x,x^*> = \sup\{<y,x^*> : y \in K\}$. In this situation, x^* is called a *tangent functional* of K at x, and the hyperplane $H = \{y \in X : <y,x^*> = <x,x^*>\}$ is called a *supporting hyperplane* of K at x. Clearly, any support point of K belongs to the (topological) boundary. It is an immediate consequence of the separation theorems of convex sets that, if $\operatorname{int} K \neq \varnothing$, then every boundary point of K is a support point of K.

Now suppose that F is a proper convex (l.s.c.) function from X into $(-\infty, \infty]$. The epigraph $\operatorname{epi}(F)$ is a (closed) convex subset of $X \times \mathbb{R}$. A hyperplane H in $X \times \mathbb{R}$, which is given by $H = \{(y, \beta) : \langle y, x^* \rangle + \alpha\beta = c\}$ for some $x^* \in X^*$ and $\alpha \neq 0$, is called non-vertical. It is easy to see that the subgradients of F (in the point x) correspond to the non-vertical supporting hyperplanes of $\operatorname{epi}(F)$ (in the point $(x, F(x))$). These observations are in fact the main point in the proof of the above proposition.

If $\operatorname{int} K = \varnothing$, then there exist, in general, boundary points of K which are not support points. However, in a Banach space the Bishop-Phelps theorem states that the support points of any closed convex set K are dense in the boundary of K. The corresponding result for convex functions is the theorem of Bronsted and Rockafellar, which says that any l.s.c. convex function F from Banach space X into $(-\infty, \infty]$ is subdifferentiable on a dense subset of $\operatorname{dom}(F)$, i.e., $\overline{\mathfrak{D}(\partial F)} = \overline{\operatorname{dom}(F)}$.

A related problem is, of course, to determine the set of tangent functionals of a given closed convex set K. Clearly, if K is weakly compact, then every $x^* \in X^*$ attains its supremum on K, and so any $x^* \neq 0$ is a tangent functional of K. Another theorem of Bishop and Phelps states that the tangent functionals of a closed convex subset of a Banach space X are dense in X^*. Finally we mention in this connection that, if every $x^* \neq 0$ in X^* is a tangent functional of the closed convex subset K of the Banach space X, then K is weakly compact, which is a deep theorem of R.C. James.

If $F : X \to (-\infty, \infty]$ and $\lambda > 0$, then it is clear that $\partial(\lambda F)(x) = \lambda \partial F(x)$ for all $x \in X$. Furthermore, if F and G are two such functions, then

$$\partial F(x) + \partial G(x) \subseteq \partial(F+G)(x)$$

for all $x \in X$. In general, this inclusion is proper, but in many cases equality occurs. The following result is sometimes useful.

Proposition A.1.18. *If $F, G : X \to (-\infty, \infty]$ are convex functions such that F is continuous at some point of $\operatorname{dom}(F) \cap \operatorname{dom}(G)$, then $\partial(F+G)(x) = \partial F(x) + \partial G(x)$ for all $x \in X$.*

Proof. Take $x_0 \in \operatorname{dom}(F) \cap \operatorname{dom}(G)$. We have to show that

$$\partial(F+G)(x_0) \subseteq \partial F(x_0) + \partial G(x_0).$$

For the sake of convenience we may assume that $x_0 = 0$, and that $F(0) = G(0) = 0$. Now take $x_0^* \in \partial(F+G)(0)$. Replacing F by the function $F - x_0^*$, we may assume $x_0^* = 0$ as well. Therefore, we have to prove that $0 \in \partial(F+G)(0)$ implies that $0 \in \partial F(0) + \partial G(0)$. Note that $0 \in \partial(F+G)(0)$ is equivalent to $0 = F(0) + G(0) = \min_{x \in X}\{F(x) + G(x)\}$, i.e., to $F(x) \geqslant -G(x)$ for all $x \in X$. Consequently, the convex set $K = \{(x,\alpha) \in X \times \mathbb{R} : \alpha \geqslant -G(x)\}$ is disjoint with $\operatorname{int}[\operatorname{epi}(F)]$. Since F is continuous at some point of $\operatorname{dom}(F)$, the interior of $\operatorname{epi}(F)$ is non-empty, and hence there exist $x^* \in X^*$ and $\beta \in \mathbb{R}$ such that

$$\sup\{<x,x^*> + \beta\alpha : (x,\alpha) \in \operatorname{epi}(F)\} \leqslant \inf\{<x,x^*> + \beta\alpha : (x,\alpha) \in K\},$$

i.e., $<x,x^*> + \beta\alpha \leqslant 0$ for all $(x,\alpha) \in \operatorname{epi}(F)$ and $<x,x^*> + \beta\alpha \geqslant 0$ for all $(x,\alpha) \in K$. Note that $\alpha \leqslant 0$. We assert that $\alpha < 0$. Indeed, $\alpha = 0$ implies that $<x,x^*> \leqslant 0$ for all $x \in \operatorname{dom}(F)$ and $<x,x^*> \geqslant 0$ for all $x \in \operatorname{dom}(G)$. By hypothesis, F is continuous at some point of $\operatorname{dom}(F) \cap \operatorname{dom}(G)$, and so $\operatorname{dom}(G)$ has a non-empty intersection with $\operatorname{int}(\operatorname{dom}(F))$, a contradiction. Hence $\alpha < 0$, and it follows that $-\alpha^{-1}x^* \in \partial F(0)$ and $\alpha^{-1}x^* \in \partial G(0)$, which shows that $0 \in \partial F(0) + \partial G(0)$. □

As before, let X be a locally convex space with dual X^*. Recall that a (multivalued) mapping A from X into its dual X^* is called *monotone* if $<x_1 - x_2, y_1^* - y_2^*> \geqslant 0$ for all $x_1, x_2 \in \mathcal{D}(A)$ and $y_1^* \in Ax_1$, $y_2^* \in Ax_2$. The following proposition can be considered as a generalization of the fact that the derivative of a convex function on $\mathbb{R}$ is monotone increasing.

PROPOSITION A.1.19. *If $F : X \to (-\infty, \infty]$ is a convex function, then ∂F is a monotone mapping from X into X^*.*

PROOF. If $y_1^* \in \partial F(x_1)$ and $y_2^* \in \partial F(x_2)$, then in particular

$$<x_2, y_1^*> - <x_1, y_1^*> \leqslant F(x_2) - F(x_1)$$

and

$$<x_1, y_2^*> - <x_2, y_2^*> \leqslant F(x_1) - F(x_2).$$

Addition of these two inequalities shows that $<x_1 - x_2, y_1^* - y_2^*> \geqslant 0$. □

In the next section we will see that much more can be said about the monotonicity of ∂F in the case that X is a Hilbert space.

Next we will discuss the relation between subdifferentials, directional derivatives and the Gâteaux derivative. Let F be a function from the locally convex space X into $(-\infty, \infty]$.

DEFINITION A.1.20. The *directional derivative* $F'(x;y)$ of F at the point $x \in \operatorname{dom}(F)$ in the direction $y \in X$ is defined by

$$F'(x;y) = \lim_{\lambda \downarrow 0} \frac{F(x + \lambda y) - F(x)}{\lambda},$$

whenever the limit exists. If there exists $x^* \in X^*$ such that $F'(x;y) = <y,x^*>$ for all $y \in X$, then F is called *Gâteaux differentiable* at x, and x^* is called the *Gâteaux derivative* of F at x, which will be denoted by $F'(x)$.

The directional derivative $F'(x;y)$ is also denoted by $D_yF(x)$, and the Gâteaux derivative of F at x is also denoted by grad $F(x)$ (or $\nabla F(x)$) and called the gradient of F at x. Recall that F is called *Fréchet differentiable* at the point x if there exists $x^* \in X^*$ such that

$$F(x+y) = F(x) + <y,x^*> + R(x;\|y\|)$$

with $R(x;\|y\|)/\|y\| \to 0$ as $\|y\| \to 0$. Then F is obviously Gâteaux differentiable at x and $x^* = F'(x)$.

Now assume that F is a convex function from X into $(-\infty,\infty]$, and take $x \in \mathrm{dom}(F)$. In the next lemma we collect some simple properties of the directional derivatives of F.

LEMMA A.1.21.
(i) *For every $y \in X$ the directional derivative $F'(x;y)$ exists (with possible values $\pm\infty$).*
(ii) *The mapping $y \mapsto F'(x;y)$ from X into $[-\infty,\infty]$, is sublinear.*
(iii) $-F'(x;-y) \leqslant F'(x;y)$ *for all* $y \in X$.
(iv) *If $x \in \mathrm{int}(\mathrm{dom}(F))$, then $F'(x;y)$ is finite for all $y \in X$.*
(v) *If $x^* \in \partial F(x)$, then $-F'(x;y) \leqslant <y,x^*> \leqslant F'(x;y)$ for all $y \in X$.*

PROOF.
(i) If $y \in X$ and $0 < \lambda_1 \leqslant \lambda_2$ then by the convexity of F we have

$$\frac{F(x+\lambda_1 y) - F(x)}{\lambda_1} \leqslant \frac{F(x+\lambda_2 y) - F(x)}{\lambda_2},$$

and hence

$$\lim_{\lambda \downarrow 0} \frac{F(x+\lambda y) - F(x)}{\lambda} = \inf_{\lambda > 0} \frac{F(x+\lambda y) - F(x)}{\lambda}$$

exists ($\pm\infty$ is possible).
(ii) Obviously $F'(x;\alpha y) = \alpha F'(x;y)$ for all $\alpha \geqslant 0$ and all $y \in X$. Now take y and z in X. For $\lambda > 0$ it follows from $x + \lambda(y+z) = (x+2\lambda y)/2 + (x+2\lambda z)/2$ that

$$\frac{F(x+\lambda(y+z)) - F(x)}{\lambda} \leqslant \frac{F(x+2\lambda y) - F(x)}{2\lambda} + \frac{F(x+2\lambda z) - F(z)}{2\lambda},$$

and hence $F'(x;y+z) \leqslant F'(x;y) + F'(x;z)$ (whenever the right-hand side makes sense).
(iii) follows immediately from the sublinearity of $F'(x;\cdot)$.
(iv) Now suppose that $x \in \mathrm{int}(\mathrm{dom}(F))$ and take $y \in X$. There exists $\lambda > 0$ such that $x + \lambda y \in \mathrm{dom}(F)$, so $F'(x;y) \leqslant \lambda^{-1}\{F(x+\lambda y) - F(x)\} < \infty$. Also we find that $F'(x;-y) < \infty$, and hence $F'(x;y) \geqslant -F'(x;-y) > -\infty$.
(v) Suppose that $x^* \in \partial F(x)$. For every $y \in X$ and $\lambda > 0$ we then have $<\lambda y,x^*> \leqslant F(x+\lambda y) - F(x)$, which implies that $<y,x^*> \leqslant F'(x;y)$. Replacing in this inequality y by $-y$, we get $-<y,x^*> \leqslant F'(x;-y)$ and hence $-F'(x;-y) \leqslant <y,x^*>$. □

COROLLARY A.1.22. *If $F:X\to(-\infty,\infty]$ is convex and Gâteaux differentiable at the point $x\in\text{dom}(F)$, then F is subdifferentiable at x and $\partial F(x)=\{F'(x)\}$.*

PROOF. For all $z\in X$ we have $F'(x;z)=\langle z,F'(x)\rangle$, hence

$$F(y)-F(x)=F(x+(y-x))-F(x)\geqslant F'(x;y-x)=\langle y-x,F'(x)\rangle$$

for all $y\in X$. This shows that $F'(x)\in\partial F(x)$. Furthermore, if $x^*\in\partial F(x)$, then $-F'(x;-y)\leqslant\langle y,x^*\rangle\leqslant F'(x;y)$ for all $y\in X$, so $x^*=F'(x)$. □

If F is a convex function from X into $(-\infty,\infty]$, and if $x\in\text{dom}(F)$, then

$$\partial F(x)=\{x^*\in X^*:\langle y,x^*\rangle\leqslant F'(x;y),\ \forall y\in X\}.$$

Indeed, suppose that $x^*\in X^*$ is such that $\langle y,x^*\rangle\leqslant F'(x;y)$ for all $y\in X$. Then

$$F(y)-F(x)=F(x+(y-x))-F(x)\geqslant F'(x;y-x)\geqslant\langle y-x,x^*\rangle$$

for all $y\in X$, hence $x^*\in\partial F(x)$. If F is continuous at a point $x\in\text{dom}(F)$, then the sublinear functional $y\mapsto F'(x;y)$ is continuous. Indeed, in this case F is bounded on a neighbourhood of x, and so it follows from the inequality $F'(x;y)\leqslant F(x+y)-F(x)$ that $F'(x;\cdot)$ is bounded on a neighbourhood of 0, and hence $F'(x;\cdot)$ is continuous.

LEMMA A.1.23. *If the convex function $F:X\to(-\infty,\infty]$ is continuous at $x\in\text{dom}(F)$ then for any $\alpha\in\mathbb{R}$ such that $-F'(x;-y)\leqslant\alpha\leqslant F'(x;y)$ for some $y\in X$, there exists $x^*\in\partial F(x)$ such that $\langle y,x^*\rangle=\alpha$.*

PROOF. Set $p(z)=F'(x;z)$ for all $y\in X$. By the above observations $p:X\to\mathbb{R}$ is a continuous sublinear functional. Now define the linear subspace $M=\{\lambda y:\lambda\in\mathbb{R}\}$ and define the element $m^*\in M^*$ by $\langle\lambda y,m^*\rangle=\lambda\alpha$. From $-p(-y)\leqslant\alpha\leqslant p(y)$ it follows that $\langle z,m^*\rangle\leqslant p(z)$ for all $z\in M$, and hence m^* has a linear extension x^* to X which satisfies $\langle z,x^*\rangle\leqslant p(z)$ for all $z\in X$. From the continuity of p it follows that $x^*\in X^*$. As observed above, $\langle z,x^*\rangle\leqslant F'(x;z)$ for all $z\in X$ now implies that $x^*\in\partial F(x)$. □

The above lemma has the following immediate consequence.

PROPOSITION A.1.24. *If the convex function $F:X\to(-\infty,\infty]$ is continuous at $x\in\text{dom}(F)$, then we have $F'(x;y)=\max\{\langle y,x^*\rangle: x^*\in\partial F(x)\}$ and $-F'(x;-y)=\min\{\langle y,x^*\rangle:x^*\in\partial F(x)\}$ for all $y\in X$.*

Furthermore, the proof of the following result is now straightforward.

PROPOSITION A.1.25. *If the convex function $F:X\to(-\infty,\infty]$ is continuous at $x\in\text{dom}(F)$, then the following statements are equivalent.*
(i) *F is Gâteaux differentiable at x.*
(ii) *F is subdifferentiable at x and $\partial F(x)=\{x^*\}$.*

(iii) $F'(x;y)=-F'(x;-y)$ *for all* $y\in X$.
(iv) *The mapping* $y\mapsto F'(x;y)$ *is linear.*
(v) $\lim\limits_{\lambda\to 0}\dfrac{F(x+\lambda y)-F(x)}{\lambda}$ *exists for all* $y\in X$.

Before discussing some examples, we make a remark about subdifferentials of sublinear functionals.

REMARK A.1.26. Suppose that $p:X\to(-\infty,\infty]$ is sublinear and take $x\in\operatorname{dom}(p)$. Now it is easy to see that the directional derivative has the properties
(i) $p'(x;y)\leqslant p(y)$ for all $y\in X$, and
(ii) $p'(x;x)=-p'(x;-x)=p(x)$.
Furthermore we claim that

$$\partial p(x)=\{x^*\in X^*:\langle x,x^*\rangle=p(x) \text{ and } \langle y,x^*\rangle\leqslant p(y) \text{ for all } y\in X\}.$$

Indeed, if $x^*\in\partial p(x)$, then $\langle y,x^*\rangle\leqslant p'(x;y)\leqslant p(y)$ for all $y\in X$ and $p(x)=-p'(x;-x)\leqslant\langle x,x'\rangle\leqslant p'(x;x)=p(x)$, so $\langle x,x^*\rangle=p(x)$. Conversely, if $x^*\in X^*$ is such that $\langle y,x^*\rangle\leqslant p(y)$ for all $y\in X$ and $\langle x,x^*\rangle=p(x)$, then $\langle y-x,x^*\rangle\leqslant p(y)-p(x)$ for all $y\in X$, hence $x^*\in\partial_\rho(x)$. Note that, if p is continuous at x, and x^* is a linear functional, then $\langle y,x^*\rangle\leqslant p(y)$ for all y implies that x^* is continuous, i.e. $x^*\in X^*$.

EXAMPLES A.1.27.
(i) Let K be a closed convex subset of the locally convex space X. The indicator function I_K of K (see Example A.1.6-(iii)) is a l.s.c. convex function on X. It follows directly from the definition that $x^*\in\partial I_K(x)$ if and only if $x\in K$ and $\langle y,x^*\rangle\leqslant\langle x,x^*\rangle$ for all $y\in K$. Note that $0\in\partial I_K(x)$ for all $x\in K$, so $\mathfrak{D}(\partial I_K)=K$.
(ii) Let K be a closed convex subset of the locally convex space X with $0\in K$. The gauge functional ρ_K of K (see Example A.1.6 -(i)) is a l.s.c. convex function on X, and $K=\{x\in X:\rho_K(x)\leqslant 1\}$. It follows from the above observations that $x^*\in\partial\rho_K(x)$ if and only if $x\in\operatorname{dom}(\rho_K)$, $\langle x,x^*\rangle=\rho_K(x)$ and $\langle y,x^*\rangle\leqslant\rho_K(y)$ for all $y\in X$. Note that $\langle y,x^*\rangle\leqslant\rho_K(y)$ for all $y\in X$ is equivalent to $\langle y,x^*\rangle\leqslant 1$ for all $y\in K$. Hence

$$\partial\rho_K(x)=\{x^*\in K^0:\langle x,x^*\rangle=\rho_K(x)\}$$

for all $x\in\operatorname{dom}(\rho_K)$. Furthermore, if $\rho_K(x)>0$, then $x^*\in\partial\rho_K(x)$ if and only if $\langle x,x^*\rangle=\rho_K(x)$ and $\sup\{\langle y,x^*\rangle:y\in K\}=1$. This shows in particular that, if $\rho_K(x)=1$, then $x^*\in\partial\rho_K(x)$ if and only if $\sup\{\langle y,x^*\rangle:y\in K\}=\langle x,x^*\rangle=1$, and so in this situation $\partial\rho_K(x)$ consists precisely of those functionals $x^*\in X^*$ for which $\{y\in X:\langle y,x^*\rangle=1\}$ is a supporting hyperplane of K at the point x.

Now suppose in addition that $0\in\operatorname{int}K$, then ρ_K is continuous on X. Let x be a point of the boundary K, i.e., $\rho_K(x)=1$. It follows from Proposition A.1.15 that ρ_K is Gâteaux-differentiable at the point x, with derivative $\rho_K{}'(x)$, if and only if K has a unique supporting hyperplane H at x, which is given by $H=\{y\in X:\langle y,\rho_K'(x)\rangle=1\}$. Such a boundary point x of K is called a *smooth point of K*. In particular, K has a unique supporting hyperplane at each of its boundary points if and only if ρ_K is Gâteaux differentiable at all points $x\neq 0$.

Now consider the case that $K=B_X$, the closed unit ball in a Banach space X. Then $\rho_K(x)=\|x\|$ for all $x\in X$, and in this situation we denote the directional derivative $\rho_K'(x;y)$ by $[x,y]_+$. By the above, for $x\neq 0$ we have $\partial\rho_K(x)=\{x^*\in X^*:\|x^*\|=1$ and $\langle x,x^*\rangle=\|x\|\}$, i.e., $\partial\rho_K(x)$ consists of all *norming functionals* of x, whereas $\partial\rho_K(0)=\{x^*\in X^*:\|x^*\|\leq 1\}$. The norm in X is called *smooth* if every point of $S_X=\{x\in X:\|x\|=1\}$ is a smooth point of B_X. Consequently, the following statements are equivalent:

(i) $\|\cdot\|$ is smooth;

(ii) B_X has a unique supporting hyperplane at each point of S_X;

(iii) the norm is Gâteaux differentiable at all points $x\neq 0$.

(iv) the functional $y\mapsto[x,y]_+$ is linear for all $x\neq 0$.

If the convex function F on X is defined by $F(x)=\|x\|^2/2$, then it is not difficult to see that $\partial F(x)=\{x^*\in X^*:\|x^*\|=\|x\|$ and $\langle x,x^*\rangle=\|x\|^2\}$ for all $x\in X$. The (multivalued) mapping ∂F from X into X^* is called the *duality map*. Now it is clear that ∂F is single valued if and only if the norm in X is smooth.

Next we consider some special classes of Banach spaces.

(a) Let (Ω,Σ,μ) be a σ-finite measure space and consider the real Banach space $X=L_p(\Omega,\mu)$ with $1<p<\infty$. Take $f,g\in L_p(\Omega,\mu)$ with $f\neq 0$. By differentiation under the sign of integration we obtain

$$\frac{d}{d\lambda}(\|f+\lambda g\|_p^p)|_{\lambda=0}=p\int_\Omega gf|f|^{p-2}d\mu,$$

and so the function $\lambda\mapsto\|f+\lambda g\|_p$ is differentiable with

$$\frac{d}{d\lambda}(\|f+\lambda g\|_p)|_{\lambda=0}=\int_\Omega g\frac{f|f|^{p-2}}{\|f\|_p^{p-1}}d\mu.$$

This shows that the norm $\|\cdot\|_p$ is Gâteaux differentiable at $f\neq 0$, with Gâteaux derivative $f|f|^{p-2}\|f\|_p^{1-p}$ (as usual, we identify the dual of L_p with L_q, $p^{-1}+q^{-1}=1$, via the canonical pairing).

(b) Consider the real Banach space $X=L_1(\Omega,\Sigma,\mu)$ for some σ-finite measure space (Ω,Σ,μ). The norm $\|\cdot\|_1$ is not smooth, but it is possible to describe the smooth points of the unit ball in L_1. Indeed, if $f\in L_1$, $\|f\|=1$, is such that $f(\omega)\neq 0$ μ-a.e., then the subdifferential of the norm at f consists only of the function $f|f|^{-1}$ (usual identification of L_1^* with L_∞). If, however, the set $N=\{\omega:f(\omega)=0\}$ has positive measure, then this subdifferential consists of all functions of the form $f|f|^{-1}+h$, with $|h|\leq\mathbf{1}_N$ (where $\mathbf{1}_N$ denotes the characteristic function of N), and so these points f are not smooth.

(c) Let $X=C(\Omega)$, the space of all real continuous functions on a compact Hausdorff space, with the supremum norm. The norm in this space is not smooth. In fact, the subdifferential of the norm at an element f consists of one functional if and only if the function $|f|$ attains its maximum at precisely one point of Ω.

(d) If $\mathcal{H}$ is a real Hilbert space with inner product $(\cdot,\cdot)$, then the norm is Gâteaux differentiable at any $x\neq 0$, and the derivative can be identified with $\|x\|^{-1}x$, i.e., $[x,y]_+=(y,\|x\|^{-1}x)$ for all $y\in\mathcal{H}$.

Finally we note that it is possible that the closed unit ball in a Banach space doesn't have any smooth points. Examples of such spaces are L_∞-spaces on non-atomic measure spaces. We note, however that if K is a subset of the *separable* Banach space X, with $\operatorname{int} K\neq\varnothing$, then by theorem of Mazur, the smooth points of A form a dense G_δ-set in the boundary of A. The corresponding theorem for convex functions states that a convex function F from a separable Banach space X into $(-\infty,\infty]$, which is continuous on $\operatorname{int}(\operatorname{dom}(F))$, is Gâteaux differentiable on a dense G_δ-subset of $\operatorname{int}(\operatorname{dom}(F))$. These results follow from a more general theorem. In fact, it was shown by ZARANTONELLO [1973], that, if A is a (multivalued) monotone mapping from a separable Banach space X into X^* (cf. Proposition A.1.19), then A is single valued on some dense G_δ-subset of $\operatorname{int}(\mathcal{D}(A))$.

5. Convex functions on Hilbert spaces

In this section H will denote a real Hilbert space with inner product $(\cdot,\cdot)$. As usual, the dual of H is identified with H itself. Recall that the (multivalued) mapping A from H into itself, i.e., $A\subseteq H\times H$, is called *monotone* if $(x_1-x_2,y_1-y_2)\geqslant 0$ for all $x_1,x_2\in\mathcal{D}(A)$ and $y_1\in Ax_1, y_2\in Ax_2$.
Furthermore, A is called *accretive* if $(I+\lambda A)^{-1}$ is a contraction for all $\lambda>0$, i.e., if $\|x_1-x_2\|\leqslant\|(x_1-x_2)+\lambda(y_1-y_2)\|$ for all elements $x_1,x_2\in\mathcal{D}(A)$, and $y_1\in Ax_1, y_2\in Ax_2$, and all $\lambda>0$ (equivalently, $-A$ is *dissipative;* see Definition 2.2). Note that A is accretive if and only if $[x_1-x_2,y_1-y_2]_+\geqslant 0$ whenever $y_1\in Ax_1$ and $y_2\in Ax_2$. Since $[x,y]_+=\|x\|^{-1}(x,y)$ for all $x\neq 0$ (and $[0,y]=\|y\|$), it is clear that A is monotone if and only if A is accretive. The monotone mapping A is called *maximal monotone* if it follows from $A\subseteq B$, B monotone, that $A=B$ (equivalently, $(x_1-x,y_1-y)\geqslant 0$ for all $x\in\mathcal{D}(A)$ and $y\in Ax$, implies that $x_1\in\mathcal{D}(A)$ and $y_1\in Ax_1$). The accretive mapping A is called *m-accretive* if $\mathcal{R}(I+\lambda A)=H$ for all $\lambda>0$ (equivalently, $\mathcal{R}(I+\lambda A)=H$ for some $\lambda>0$). It is easy to see that any m-accretive mapping is maximal monotone, and it can be shown that the converse holds as well. If A is m-accretive, then $(I+\lambda A)^{-1}$ is a contraction from X onto $\mathcal{D}(A)$ for all $\lambda>0$. Then $J_\lambda^A=(I+\lambda A)^{-1}$ for $\lambda>0$, is called the *resolvent* of A (note that $J_\lambda^A=E_\lambda^{-A}$; see Section 2.2), and the mapping $A_\lambda=\lambda^{-1}(I-J_\lambda^A)$, $\lambda>0$, is called the *Yosida approximation* of A.

For any convex function $F:H\to(-\infty,\infty]$, the subdifferential mapping ∂F is now a mapping from H into itself, i.e., $\partial F\subseteq H\times H$. As observed in Proposition A.1.19, ∂F is monotone. Note that if F is the function defined by $F(x)=\frac{1}{2}\|x\|^2$ for all $x\in H$, then $\partial F=I$.

PROPOSITION A.1.28. *If $F \in \Gamma(H)$, then ∂F is m-accretive.*

PROOF. Let $\lambda > 0$ be given. We have to show that $\mathcal{R}(I+\lambda\partial F)=H$. Take $y_0 \in H$ and consider the convex function G defined by $G(x)=(2\lambda)^{-1}\|x-y_0\|^2+F(x)$. It follows from Proposition A.1.18 that $\partial G(x)=\lambda^{-1}(x-y_0)+\partial F(x)$, and hence $0\in\partial G(x_0)$ if and only if $y_0=x_0+\lambda\partial F(x_0)$. Since $0\in\partial G(x_0)$ if and only if $G(0)=\inf_{x\in H}G(x)$, it remains to show that G is bounded from below and attains its infimum. To this end, first note that, by Proposition A.1.4, F has an affine minorant $g\in A(H)$, so $G(x)\geqslant(2\lambda)^{-1}\|x-y_0\|^2+g(x)$ for all $x\in H$, and hence G is bounded from below. Put $\alpha=\inf_{x\in H}G(x)$, and let $\{x_n\}$ be a sequence such that $G(x_n)\downarrow\alpha$. Using once more that F has an affine minorant, it follows that $\{x_n\}$ is bounded, and hence $\{x_n\}$ has a subsequence $\{x_{n_k}\}$ which is weakly convergent to an element $x_0\in H$. Since G is l.s.c., G is weakly l.s.c. as well (see section A.1.1), and so $G(x_0)=\liminf G(x_{n_k})=\alpha$, and we are done. □

Note that it follows from the proof of the above proposition that the function $G(x)=(2\lambda)^{-1}\|x-y_0\|^2+F(x)$ attains its minimum at the unique point $J_\lambda^{\partial F}y_0=(I+\lambda\partial F)^{-1}y_0$. Moreover, it is implicit in the above result that $\mathcal{D}(\partial F)$ is non-empty. Next we will consider the Yosida approximation of ∂F.

PROPOSITION A.1.29. *Let $F\in\Gamma(H)$, and define for $\lambda>0$ the convex function F_λ on H by*

$$F_\lambda(x)=\min_{y\in H}\{\|y-x\|^2/2\lambda+F(y)\}$$

for all $x\in H$. Then the following statements hold.

(i) $\mathrm{dom}(F_\lambda)=H$, *$F_\lambda$ is Fréchet differentiable with derivative $F_\lambda{}'=\partial F_\lambda=(\partial F)_\lambda$, where $(\partial F)_\lambda$ is the Yosida approximation of F.*

(ii) *The domain of ∂F is dense in the effective domain of F, i.e., $\overline{\mathcal{D}(\partial F)}=\overline{\mathrm{dom}(F)}$.*

(iii) *$F_\lambda(x)\uparrow F(x)$ as $\lambda\downarrow 0$ for all $x\in H$.*

PROOF. For the sake of convenience we set $A=\partial F$, which is an m-accretive operator by the above proposition. As above, J_λ^A denotes the resolvent, and A_λ denotes the Yosida approximation of A, for all $\lambda>0$. We recall from Section 2.2 that $A_\lambda\subseteq AJ_\lambda^A$ and that $\|A_\lambda x-A_\lambda y\|\leqslant(2/\lambda)\|x-y\|$ for all $x,y\in H$. Moreover, since $A_\lambda=\lambda^{-1}(I-J_\lambda^A)$, and J_λ^A is a contraction, the mappings A_λ are m-accretive.

(i) As observed above, the minimum in the definition of $F_\lambda(x)$ is attained in the unique point $J_\lambda^A x$ for all $x\in H$, and so $\mathrm{dom}(F_\lambda)=H$ and

$$F_\lambda(x)=(2\lambda)^{-1}\|J_\lambda^A x-x\|+F(J_\lambda^A x)=(\lambda/2)\|A_\lambda x\|^2+F(J_\lambda^A x).$$

We claim that $A_\lambda x\in\partial F_\lambda(x)$ for all $x\in H$. Indeed, first note that it follows from the inclusion $A_\lambda\subseteq AJ_\lambda^A$ that $A_\lambda x\in\partial F_\lambda(J_\lambda^A x)$, and so

$$(J_\lambda^A y-J_\lambda^A x,A_\lambda x)\leqslant F(J_\lambda^A y)-F(J_\lambda^A x),$$

for all $y \in H$. Now

$$F_\lambda(y) - F_\lambda(x) = (\lambda/2)(\|A_\lambda y\|^2 - \|A_\lambda x\|^2) + (F(J_\lambda^A y) - F(J_\lambda^A))\geqslant$$
$$(\lambda/2)(\|A_\lambda y\|^2 - \|A_\lambda x\|^2) + (J_\lambda^A y - J_\lambda^A x\,, A_\lambda x) =$$
$$(y - x, A_\lambda x) + (\lambda/2)(\|A_\lambda y\|^2 - \|A_\lambda x\|^2) + \lambda(A_\lambda x - A_\lambda y, A_\lambda x) =$$
$$(y - x, A_\lambda x) + (\lambda/2)\|A_\lambda y - A_\lambda x\|^2 \geqslant (y - x, A_\lambda x),$$

for all $y \in H$, which shows that $A_\lambda x \in \partial F_\lambda(x)$. We thus have proved that $A_\lambda \subseteq \partial F_\lambda$. Note that, since A_λ and ∂F_λ are *m*-accretive, it follows that $A_\lambda = \partial F_\lambda$, which shows already that F_λ is Gâteaux differentiable with derivative $F_\lambda{}' = A_\lambda$. In order to show that F_λ if Fréchet differentiable, take $x, y \in H$, and note that if follows from $A_\lambda y \in \partial F_\lambda(y)$ that $F_\lambda(x) - F_\lambda(y) \geqslant (x - y, A_\lambda y)$. Hence

$$0 \leqslant F_\lambda(y) - F_\lambda(x) - (y - x, A_\lambda x) \leqslant (x - y, A_\lambda x - A_\lambda y) \leqslant (\lambda/2)\|x - y\|^2,$$

and so F_λ is Fréchet differentiable.
(ii) We will show that $J_\lambda^A x \to x$ as $\lambda \downarrow 0$ for all $x \in H$ for which $\sup_{\lambda>0} F_\lambda(x) = M < \infty$. Let such an x be fixed. By Proposition A.1.4 there exist $z \in H$ and $\alpha \in \mathbb{R}$ such that $F(y) \geqslant <y, z> + \alpha$ for all $y \in H$. Now

$$\frac{1}{2\lambda}\|J_\lambda^A x - x\|^2 \leqslant M - F(J_\lambda^A x)$$
$$\leqslant M - <J_\lambda^A x, z> - \alpha$$
$$= M + <x, z> - \alpha + <J_\lambda^A x, z>$$
$$\leqslant M + <x, z> - \alpha + \frac{1}{2}\|z\|^2 + \frac{1}{2}\|J_\lambda^A x - x\|^2,$$

which implies that $\|J_\lambda^A x - x\| \to 0$ as $\lambda \downarrow 0$.
Since $F_\lambda(x) \leqslant F(x)$ for all x, the above implies that $J_\lambda^A x \to x$ as $\lambda \downarrow 0$ for all $x \in \operatorname{dom}(F)$, and since $J_\lambda^A x \in \mathfrak{D}(\partial F)$ for all $\lambda > 0$, we may conclude that $\mathfrak{D}(\partial F)$ is dense in $\operatorname{dom}(F)$.
(iii) Evidently, $\sup_{\lambda>0} F_\lambda(x) \leqslant F(x)$ for all $x \in H$. For the proof of the converse inequality we may assume that $\sup F_\lambda(x) < \infty$. By the proof of (ii) we then have $J_\lambda^A x \to x$ as $\lambda \downarrow 0$. Now it follows from the lower semi-continuity of F that

$$F(x) = \liminf_{\lambda \downarrow 0} F(J_\lambda^A x) \leqslant \liminf_{\lambda \downarrow 0} F_\lambda(x) = \sup_{\lambda > 0} F_\lambda(x),$$

and the proof is complete. □

It was mentioned in Remark A.1.17, without proof, that $\mathfrak{D}(\partial F)$ is dense in $\operatorname{dom}(F)$ for any $F \in \Gamma(X)$, with X a Banach space. Statement (ii) in the above proposition is a special case of this general result.

Appendix 2

Ordered Banach Spaces

1. Ordered Banach spaces

The purpose of this appendix is to list some terminologies and results of the theory of partially ordered Banach spaces, which is used in the sections on positive semigroups. The real vector space X is called an *ordered vector space* if a partial ordering "$\leqslant$" is defined in X with the additional property that $x \leqslant y$ in X implies that $x+z \leqslant y+z$ for all $z \in X$, and $\lambda x \leqslant \lambda y$ for all $0 \leqslant \lambda \in \mathbb{R}$. Given such a partial ordering, the *positive cone* of X is defined by $X^+ = \{x \in X : x \geqslant 0\}$. Clearly, X^+ is a cone, i.e., $\alpha x + \beta y \in X^+$ whenever $x, y \in X^+$ and $0 \leqslant \alpha, \beta \in \mathbb{R}$. Moreover, $X^+ \cap (-X^+) = \{0\}$, so X^+ is a *proper cone.* Conversely, given the proper cone K in X, a partial order in X is defined by setting $x \leqslant y$ whenever $y - x \in K$, and then $(X, \leqslant)$ is an ordered vector space with positive cone $X^+ = K$.

The real Banach space $(X, \|\cdot\|)$ is called an *ordered Banach space* if X is an ordered vector space such that X^+ is norm closed. Note that any ordered Banach space is Archimedean, i.e., if $0 \leqslant nx \leqslant y$ for all $n = 1, 2, \ldots,$ then $x = 0$. Indeed, if $0 \leqslant nx \leqslant y$ for all n, then $n^{-1}y - x \in X^+$ for all n, and so $-x \in X^+$, since X^+ is closed. Since X^+ is proper, we may conclude that $x = 0$.

From now on we assume that X is an ordered Banach space with positive cone X^+. The cone X^+ is called *generating* if $X = X^+ - X^+$ (and X^+ is called *weakly generating* if $X^+ - X^+$ is norm dense in X). From the Baire category theorem it follows that, if X^+ is generating, then there exists a constant $C \geqslant 1$ such that every $x \in X$ has a decomposition $x = y - z$ with $y, z \in X^+$ and $\max(\|y\|, \|z\|) \leqslant C\|x\|$. Given $x, y \in X$ we denote by $[x, y]$ the *order interval* determined by x and y, i.e., $[x, y] = \{z \in X : x \leqslant z \leqslant y\}$. A subset S of X is called *solid* if $x,y \in S$ implies that $[x,y] \subseteq S$. A subset D of X is called *order bounded* if D is contained in some order interval. Note that order intervals are closed in X.

Now we introduce some properties which relate the norm in X with the order.

DEFINITION A.2.1.

(i) The positive cone X^+ is called normal if there exists a neighbourhood base at 0 (for the norm topology) consisting of solid sets (i.e. if there exist $\epsilon > 0$ and a solid set S such that $\epsilon B \subseteq S \subseteq B$, where $B = \{x \in X : \|x\| \leqslant 1\}$).

(ii) The norm in X is called M-monotone if there exists a constant $M \geqslant 1$ such that $0 \leqslant x \leqslant y$ implies that $\|x\| \leqslant M\|y\|$. If $M = 1$, then the norm is simply called monotone.

PROPOSITION A.2.2. *In an ordered Banach space X the following conditions are equivalent.*

(i) *X^+ is normal,*

(ii) *order bounded sets are norm bounded,*

(iii) *the norm is M-monotone for some $M \geqslant 1$.*

PROOF. (i)$\Rightarrow$(ii). It is sufficient to show that $[0,x]$ is norm bounded for $0 < x \in X$. By hypothesis, there exist an $\epsilon > 0$ and a solid set S such that $\epsilon B \subseteq S \subseteq B$. Then $\epsilon\|x\|^{-1}x \in \epsilon B \subseteq S$, and so $[0, \epsilon\|x\|^{-1}x] \subseteq S \subseteq B$, which implies that $\|z\| \leqslant \epsilon^{-1}\|x\|$ for all $z \in [0,x]$.
(ii)$\Rightarrow$(iii). Suppose that (iii) does not hold. Then we can find elements $0 < x_n \leqslant y_n$ in X with $\|x_n\| > 4^n\|y_n\|$ $(n = 1,2, \cdots)$. Replacing x_n and y_n by $\|y_n\|^{-1}$ x_n and $\|y_n\|^{-1}y_n$ respectively, we may assume that $0 \leqslant x_n \leqslant y_n$, $\|y_n\| \leqslant 1$ and $\|x_n\| > 4^n$ $(n = 1,2 \cdots)$. Define $y = \Sigma_{n=1}^{\infty} 2^{-n} y_n$, then $0 \leqslant 2^{-n}x_n \leqslant 2^{-n}y_n \leqslant y$ for all n. Since $\|2^{-n}x_n\| > 2^n$, this implies that $[0,y]$ is not norm bounded, which is a contradiction.
(iii)$\Rightarrow$(i). As before denote by $B = \{x \in X : \|x\| \leqslant 1\}$. Let S be the solid hull of B, i.e. $S = \bigcup \{[x,y] : x,y \in B\}$. Clearly, S is a solid set and $B \subseteq S$. Now suppose that $x \leqslant z \leqslant y$ with $x,y \in B$. Then $z = x + (z - x)$, and since $0 \leqslant z - x \leqslant y$, it follows from the assumption on the norm that $\|z\| \leqslant \|x\| + \|z - x\| \leqslant \|x\| + M\|y\| \leqslant 1 + M$. This shows that $S \subseteq (1+M)B$, and therefore X^+ is normal. □

In connection with the above proposition we note that it follows from Proposition 7.9 that X^+ is normal if and only if there exists a monotone norm in X which is equivalent to the given norm.

Suppose that X is an ordered Banach space with positive cone X^+. The *dual cone* in the Banach dual X^* is defined by

$$(X^*)^+ = \{\phi \in X^* : \langle x, \phi \rangle \geqslant 0 \text{ for all } x \geqslant 0\}.$$

Evidently, $(X^*)^+$ is a closed cone in X^*, which is proper if and only if X^+ is weakly generating. The elements of $(X^*)^+$ are called *norm bounded positive* *linear functionals* on X. The following proposition shows that, in some sense, there exist sufficiently many positive linear functionals on X.

PROPOSITION A.2.3. *If* $x \in X$, *then* $x \in X^+$ *if and only if* $<x,\phi> \geq 0$ *for all* $\phi \in (X^*)^+$.

PROOF. By definition of $(X^*)^+$, if $x \in X^+$ then $<x,\phi> \geq 0$ for all $\phi \in (X^*)^+$. Now assume that $x \in X \setminus X^+$. Since X^+ is a closed convex subset, it follows from the Hahn-Banach theorem that there exist $\phi \in X^*$ and $\alpha \in \mathbb{R}$ such that $<x,\phi> < \alpha < <y,\phi>$ for all $y \in X^+$. Since $y \in X^+$ implies that $\lambda y \in X^+$ for all $0 \leq \lambda \in \mathbb{R}$, it follows that $\alpha < 0$ and $<y,\phi> \geq 0$ for all $y \in X^+$. Hence, $\phi \in (X^*)^+$ and $<x,\phi> < 0$, and we are done. □

PROPOSITION A.2.4. *If the positive cone* X^+ *is normal and generating, then the dual cone* $(X^*)^+$ *is normal and generating.*

PROOF. First we show that $(X^*)^+$ is normal. As observed at the beginning of this section, since X^+ is generating, there exists a constant $C \geq 1$ such that every $x \in X$ has a decomposition $x = x_1 - x_2$ with $x_1, x_2 \in X^+$ and $\max(\|x_1\|, \|x_2\|) \leq C\|x\|$. Suppose $0 \leq \psi \leq \phi$ in X^* (note that, since X^+ is generating, $(X^*)^+$ is proper, and so X^* is an ordered Banach space with positive cone $(X^*)^+$). Take $x \in X$ and write $x = x_1 - x_2$ as above. Then

$$|<x,\psi>| \leq <x_1,\psi> + <x_2,\psi> \leq <x_1 + x_2,\phi>$$
$$\leq \|\phi\|(\|x_1\| + \|x_2\|) \leq 2C\|\phi\| \cdot \|x\|,$$

and hence $\|\psi\| \leq 2C\|\phi\|$. This shows that the norm in X^* is $2C$-monotone, and so $(X^*)^+$ is normal.

Now we shall prove that $(X^*)^+$ is generating. Since X^+ is normal, it follows from Proposition A.2.2 that the norm in X is M-monotone for some $M \geq 1$. Let $\phi \in X^*$ be given. For $x \in X^+$ define $p(x) = \sup\{<z,\phi> : z \in [0,x]\}$. Note that $p(x) < \infty$. In fact, if $z \in [0,x]$, then $<z,\phi> \leq \|\phi\| \cdot \|z\| \leq M\|\phi\| \cdot \|x\|$, so $p(x) \leq M\|\phi\| \cdot \|x\|$. It is easy to see that $p(x+y) \geq p(x) + p(y)$ for all $x,y \in X^+$ and $p(\alpha x) = \alpha p(x)$ for all $x \in X^+$ and $0 \leq \alpha \in \mathbb{R}$. Consider the product space $\mathbb{R} \times X$ with norm $\|(\alpha,x)\| = |\alpha| + \|x\|$, and define

$$K = \{(\alpha,x) \in \mathbb{R} \times X : x \in X^+ \text{ and } 0 \leq \alpha \leq p(x)\},$$

which is clearly a cone in $\mathbb{R} \times X$. Observe that, if $(\alpha_n, x_n) \in K$ $(n = 1,2, \cdots)$ such that $x_n \to 0$ (norm) in X, then it follows from $0 \leq \alpha_n \leq p(x_n) \leq M\|\phi\| \cdot \|x_n\|$, that $\alpha_n \to 0$. This shows that $(1,0) \notin \overline{K}$, and so, by the Hahn-Banach separation theorem, there exists an $\eta \in (\mathbb{R} \times X)^*$ and a constant c such that $<(1,0),\eta> < c < <(\alpha,x),\eta>$ for all $(\alpha,x) \in K$. Since K is a cone it follows that $<(1,0),\eta)> < 0$ and $<(\alpha,x),\eta> \geq 0$ for all $(\alpha,x) \in K$. We may assume that $<(1,0),\eta> = -1$. Since $(\mathbb{R} \times X)^* = \mathbb{R} \times X^*$, there exist $\psi \in X^*$ and $\lambda \in \mathbb{R}$ such that $<(\alpha,x),\eta> = \lambda\alpha + <x,\psi>$ for all $(\alpha,x) \in \mathbb{R} \times X$. It follows that $\lambda = -1$, and so $<(\alpha,x),\eta> = -\alpha + <x,\psi>$. We claim that $\psi \in (X^*)^+$. Indeed, if

$x \in X^+$, then $(0,x) \in K$ and so $<x,\psi> = <(0,x),\eta> \geq 0$. Furthermore, if $x \in X^+$ then $(p(x),x) \in K$, so $-p(x) + <x,\psi> \geq 0$, which implies that $<x,\phi> \leq p(x) \leq <x,\psi>$. Therefore $<x,\psi-\phi> \geq 0$ for all $x \in X^+$, so that $\psi - \phi \in (X^*)^+$. Now $\phi = \psi - (\psi - \phi)$, and hence $(X^*)^+$ is generating. □

REMARK. If the ordered Banach space X possesses the so called *Riesz decomposition property* (i.e., if $0 \leq x,y \in X$ and $0 \leq z \leq x+y$, then z can be decomposed as $z = u + v$ with $0 \leq u \leq x$ and $0 \leq v \leq y$), then the second half of the proof of the above proposition can be simplified considerably. Indeed, in that situation it is easily seen that the functional p is not only superadditive, but even additive on X^+, and therefore there exists a $\psi_0 \in (X^*)^+$ such that $p(x) = <x,\psi_0>$ for all $x \in X^+$. The decomposition $\phi = \psi_0 - (\psi_0 - \phi)$ is now the desired one, as $\phi \leq \phi_0$. Moreover, it is not difficult to see that ψ_0 is in fact the supremum of the linear functionals ϕ and 0 in the ordered Banach space X^*. This shows that, in this case, X^* is a vector lattice (see section A.2.4). In particular, any vector lattice has the Riesz decomposition property.

We end this section with some remarks concerning the complexification of real Banach spaces. Let $(X, \|\cdot\|)$ be a real Banach space. The product space $X \times X$ is a complex vector space with the linear structure defined by $(x,y)+(u,v) = (x+u, y+v)$ and $(\alpha + i\beta)(x,y) = (\alpha x - \beta y, \beta x + \alpha y)$ for all $x,y,u,v \in X$ and all $\alpha, \beta \in \mathbb{R}$. The mapping $x \rightarrow (x, 0)$ is a real-linear isomorphism from X onto a real-linear subspace of $X \times X$. Via this isomorphism, X is identified with a real-linear subspace of $X \times X$, and now it is clear that $X \times X = X \oplus iX$. This space $X_{\mathbb{C}} = X \oplus iX$ is called the *complexification* of the real space X. Equip $X_{\mathbb{C}}$ with the norm defined by

$$\|x + iy\|_{\mathbb{C}} = \sup_{0 \leq \theta \leq 2\pi} \|(\cos\theta)x + (\sin\theta)y\|, \quad x,y \in X.$$

Since $\|x\|, \|y\| \leq \|x+iy\|_{\mathbb{C}} \leq \|x\| + \|y\|$, it is evident that $X_{\mathbb{C}}$ is a Banach space. Note that the embedding of X into $X_{\mathbb{C}}$ is an isometry. If $z = x + iy \in X_{\mathbb{C}}$, with $x,y \in X$, then we write $x = \operatorname{Re} z$ and $y = \operatorname{Im} z$. Similarly, $X = \operatorname{Re} X_{\mathbb{C}}$. In many situations there exists already a norm on $X_{\mathbb{C}}$ which makes $X_{\mathbb{C}}$ into a complex Banach space and the restriction of which to X coincides with the given norm in X. Such a norm need not be equal to $\|\cdot\|_{\mathbb{C}}$, but is equivalent to $\|\cdot\|_{\mathbb{C}}$.

Assume that X and Y are real Banach spaces with complexifications $X_{\mathbb{C}}$ and $Y_{\mathbb{C}}$ respectively. As usual, $\mathfrak{L}(X,Y)$ denotes the Banach space of all (real) linear bounded operators from X into Y, and $\mathfrak{L}(X_{\mathbb{C}}, Y_{\mathbb{C}})$ is the space of all (complex) linear bounded operators from $X_{\mathbb{C}}$ into $Y_{\mathbb{C}}$. Any $T \in \mathfrak{L}(X,Y)$ has a unique extension to a complex linear $T_{\mathbb{C}} \in \mathfrak{L}(X_{\mathbb{C}}, Y_{\mathbb{C}})$, given by $T_{\mathbb{C}}(x+iy) = Tx + iTy$ for all $x,y \in X$. The mapping $T \mapsto T_{\mathbb{C}}$ is a real-linear isomorphism from $\mathfrak{L}(X,Y)$ into $\mathfrak{L}(X_{\mathbb{C}}, Y_{\mathbb{C}})$, and we can consider $\mathfrak{L}(X,Y)$ as a real-linear subspace of $\mathfrak{L}(X_{\mathbb{C}}, Y_{\mathbb{C}})$. Conversely, any $T \in \mathfrak{L}(X_{\mathbb{C}}, Y_{\mathbb{C}})$ can be written uniquely in the form $T = T_1 + iT_2$ for some $T_1, T_2 \in \mathfrak{L}(X,Y)$. Indeed, define $T_1 x = \operatorname{Re}(Tx)$ and $T_2 x = \operatorname{Im}(Tx)$ for all $x \in X$. Furthermore, if $T \in \mathfrak{L}(X,Y)$ with operator norm $\|T\|$, then $\|T\| = \|T_{\mathbb{C}}\|$, the operator norm in $\mathfrak{L}(X_{\mathbb{C}}, Y_{\mathbb{C}})$ (in $X_{\mathbb{C}}$ and $Y_{\mathbb{C}}$ we consider the complexifications of the given norms). It follows from the above observations that $\mathcal{L}(X_{\mathbb{C}}, Y_{\mathbb{C}})$ is the complexification of $\mathcal{L}(X, Y)$ as a vector space. In general, however, the operator norm in $\mathcal{L}(X_{\mathbb{C}}, Y_{\mathbb{C}})$ is not the complexification of the operator norm in $\mathcal{L}(X, Y)$ (but these norms are equivalent).

DEFINITION A.2.5. A complex Banach space is called an ordered Banach space if it is the complexification of a real ordered Banach space.

2. ORDERED BANACH SPACES WITH ORDER UNIT

Let $(X, \|\cdot\|)$ be an ordered Banach space with positive cone X^+.

DEFINITION A.2.6. The element $u \in X^+$ is called an order unit in X if for every $x \in X$ there exists $0 \leqslant \lambda \in \mathbb{R}$ such that $-\lambda u \leqslant x \leqslant \lambda u$.

If X possesses an order unit, then X^+ is generating. Indeed if $-\lambda u \leqslant x \leqslant \lambda u$, then $x = (x + \lambda u)/2 - (\lambda u - x)/2$, and $x + \lambda u$, $\lambda u - x \geqslant 0$. The next proposition gives a topological characterization of order units.

PROPOSITION A.2.7. *The element* $u \in X^+$ *is an order unit if and only if* $u \in \operatorname{int} X^+$.

PROOF. First suppose that $u \in \operatorname{int} X^+$. Then there exists an $\epsilon > 0$ such that $B(u, \epsilon) = \{x \in X : \|x - u\| \leqslant \epsilon\} \subseteq X^+$. Take $x \in X$, $x \neq 0$. Then it is clear that $u \pm \epsilon x / \|x\| \in B(u, \epsilon)$, so $u \pm \epsilon x / \|x\| \geqslant 0$ which implies that $-\epsilon^{-1} \|x\| u \leqslant x \leqslant \epsilon^{-1} \|x\| u$. Hence u is an order unit.

Now assume that $u \in X^+$ is an order unit, so $X = \cup_{\lambda > 0} [-\lambda u, \lambda u]$. By the Baire category theorem, there exists a $\lambda_0 > 0$ and a $w \in X$ such that $w \in \operatorname{int}[-\lambda_0 u, \lambda_0 u]$. Then $0 \in \operatorname{int}([-\lambda_0 u, \lambda_0 u] - w) \subseteq \operatorname{int}[-\lambda_1 u, \lambda_2 u]$ for some $\lambda_1 > 0$, and so $B(0,1) \subseteq [-\alpha u, \alpha u]$, for an appropriate $\alpha > 0$. This implies that $B(u, \alpha^{-1}) \subseteq X^+$, so $u \in \operatorname{int} X^+$. □

As observed in Proposition A.2.2, X^+ is normal if and only if all order intervals in X are norm bounded. This result can be improved in spaces with an order unit.

PROPOSITION A.2.8. *If X is an ordered Banach space with* $\operatorname{int} X^+ \neq \varnothing$, *then* X^+ *is normal if and only if there exists* $u \in \operatorname{int} X^+$ *such that* $[-u, u]$ *is norm bounded.*

PROOF. Suppose that $u \in \operatorname{int} X^+$ is such that $\|x\| \leqslant M$ for all $x \in [-u, u]$. As in the proof of the above proposition, there exists an $\epsilon > 0$ such that $-\epsilon^{-1} \|x\| u \leqslant x \leqslant \epsilon^{-1} \|x\| u$ for all $x \in X$. Now assume that $0 < z \leqslant y$ in X, then $0 \leqslant \epsilon \|y\|^{-1} y \leqslant u$, and so $0 \leqslant \epsilon \|y\|^{-1} z \leqslant u$. This implies that $\|z\| \leqslant (\epsilon^{-1} M) \|y\|$, hence the norm is $\epsilon^{-1} M$-monotone. The converse is evident from Proposition A.2.2. □

In general, norm bounded subsets of an ordered Banach space X are not order bounded. In fact, a glance at the proof of Proposition A.2.7 shows that all norm bounded sets in X are order bounded if and only if X has an order unit.

Let X be an ordered Banach space with $u \in \operatorname{int} X^+$. The *order unit norm* with respect to u in X is defined by

$$\|x\|_u = \inf\{\lambda \geqslant 0 : -\lambda u \leqslant x \leqslant \lambda u\}, \quad x \in X.$$

The verification that $\|\cdot\|_u$ is a norm is left to the reader. Note that $-\|x\|_u u \leqslant x \leqslant \|x\|_u u$, and so the above infimum is, in fact, a minimum. As observed in the proof of Proposition A.2.7, there exists $\epsilon > 0$ such that $-\epsilon^{-1}\|x\|u \leqslant x \leqslant \epsilon^{-1}\|x\|u$ for all $x \in X$, and hence $\|x\|_u \leqslant \epsilon^{-1}\|x\|$ for all $x \in X$. Furthermore, $\|\cdot\|_u$ is an *absolutely monotone norm*, i.e., if $-y \leqslant x \leqslant y$ in X, then $\|x\|_u \leqslant \|y\|_u$. In particular, $\|\cdot\|_u$ is monotone. From the above observations it follows easily that X^+ is normal if and only if $\|\cdot\|_u$ and $\|\cdot\|$ are equivalent norms. The ordered Banach space X is called an *order unit space* if there exists $u \in \operatorname{int} X^+$ such that $\|\cdot\| = \|\cdot\|_u$.

There is a notion of interior point which is weaker than order unit.

DEFINITION A.2.9. Let X be an ordered Banach space. The element $u \in X^+$ is called a quasi-interior point if $\langle u,\phi\rangle > 0$ for all $0 < \phi \in X^*$.

If $u \in X^+$ is an order unit, then u is a quasi-interior point. Indeed, take $0 \leqslant \phi \in X^*$ and suppose that $\langle u,\phi\rangle = 0$. For $x \in X$ there exists $0 \leqslant \lambda \in \mathbb{R}$ such that $-\lambda u \leqslant x \leqslant \lambda u$, which implies that $-\lambda\langle u,\phi\rangle \leqslant \langle x,\phi\rangle \leqslant \lambda\langle u,\phi\rangle$, and so $\langle x,\phi\rangle = 0$. Hence $\phi = 0$.

Furthermore, it should be observed that $u \in X^+$ is a quasi-interior point if and only if for every $x \in X$ and $\epsilon > 0$ there exists $y \in X$ such that $\|x - y\| < \epsilon$ and $y \leqslant \lambda u$ for some $0 \leqslant \lambda \in \mathbb{R}$ (this last statement simply says that $\cup_{\lambda > 0}(\lambda u - X^+)$ is norm dense in X; the equivalence is a straightforward application of the Hahn-Banach separation theorem).

PROPOSITION A.2.10. *If X is an ordered Banach space with $\operatorname{int} X^+ \neq \varnothing$, then $u \in X^+$ is an order unit if and only if u is a quasi-interior point.*

PROOF. As observed above, if u is an order unit, then u is a quasi-interior point. Now assume that $u \in X^+$ is a quasi-interior point, and suppose that $u \notin \operatorname{int} X^+$. Since $\operatorname{int} X^+ \neq \varnothing$, it follows from the Hahn-Banach separation theorem that there exists $0 \neq \phi \in X^*$ such that $\langle u,\phi\rangle \leqslant \langle x,\phi\rangle$ for all $x \in X^+$. Since $x \in X^+$ implies $\alpha x \in X^+$ for all $\alpha \geqslant 0$, it follows that $\langle x,\phi\rangle \geqslant 0$ for all $x \in X^+$. In particular, $\langle u,\phi\rangle \geqslant 0$. On the other hand $\langle u,\phi\rangle \leqslant 0$, as $0 \in X^+$, and so $\langle u,\phi\rangle = 0$. Since u is quasi-interior point, this implies that $\phi = 0$, which is a contradiction. □

3. POSITIVE OPERATORS

Suppose that X and Y are ordered Banach spaces. A linear operator T from X into Y is called *positive* if $T(X^+) \subseteq Y^+$ (equivalently, $Tx \leqslant Ty$ in Y whenever $x \leqslant y$ in X). In many situations, positive linear operators are automatically norm bounded.

PROPOSITION A.2.11. *If X^+ is generating and Y^+ is normal, then every positive linear operator T from X into Y is norm bounded.*

PROOF. As observed at the beginning of section A.2.1, since X^+ is generating, there exists a constant $C \geqslant 1$ such that every $x \in X$ can be decomposed as $x = x' - x''$ with $x', x'' \geqslant 0$ and $\max(\|x'\|, \|x''\|) \leqslant C\|x\|$. Furthermore, since Y^+ is normal, there exists (by Proposition A.2.2) a constant $M \geqslant 1$ such that $0 \leqslant y \leqslant y'$ in Y implies that $\|y\| \leqslant M\|y'\|$. Now assume that T is a positive linear operator from X into Y, which is not norm bounded. Then there exist $x_n \in X$ such that $\|x_n\| \leqslant 1$ and $\|Tx_n\| \geqslant 2n^3$ $(n = 1, 2, \cdots)$. Take $0 \leqslant x_n', x_n'' \in X$ such that $x_n = x_n' - x_n''$ and $\max(\|x_n'\|, \|x_n''\|) \leqslant C\|x_n\|$ for all n. Since $\|Tx_n'\| + \|Tx_n''\| \geqslant \|Tx_n\| \geqslant 2n^3$, we have $\|Tx_n'\| \geqslant n^3$ or $\|Tx_n''\| \geqslant n^3$. Therefore we can find $0 \leqslant z_n \in X$ such that $\|z_n\| \leqslant C$ and $\|Tz_n\| \geqslant n^3$ for all n. Now define $z = \Sigma_{n=1}^{\infty} n^{-2} z_n$, then $0 \leqslant n^{-2} z_n \leqslant z$, so $0 \leqslant n^{-2} Tz_n \leqslant Tz$ for all n. This implies that $n \leqslant n^{-2}\|Tz_n\| \leqslant M\|Tz\|$ for $n = 1, 2, \ldots,$ which is absurd, and we are done. □

For the special case that $Y = \mathbb{R}$ we obtain immediately the following result.

COROLLARY A.2.12. *If X is an ordered Banach space with generating positive cone, then every positive linear functional on X is norm bounded.*

If we combine the above corollary with Proposition A.2.4, then we see that for an ordered Banach space X with generating, normal positive cone, the Banach dual X^* consists precisely of those linear functionals which can be written as the difference of two positive linear functionals.

Given the ordered Banach spaces X and Y, the collection of all bounded positive linear operators from X into Y is denoted by $\mathfrak{L}(X, Y)^+$, which clearly is a (strongly) closed cone in the Banach space $\mathfrak{L}(X, Y)$. If X^+ is generating, then $\mathfrak{L}(X, Y)^+$ is a proper cone and so $\mathfrak{L}(X, Y)$ is an ordered Banach space. It should be noted that only in some special situations $\mathfrak{L}(X, Y)^+$ happens to be generating. It is not difficult to see that, if X^+ is generating and Y^+ is normal, then $\mathfrak{L}(X, Y)^+$ is normal as well. The operator norm in $\mathfrak{L}(X, Y)$ is said to be *positively attained* if

$$\|T\| = \sup\{\|Tx\| : 0 \leqslant x \in X, \|x\| \leqslant 1\},$$

for all $T \in \mathfrak{L}(X, Y)^+$. Note that if the operator norm is positively attained, then $\mathfrak{L}(X, Y)^+$ is proper. Furthermore, if the operator norm is positively attained and if the norm in Y is monotone, then the operator norm is likewise monotone.

Now we will discuss some sufficient conditions on X and Y in order that the operator norm is positively attained. To this end we introduce the following properties of an ordered Banach space $(X, \|\cdot\|)$:

(i) $\|\cdot\|$ is called *absolutely monotone* if $-y \leqslant x \leqslant y$ in X implies that $\|x\| \leqslant \|y\|$ (note that 'absolutely monotone' implies 'monotone');

(ii) X^+ is called *approximately absolutely dominating* if for every $x \in X$ and all $\alpha > 1$ there exists $y \in X^+$ such that $-y \leqslant x \leqslant y$ and $\|y\| \leqslant \alpha \|x\|$ (note that such a cone is generating);

(iii) X^+ is called *absolutely dominating* if for every $x \in X$ there exists $y \in X^+$ such that $-y \leqslant x \leqslant y$ and $\|y\| \leqslant \|x\|$;

(iv) $\|\cdot\|$ is called a *Riesz norm* if $\|\cdot\|$ is absolutely monotone and X^+ is approximately absolutely dominating ; furthermore, $\|\cdot\|$ is called a *strong Riesz norm* if $\|\cdot\|$ is absolutely monotone and X^+ is absolutely dominating.

From the above definitions it is clear that:

(1) $\|\cdot\|$ is a Riesz norm if and only if $\|x\| = \inf\{\|y\| : -y \leqslant x \leqslant y,\ y \in X^+\}$ for all $x \in X$,

(2) $\|\cdot\|$ is a strong Riesz norm if and only if $\|x\| = \min\{\|y\| : -y \leqslant x \leqslant y, y \in X^+\}$.

Observe that, if $-y \leqslant x \leqslant y$ in X, then $(x+y)/2, (y-x)/2 \geqslant 0$. Since, obviously, $x = (x+y)/2 - (y-x)/2$ and $y = (y+x)/2 + (y-x)/2$, this shows that $\|\cdot\|$ is a Riesz norm if and only if

$$\|x\| = \inf\{\|u+v\| : x = u - v,\ 0 \leqslant u, v \in X\},$$

and $\|\cdot\|$ is a strong Riesz norm if and only if the infimum is attained.

PROPOSITION A.2.13. *If X and Y are ordered Banach spaces such that X^+ is approximately absolutely dominating and the norm in Y is absolutely monotone, then the norm in $\mathfrak{L}(X,Y)$ is positively attained. In particular, if X is an ordered Banach space with a Riesz norm then the operator norm in $\mathfrak{L}(X)$ is positively attained.*

PROOF. Take $0 \leqslant T \in \mathfrak{L}(X,Y)$ and set $\|T\|_+ = \sup\{\|Tx\| : 0 \leqslant x \in X, \|x\| \leqslant 1\}$. Clearly, $\|Tx\| \leqslant \|T\|_+ \|x\|$ for all $0 \leqslant x \in X$, and $\|T\|_+ \leqslant \|T\|$. Take $\alpha > 1$. Given $x \in X$ there exists $u \in X^+$ such that $-u \leqslant x \leqslant u$ and $\|u\| \leqslant \alpha \|x\|$. Since T is positive, we obtain that $-Tu \leqslant Tx \leqslant Tu$ and hence $\|Tx\| \leqslant \|Tu\|$, so that $\|Tx\| \leqslant \|Tu\| \leqslant \|T\|_+ \|u\| \leqslant \alpha \|T\|_+ \|x\|$. This shows that $\|T\| \leqslant \alpha \|T\|_+$. Since $\alpha > 1$ is arbitrary, we get $\|T\| \leqslant \|T\|_+$, and therefore $\|T\| = \|T\|_+$. □

COROLLARY A.2.14. *If X is an ordered Banach space such that X^+ is approximately absolutely dominating, then the norm in X^* is positively attained and monotone.*

REMARK.

1) If $(X, \|\cdot\|)$ is an ordered Banach space, and if there exists an equivalent

Riesz norm in X, then it is clear that X^+ is generating and normal. Conversely, if X^+ is generating and normal, then it is not difficult to see that

$$|||x||| = \inf\{\|u+v\| : 0 \leqslant u, v \in X, x = u - v\}, \quad x \in X,$$

defines a Riesz norm in X, which is equivalent to the given norm.

2) Suppose that X is an ordered Banach space with $u \in \operatorname{int} X^+$. We claim that the order unit norm $\|\cdot\|_u$ is a strong Riesz norm in X. Indeed, as observed already in Section A.2.2, $\|\cdot\|_u$ is monotone and $-\|x\|_u u \leqslant x \leqslant \|x\|_u u$ for all $x \in X$, and so X^+ is absolutely dominating (with respect to $\|\cdot\|_u$). In other words, the norm in any order unit space X is a strong Riesz norm, and in particular, the norm in $\mathfrak{L}(X)$ is positively attained.

4. Banach lattices

Let X be an ordered vector space (see the beginning of section A.2.1). If for any two elements $x,y \in X$ the $\sup(x,y)$ and $\inf(x,y)$ exist, then X is called a *vector lattice* (or *Riesz space*). As usual we denote $\sup(x,y)$ by $x \vee y$, and $\inf(x,y)$ by $x \wedge y$. Furthermore we denote $x^+ = x \vee 0$ and $x^- = (-x) \vee 0$. Then $x = x^+ - x^-$ and $x^+ \wedge x^- = 0$. In particular, the positive cone in a vector lattice is generating. Moreover, for any $x \in X$ the *absolute value* is defined by $|x| = x \vee (-x)$. Then $|x| = x^+ + x^- = x^+ \vee x^-$.

A norm $\|\cdot\|$ on the vector lattice X is called a *Riesz norm* if $|x| \leqslant |y|$ in X implies that $\|x\| \leqslant \|y\|$ (it should be observed that this terminology is consistent with the one introduced in the preceding section). If the vector lattice X is a Banach space with respect to the Riesz norm $\|\cdot\|$, then $(X, \|\cdot\|)$ is called a *Banach lattice*. Note that it follows from Proposition A.2.11 that, if $\|\cdot\|_1$ and $\|\cdot\|_2$ are two Riesz norms on the vector lattice X, such that $(X, \|\cdot\|_1)$ and $(X, \|\cdot\|_2)$ are both Banach spaces, then $\|\cdot\|_1$ and $\|\cdot\|_2$ are equivalent norms.

From now on we assume that $(X, \|\cdot\|)$ is a Banach lattice. We introduce some further terminology and notation. A sequence $\{x_n\}_{n=1}^{\infty}$ in X is called *increasing* if $x_n \leqslant x_{n+1}$ for all n, which is denoted by $x_n \uparrow$. If, in addition $\sup_n x_n = x$ exists in X, then we set $x_n \uparrow x$. An indexed subset $\{x_\alpha\}$ of X is called *upwards directed* if for any pair α_1, α_2 there exists α_3 such that $x_{\alpha_1}, x_{\alpha_2} \leqslant x_{\alpha_3}$, which is denoted by $x_\alpha \uparrow$, and if $\sup_\alpha x_\alpha = x$ exists in X, then we set $x_\alpha \uparrow x$. Similarly for decreasing sequences and downwards directed systems. Note that if $\{x_n\}_{n=1}^{\infty}$ is such that $x_n \uparrow$, and $x_n \to x$ (norm), then $x_n \uparrow x$. Next we introduce some special classes of subspaces of a Banach lattice X. If Y is a linear subspace of X, then:

(i) Y is called a *vector sublattice* (or *Riesz subspace*) of X if $x,y \in Y$ implies that $x \vee y$, $x \wedge y \in Y$ (equivalently, $|x| \in Y$ whenever $x \in Y$),

(ii) Y is called an *ideal* if it follows from $y \in Y, x \in X$ and $|x| \leqslant |y|$ that $x \in Y$,

(iii) Y is called a *band* if Y is an ideal with the additional property that, if $D \subseteq Y$ and $\sup D = x \in X$ exists, then $x \in Y$ (equivalently, Y is an ideal and it follows from $0 \leqslant x_\alpha \uparrow x$, $x_\alpha \in Y$ that $x \in Y$).

It is easy to see that an ideal is a vector sublattice. Furthermore, the norm closure of any ideal (vector sublattice) is likewise an ideal (vector sublattice). Any band in X is norm closed. For an element $z \in X$ we denote by X_z the ideal generated by z in X. Such an ideal is called a *principal ideal,* and it is easily seen that

$$X_z = \{x \in X : |x| \leqslant n|z| \text{ for some } n \in \mathbb{N}\}.$$

If $0 \leqslant u \in X$, then u is an order unit in X (Definition A.2.6.) if and only if $X = X_u$. Moreover, u is a quasi-interior point in X (Definition A.2.9) if and only if $X = \overline{X_u}$.

Two elements $x, y \in X$ are called *disjoint* if $|x| \wedge |y| = 0$, which is denoted by $x \perp y$. For any subset $D \subseteq X$ we put

$$D^d = \{x \in X : x \perp y \text{ for all } y \in D\},$$

the *disjoint complement* of D. Any disjoint complement D^d is a band in X. Clearly $D \subseteq D^{dd}$, and so it is clear that the band generated by D is contained in D^{dd}. However, it can be shown (using the Archimedean property) that $B^{dd} = B$ holds for any band B in X. This implies that for any $D \subseteq X$, the band generated by D precisely equals D^{dd}. For any element $z \in X$ we denote by B_z the band generated by z (equivalently, the band generated by X_z). It follows from the above observations that $B_z = \{z\}^{dd}$, and such a band is called a *principal band.* Clearly, for any $z \in X$ we have

$$X_z \subseteq \overline{X}_z \subseteq B_z = \{z\}^{dd}.$$

If $0 \leqslant u \in X$ is such that $\{u\}^{dd} = X$, then u is called a *weak order unit* in X. Note that $0 \leqslant u \in X$ is a weak order unit if and only if it follows from $x \in X$ and $|x| \wedge u = 0$ that $x = 0$. Furthermore, it is clear from the above inclusions that any quasi-interior point is a weak order unit.

For the sake of convenience we mention another useful description of $\overline{X}_z$ and B_z. For any $0 \leqslant z \in X$ we have

$$(\overline{X}_z)^+ = \{0 \leqslant x \in X : x \wedge (nz) \to x \text{ (norm) as } n \to \infty\},$$

$$B_z^+ = \{0 \leqslant x \in X : x \wedge (nz) \uparrow x\}.$$

We leave it to the reader to formulate the corresponding descriptions of quasi-interior point and weak order unit.

A band B in X is called a *projection band* if $X = B \oplus B^d$. The projection P onto B along B^d is called the *band projection* onto B. This projection P satisfies $0 \leqslant Px \leqslant x$ for all $0 \leqslant x \in X$ (i.e., $0 \leqslant P \leqslant I$ in $\mathfrak{L}(X)$). Conversely, if P is a projection in X and $0 \leqslant P \leqslant I$, then $P(X)$ is a band and P is the band projection onto $P(X)$. If all bands in X are projection bands, then we say that X has the *projection property.*

The Banach lattice X is called *Dedekind complete* if every order bounded subset of X has a supremum in X; equivalently, if $0 \leqslant x_\alpha \uparrow \leqslant x$ in X, then there exists $y \in X$ such that $x_\alpha \uparrow y$. If X is Dedekind complete, then X has the projection property. Indeed, if B is a band in X, then the band projection P_B onto B

is given by

$$P_B x = \sup\{y \in B : 0 \leqslant y \leqslant x\}, \quad 0 \leqslant x \in X,$$

where the existence of the supremum is guaranteed by the Dedekind completeness of X. In fact, it can be shown that Dedekind completeness of the Banach lattice X is equivalent to the projection property.

If every principal band in X is a projection band, then we say that X has the *principal projection property,* which is in fact equivalent to *Dedekind σ-completeness* of X (i.e., every countable order bounded subset of X has a supremum). In this situation, the band projection P_z onto the principal band B_z generated by $0 \leqslant z \in X$ is given by

$$P_z x = \sup\{x \wedge (nz) : n \in \mathbb{N}\}, \quad 0 \leqslant x \in X.$$

A remarkable and important property of band projections is that these projections mutually commute. In fact, if P_A and P_B are band projections onto the band A and B respectively, then $A \cap B$ is likewise a projection band and the corresponding band projection is given by $P_{A \cap B} = P_A P_B = P_B P_A$. Furthermore, the sum $A + B$ is a projection band as well, and the corresponding band projection is given by $P_{A+B} = P_A + P_B - P_A P_B$ (it should be noted that, in general, the sum of two bands need not be a band).

As before, X^* denotes the Banach dual of the Banach lattice X. In X^* we consider the ordering induced by the dual cone $(X^*)^+$, so, if $\phi, \psi \in X^*$, then $\phi \leqslant \psi$ if and only if $<x,\phi> \leqslant <x,\psi>$ for all $0 \leqslant x \in X$. As observed already in the remark following Proposition A.2.4, X^* is a vector lattice, and for each $\phi \in X^*$ the positive part $\phi^+ = \sup(\phi, 0)$ is given by

$$<x,\phi^+> = \sup\{<z,\phi> : z \in X, 0 \leqslant z \leqslant x\}, \quad 0 \leqslant x \in X.$$

This implies that for $\phi, \psi \in X^*$ the functionals $\phi \vee \psi$ and $\phi \wedge \psi$ are given by

$$<x,\phi \vee \psi> = \sup\{<y,\phi> + <z,\psi> : 0 \leqslant y, z \in X \text{ such that } y + z = x\}$$

$$<x,\phi \wedge \psi> = \inf\{<y,\phi> + <z,\psi> : 0 \leqslant y, z \in X \text{ such that } y + z = x\}$$

for all $0 \leqslant x \in X$. In particular,

$$<x,|\phi|> = \sup\{|<y,\phi>| : y \in X, |y| \leqslant x\},$$

for all $0 \leqslant x \in X$ and all $\phi \in X^*$. It should be noted that $|<x,\phi>| \leqslant <|x|,|\phi|>$ for all $x \in X$ and $\phi \in X^*$. Now it is not difficult to show that with respect to the usual dual norm, X^* is a Banach lattice, and

$$\|\phi\| = \sup\{<x,\phi> : 0 \leqslant x \in X, \|x\| \leqslant 1\}$$

for all $0 \leqslant \phi \in X^*$. Moreover, X^* is Dedekind complete. In fact, if $0 \leqslant \phi_\alpha \uparrow \leqslant \psi$ in X^*, then $\phi_0 = \sup_\alpha \phi_\alpha$ is given by

$$<x,\phi_0> = \sup_\alpha <x,\phi_\alpha>, \quad 0 \leqslant x \in X.$$

We introduce some more terminology. The norm $\|\cdot\|$ in the Banach lattice X is called *σ-order continuous* if it follows from $0 \leqslant x_n \in X$ ($n=1,2,\cdots$) and $x_n \downarrow 0$ that $\|x_n\| \downarrow 0$. Furthermore, if $x_\alpha \downarrow 0$ implies $\|x_\alpha\| \downarrow 0$ for arbitrary downwards directed systems in X, then $\|\cdot\|$ is called *order continuous* (a similar terminology is used for positive linear functionals). It can be shown that a Banach lattice with order continuous norm is Dedekind complete. Moreover, the following statements are equivalent in a Banach lattice X:

(1) the norm in X is order continuous;

(2) the norm in X is σ-order continuous and X is Dedekind σ-complete.

If X has order continuous norm, then the bands in X are precisely the closed ideals in X (in fact, the order continuity of the norm is equivalent to this latter property). In particular, $B_z = \overline{X_z}$, for all $z \in X$, and so, in that situation, the notions of quasi-interior point and weak order unit coincide.

Suppose that K is a compact Hausdorff space, and denote by $C(K)$ the vector space of all real continuous functions on K. With respect to the pointwise ordering, $C(K)$ is clearly a vector lattice. Moreover, equipped with the sup-norm $\|\cdot\|_\infty$, $C(K)$ is a Banach lattice. Evidently, the function $\mathbf{1}$, identically equal to one, is an order unit in $C(K)$ and $\|\cdot\|_\infty$ is equal to the order unit norm $\|\cdot\|_\mathbf{1}$. In other words, $C(K)$ is an order unit space with respect to the order unit $\mathbf{1}$ (see Section A.2.2). For Banach lattices which are order unit spaces, there is a representation theorem analogous to the Gelfand representation for commutative Banach algebras.

THEOREM A.2.15 (KAKUTANI'S REPRESENTATION THEOREM). *Suppose that $(X, \|\cdot\|)$ is a Banach lattice which is an order unit space with respect to the order unit $0 \leqslant u \in X$ (so $\|\cdot\| = \|\cdot\|_u$). Then there exists a unique compact Hausdorff space K and a linear isometry $x \mapsto \hat{x}$, which preserves the lattice operations, from X onto $C(K)$, such that $\hat{u} = \mathbf{1}$.*

Let $(X, \|\cdot\|)$ be a Banach lattice and take $0 \leqslant u \in X$. Clearly, u is an order unit in the principal ideal X_u generated by u. Furthermore, it is not difficult to see that X_u is Banach with respect to the order unit norm $\|\cdot\|_u$. Therefore, $(X_u, \|\cdot\|_u)$ is a Banach lattice which is an order unit space, and so by the above theorem, $(X_u, \|\cdot\|_u)$ is isometrically isomorphic to the Banach lattice $C(K_u)$ for some compact Hausdorff space K_u. It should be noted that, in general, the order unit norm $\|\cdot\|_u$ is not equivalent to the given norm $\|\cdot\|$ on X_u.

Next we discuss some special classes of linear operators on Banach lattices. Given the Banach lattices X and Y, the linear operator T from X into Y is called a *Riesz homomorphism* (or *lattice homomorphism*) if $T(x \vee y) = (Tx) \vee (Ty)$ and $T(x \wedge y) = (Tx) \wedge (Ty)$ for all $x, y \in X$. Clearly, any Riesz homomorphism is a positive operator. The linear operator T from X into Y is a Riesz homomorphism if and only if $(Tx) \wedge (Ty) = 0$ in Y whenever $x \wedge y = 0$ in X, which is in its turn equivalent to saying that $|Tx| = T|x|$ for all $x \in X$. Furthermore, it is evident that the kernel $\mathfrak{N}(T)$ of a Riesz homomorphism T is a closed ideal, and that the range $\mathfrak{R}(T)$ is a vector sublattice.

As usual, there is a canonical way to produce Riesz homomorphisms. Suppose that J is a closed ideal in the Banach lattice X. The quotient space X/J is a Banach space with respect to the quotient norm, given by

$$\|[x]\| = \inf\{\|y\| : y \in [x]\} = \inf\{\|y\| : x - y \in J\},$$

where $[x] = x + J$. Denote the canonical map by T, i.e., $Tx = [x]$ for all $x \in X$. Now define $(X/J)^+ = T(X^+)$, which is clearly a cone in X/J. Furthermore, it is a proper cone. Indeed, if $[x] \in X/J$ is such that there exist $0 \leqslant x_1, x_2 \in X$ with $x_1 \in [x]$ and $-x_2 \in [x]$, then $x - x_1, x + x_2 \in J$ and $x - x_1 \leqslant x \leqslant x + x_2$, which implies that $x \in J$, i.e., that $[x] = [0]$. Therefore, X/J is a partially ordered vector space with $(X/J)^+$ as positive cone. Moreover, it is not difficult to show that X/J is in fact a vector lattice, with

$$[x] \vee [y] = [x \vee y],$$

$$[x] \wedge [y] = [x \wedge y],$$

for all $[x], [y] \in X/J$. In particular $|[x]| = [|x|]$ for all $[x] \in X/J$. Now suppose that $|[x]| \leqslant |[y]|$. Put $z = (x \wedge |y'|) \vee (-|y'|)$, then $[z] = [x]$ and $|z| \leqslant |y'|$ for all $y' \in [y]$, and so $\|[x]\| \leqslant \|z\| \leqslant \|y'\|$ for all $y' \in [y]$. This shows that $\|[x]\| \leqslant \|[y]\|$, and hence X/J is a Banach lattice. The canonical mapping from X onto X/J is clearly a Riesz homomorphism.

Another example of a Riesz homomorphism is the embedding of a Banach lattice into its second dual space. Let X be a Banach lattice, then the Banach dual X^* is a Dedekind complete Banach lattice, and hence the bidual X^{**} is a Dedekind complete Banach lattice as well. Let $j : X \to X^{**}$ be the canonical embedding of X into X^{**}, so $<x, \phi> = <jx, \phi>$ for all $x \in X$ and all $\phi \in X^*$. The fact that j is a Riesz homomorphism follows from the so-called Namioka-formulas

$$<x \vee y, \phi> = \sup\{<x, \psi_1> + <y, \psi_2> : 0 \leqslant \psi_1, \psi_2 \in X^*, \psi_1 + \psi_2 = \phi\},$$

$$<x \wedge y, \phi> = \inf\{<x, \psi_1> + <y, \psi_2> : 0 \leqslant \psi_1, \psi_2 \in X^*, \psi_1 + \psi_2 = \phi\},$$

for all $x, y \in X$ and all $0 \leqslant \phi \in X^*$. This implies in particular that the image of X, under the canonical embedding, is a Banach sublattice of X^{**}.

An important and useful class of operators in a Banach lattice X is the *center* $Z(X)$ of X. A linear operator π from X into itself belongs to $Z(X)$ if there exists $0 \leqslant \lambda \in \mathbb{R}$ such that $-\lambda x \leqslant \pi x \leqslant \lambda x$ for all $0 \leqslant x \in X$. Clearly, $Z(X)$ is a subalgebra of $\mathfrak{L}(X)$, and $Z(X)$ is an ordered vector space with positive cone $Z(X)^+ = Z(X) \cap \mathfrak{L}(X)^+$. Furthermore, it is evident that the identity I in X belongs to $Z(X)$, and that I is an order unit in $Z(X)$. It can be shown that $Z(X)$ is a vector lattice, with

$$(\pi_1 \vee \pi_2)(x) = (\pi_1 x) \vee (\pi_2 x), \quad (\pi_1 \wedge \pi_2)(x) = (\pi_1 x) \wedge (\pi_2 x)$$

for all $\pi_1, \pi_2 \in Z(X)$ and all $0 \leqslant x \in X$. In particular, $|\pi| x = |\pi x|$, $\pi^+(x) = (\pi x)^+$ and $\pi^-(x) = (\pi x)^-$ for all $\pi \in Z(X)$ and $0 \leqslant x \in X$. Moreover, $Z(X)$ is a commutative subalgebra of $\mathfrak{L}(X)$. For any $\pi \in Z(X)$ the operator norm $\|\pi\|$ is equal to the order in norm $\|\pi\|_I$. This shows in particular that $Z(X)$ is norm closed in $\mathfrak{L}(X)$. It should be noted that any $\pi \in Z(X)$ is *band preserving*, i.e., $\pi(B) \subseteq B$ for all bands $B \subseteq X$, equivalently, it follows from $x \perp y$ in X that $\pi x \perp y$ (i.e., π is an *orthomorphism*). It is a by no means trivial result that any band preserving linear operator in the Banach lattice X is automatically norm bounded and belongs to $Z(X)$. From this last remark it follows that $Z(X)$ is, in fact, a maximal abelian subalgebra of $\mathfrak{L}(X)$.

If $X = C(K)$ for some compact Hausdorff space K, then $Z(X)$ can be identified with the space $C(K)$, acting on itself by multiplication. In general, the center can be considered as the algebra of 'abstract multiplication operators' in X. It is possible that $Z(X)$ is trivial, i.e., consists of multiples of the identity only. However, any band projection in X belongs to $Z(X)$, and so in Dedekind complete Banach lattices the center has a rich structure.

Also in Banach lattices with a quasi-interior point the center contains sufficiently many elements. To be more precise, suppose that $0 < e \in X$ is a quasi-interior point, so $X = \overline{X}_e$. Since, by continuity, any $\pi \in Z(X_e)$ extends uniquely to an element $\pi \in Z(X)$, it follows easily that $Z(X_e)$ and $Z(X)$ are isomorphic. By the Kakutani Representation Theorem A 2.15, there exists a compact Hausdorff space K_e and a norm and Riesz isomorphism $x \mapsto \hat{x}$ from X_e onto $C(K_e)$. Collecting the above observations we see that $Z(X)$ is isomorphic to $C(K_e)$ and for any $x \in X_e$ there exists a unique $\pi \in Z(X)$ such that $\pi e = x$.

This last result can also be applied locally, i.e., to the closure of principal ideals in X, which shows that for any $u \in X$ there exists a unique $\pi \in Z(\overline{X}_{|u|})$ such that $\pi|u| = u$.

We end this section with some remarks concerning the complexification of a Banach lattice. Let $(X, \|\cdot\|)$ be a real Banach lattice, and let $X_{\mathbf{C}} = X \oplus iX$ be the complexification of the vector space X (see the last part of section A.2.1). The absolute value in X can be extended in a natural way to $X_{\mathbf{C}}$. In fact, if $z \in X_{\mathbf{C}}$, $z = x + iy$ with $x, y \in X$, then

$$|z| = \sup\{(\cos\theta)x + (\sin\theta)y : 0 \leqslant \theta \leqslant 2\pi\}$$

exists in X (which can be seen, for example, by using A.2.15). Observe that $|x| \leqslant |z|$ and $|z| \leqslant |x| + |y|$. Furthermore, $|z_1 + z_2| \leqslant |z_1| + |z_2|$ for all $z_1, z_2 \in X_{\mathbf{C}}$. If $z = x + iy$, with $x, y \in X$ and $\bar{z} = x - iy$, then $|\bar{z}| = |z|$.

Now the norm in X can be extended to $X_{\mathbf{C}}$ by setting $\|z\|_{\mathbf{C}} = \||z|\|$ for all $z \in X_{\mathbf{C}}$, and then $(X_{\mathbf{C}}, \|\cdot\|_{\mathbf{C}})$ is a complex Banach space, which is called a *complex Banach lattice.* All of the above results extend straightforwardly to complex Banach lattices. We note in particular that the dual $X_{\mathbf{C}}^*$ is the complexification of X^*, and the dual norm in $X_{\mathbf{C}}^*$ is precisely the complexification of the dual norm in X^*. As in the real case, the center of $X_{\mathbf{C}}$ is the set of all linear operators from $X_{\mathbf{C}}$ into itself such that $|\pi z| \leqslant \lambda |z|$ for all

$z \in X_{\mathbb{C}}$ and all $0 \leqslant \lambda \in \mathbb{R}$. It can be shown that $Z(X_{\mathbb{C}})$ is the complexification of $Z(X)$. For the sake of convenience we explicitly state the following result, the proof of which is the same as for real spaces.

PROPOSITION A.2.16. *Let $X_{\mathbb{C}}$ be a complex Banach lattice. For any $z \in X_{\mathbb{C}}$ there exists a unique $\pi \in Z(\overline{X_{|z|}})$ such that $\pi|z| = z$. Moreover, π is invertible in $Z(\overline{X_{|z|}})$ with $\pi^{-1} = \overline{\pi}$.*

Appendix 3

Some Results from Spectral Theory

1. The Browder essential spectrum

Let $X \neq \{0\}$ be a complex Banach space with norm $\|\cdot\|$. We denote its (topological dual) with X^*, and for two elements $x^* \in X^*$ and $x \in X$ their duality pairing is denoted by $<x,x^*>$. Let L be a closed linear operator on X with domain $\mathfrak{D}(L)$. We denote the null space and range of L by $\mathfrak{N}(L)$ and $\mathfrak{R}(L)$ respectively. The spectrum of L is denoted by $\sigma(L)$, the point spectrum by $\sigma_p(L)$, the residual spectrum by $\sigma_r(L)$, the continuous spectrum by $\sigma_c(L)$, and, finally, the approximate point spectrum by $\sigma_{ap}(L)$. The resolvent set $\rho(L)$ is by definition the complement of the spectrum, so $\rho(L) = \mathbb{C} \setminus \sigma(L)$. We denote the set of all bounded linear operators on X with $\mathfrak{L}(X)$, which is a Banach algebra. For every element $L \in \mathfrak{L}(X)$ we can define its spectral radius $r(L)$:

$$r(L) = \sup \{|\lambda| : \lambda \in \sigma(L)\}.$$

A well-known result from spectral theory says that for any $L \in \mathfrak{L}(X)$

$$r(L) = \lim_{n \to \infty} \|L^n\|^{1/n}. \tag{A.3.1}$$

We denote the boundary of a subset V of $\mathbb{C}$ by ∂V. For every closed linear operator on X the following inclusion holds:

$$\partial\sigma(L) \subseteq \sigma_{ap}(L). \tag{A.3.2}$$

For every $\lambda \in \rho(L)$ the resolvent operator $R(\lambda, L)$ is defined by

$$R(\lambda, L) = (\lambda I - L)^{-1}.$$

In stead of $\lambda I - L$ we shall often write $\lambda - L$.

Throughout the remainder of this section we assume that L is a closed linear operator on X. We are particularly interested in the isolated elements of the spectrum $\sigma(L)$. Let λ_0 be an isolated point in $\sigma(L)$. Then the resolvent $R(\lambda, L)$ can be expanded in a Laurent series about λ_0:

$$R(\lambda,L) = \sum_{k=-\infty}^{\infty} (\lambda-\lambda_0)^k B_k, \qquad \text{(A.3.3)}$$

where for each integer k,

$$B_k = \frac{1}{2\pi i}\int_\Gamma (\lambda-\lambda_0)^{-k-1} R(\lambda,L)d\lambda, \qquad \text{(A.3.4)}$$

and Γ is a positively oriented circle of sufficiently small radius such that no other points than λ_0 of $\sigma(L)$ lie on or inside Γ. The following relations hold:

$$B_{-k-1}B_{-l-1} = B_{-k-l-1}\,, k\,, l \geqslant 0. \qquad \text{(A.3.5)}$$

Our next result is stated without proof.

THEOREM A.3.1.
a) B_{-1} *is a projection in* X, $\mathcal{R}(B_{-1})$ *and* $\mathcal{R}(I-B_{-1})$ *are closed, and the restriction of* L *to* $\mathcal{R}(B_{-1})$ *is bounded and has spectrum* $\{\lambda_0\}$.
b) *If* $\mathcal{R}(B_{-1})$ *has finite dimension, then* λ_0 *is a pole of* $R(\lambda,L)$.
c) *If* λ_0 *is a pole of* $R(\lambda,L)$ *of order* p, *then* λ_0 *is an eigenvalue of* L,

$$\mathcal{R}(B_{-1}) = \mathcal{N}((\lambda_0-L)^p) = \mathcal{N}((\lambda_0-L)^{p+1}) = ...,$$

$$\mathcal{R}(I-B_{-1}) = \mathcal{R}((\lambda_0-L)^p) = \mathcal{R}((\lambda_0-L)^{p+1}) = ...,$$

and

$$X = \mathcal{N}((\lambda_0-L)^p)\oplus\mathcal{R}((\lambda_0-L)^p).$$

In the sequel we use the following notation:

$$M_{\lambda_0}(L) = \bigcup_{k=1}^{\infty} \mathcal{N}((\lambda_0-L)^k).$$

We call $M_{\lambda_0}(L)$ the *generalized eigenspace* of L associated with the value λ_0, and $\dim M_{\lambda_0}$ the *algebraic multiplicity* of λ_0 (which is zero if λ_0 is not an eigenvalue of L).

REMARK A.3.2. If the eigenvalue λ_0 of L has finite algebraic multiplicity, then there exists a finite integer $p \geqslant 0$ such that

$$M_{\lambda_0}(L) = \mathcal{N}((\lambda_0-L)^p).$$

A subset of the spectrum which turns out to be quite useful in applications is the so-called *(Browder) essential spectrum* $\sigma_{ess}(L)$.

DEFINITION. The complex number λ belongs to $\sigma_{ess}(L)$ if at least one of the following conditions is satisfied:
(C_1) λ is a limit point of $\sigma(L)$,
(C_2) $\mathfrak{R}(\lambda - L)$ is not closed,
(C_3) $M_\lambda(L)$ is infinite-dimensional.

The following result provides a very useful characterization of the non-essential elements of the spectrum. These are sometimes called the *normal eigenvalues.*

THEOREM A.3.3. *Let $\lambda_0 \in \sigma(L) \setminus \sigma_{ess}(L)$. Then λ_0 is a pole of $R(\lambda, L)$ and the residue B_{-1} is an operator of finite rank.*

PROOF. Let $\lambda_0 \in \sigma(L) \setminus \sigma_{ess}(L)$, then λ_0 is an isolated point of $\sigma(L)$ and $M_0 := M_{\lambda_0}(L)$ has finite dimension. By virtue of Theorem A.3.1 it suffices to show that $M_0 = X_0$ where $X_0 = \mathfrak{R}(B_{-1})$. We note that $M_0 = \mathfrak{N}((\lambda_0 - L)^p)$ for some infinite integer $p \geqslant 1$, since M_0 is finite-dimensional (see Remark A.3.2). First we show that $M_0 \subseteq X_0$. Obviously $\mathfrak{N}((\lambda_0 - L)^0) = \{0\} \subseteq X_0$. Let $m \geqslant 0$ and suppose that $\mathfrak{N}((\lambda_0 - L)^m) \subseteq X_0$. Let $x \in \mathfrak{N}((\lambda_0 - L)^{m+1})$ and $y = (\lambda_0 - L)x$, then $y \in \mathfrak{N}((\lambda_0 - L)^m) \subseteq X_0$. Let Γ be the contour of (A.3.4). If $\lambda \in \Gamma$, then

$$y = \lambda_0 x - Lx = (\lambda_0 - \lambda)x + (\lambda - L)x,$$

or equivalently

$$(\lambda - L)^{-1}x = (\lambda_0 - \lambda)^{-1}(\lambda - L)^{-1}x - (\lambda_0 - \lambda)^{-1}x.$$

Since $y \in X_0 = \mathfrak{R}(B_{-1})$ we have $y = B_{-1}z$ for some $z \in X$, hence

$$(\lambda - L)^{-1}y = (\lambda - L)^{-1}B_{-1}z = B_{-1}(\lambda - L)^{-1}z.$$

Now integration along the contour Γ yields

$$\begin{aligned} B_{-1}x &= \frac{1}{2\pi i}\int_\Gamma (\lambda - L)^{-1}x\,d\lambda \\ &= \frac{1}{2\pi i}\int_\Gamma (\lambda_0 - \lambda)^{-1}(\lambda - L)^{-1}y\,d\lambda - \frac{1}{2\pi i}\int_\Gamma (\lambda_0 - \lambda)^{-1}x\,d\lambda, \end{aligned}$$

or equivalently,

$$\begin{aligned} x &= \frac{1}{2\pi i}\int_\Gamma (\lambda_0 - \lambda)^{-1}(\lambda - L)^{-1}y\,d\lambda - B_{-1}x \\ &= B_{-1}\left(\frac{1}{2\pi i}\int_\Gamma (\lambda_0 - \lambda)^{-1}(\lambda - L)^{-1}z\,d\lambda - x\right), \end{aligned}$$

which gives us that $x \in \mathfrak{R}(B_{-1}) = X_0$. Therefore $M_0 \subseteq X_0$. We assert that $M_0 = X_0$. Define $L_0 = (L - \lambda_0)|_{X_0}$. We know from Theorem A.3.1 that L_0 is bounded and $\sigma(L_0) = \{0\}$. Therefore L_0 is quasinilpotent. Let $\tilde{X}_0 = X_0/M_0$, with the usual Banach space structure, and let $\tilde{L}_0$ be the operator on $\tilde{X}_0$ induced by L_0, i.e. $\tilde{L}_0(x + M_0) = L_0 x + M_0$. For every $k \geqslant 1$, $\tilde{L}_0^k$ is the operator on $\tilde{X}_0$ induced by L_0^k, and $\|\tilde{L}_0^k\| \leqslant \|L_0^k\|$, whence we get $r(\tilde{L}_0) \leqslant r(L_0) = 0$. On the other hand, $\mathcal{R}(\tilde{L}_0)$ is a closed subspace of $\tilde{X}_0$, as its inverse image in X_0 under the quotient mapping is $\mathcal{R}(L_0) + M_0$, $\mathcal{R}(L_0)$ is a closed subspace in X_0, and the sum of a closed subspace and a finite-dimensional subspace is closed. Finally $\tilde{L}_0$ has a trivial kernel, for $\tilde{L}_0(x + M_0) = M_0$ implies that $L_0 x \in M_0$, and so $x \in M_0$. This proves that $\tilde{L}_0$ is a one-to-one mapping of $\tilde{X}_0$ onto the closed subspace $\mathcal{R}(\tilde{L}_0)$ of $\tilde{X}_0$. If $\tilde{X}_0$ is not trivial, then by the open mapping theorem, there is a constant $c > 0$ such that

$$\|\tilde{L}_0 x\| \geqslant c\|x\|, \quad x \in \tilde{X}_0.$$

This, however, implies that $\|\tilde{L}_0^k x\| \geqslant c^k \|x\|$, $x \in \tilde{X}_0$, hence $r(\tilde{L}_0) \geqslant c$ contradicting $r(\tilde{L}_0) = 0$. We may conclude therefore that $\tilde{X}_0 = \{0\}$, hence $X_0 = M_0$. □

This result implies that for $\lambda \notin \sigma_{ess}(L)$ there exists an integer $p \geqslant 0$ such that
$(C_1{}')$ $\mathcal{N}(\lambda - L) \subset \mathcal{N}((\lambda - L)^2) \subset \cdots \subset \mathcal{N}((\lambda - L)^p) = \mathcal{N}((\lambda - L)^{p+1}) = \cdots$
$(C_2{}')$ $\mathcal{R}(\lambda - L) \supset \mathcal{R}((\lambda - L)^2) \supset \cdots \supset \mathcal{R}((\lambda - L)^p) = \mathcal{R}((\lambda - L)^{p+1}) = \cdots$
$(C_3{}')$ $\mathcal{N}(\lambda - L)$ has finite dimension.
The converse result holds under some additional hypotheses.

Proposition A.3.4. *Let L be a closed operator with a nonempty resolvent set. Let $\lambda \in \mathbb{C}$ be such that the conditions $(C_1{}')-(C_3{}')$ hold for some integer $p \geqslant 0$. Then $\lambda \notin \sigma_{ess}(L)$.*

We shall not present a proof of this result. Below we shall prove a spectral mapping theorem for the Browder essential spectrum. For that purpose we need the following characterization of $\sigma_{ess}(L)$ with L a norm bounded operator. Let $\mathcal{K}(X)$ be the ideal of compact operators.

Lemma A.3.5. *Let $L \in \mathcal{L}(x)$ and let $\mathcal{U}_L$ be a maximal abelian subalgebra of $\mathcal{L}(X)$ containing L. Let*

$$\pi : \mathcal{U}_L \to \mathcal{U}_L / (\mathcal{U}_L \cap \mathcal{K}(X))$$

be the canonical quotient homomorphism (note that $\mathcal{U}_L / (\mathcal{U}_L \cap \mathcal{K}(X))$ is again a Banach algebra). Then

$$\sigma_{ess}(L) = \sigma(\pi(L)).$$

Proof. If the complex-valued function f is locally holomorphic on $\sigma(L)$, then $f(L) \in \mathcal{U}_L$. Now suppose $\lambda \notin \sigma_{ess}(L)$. Then $X = \mathcal{N}((\lambda - L)^p) \oplus \mathcal{R}((\lambda - L)^p)$ for some integer $p \geqslant 0$. If P is the projection on $\mathcal{N}((\lambda - L)^p)$ along $\mathcal{R}((\lambda - L)^p)$, then P has finite dimensional range, hence $P \in \mathcal{K}(X)$. Since P is of the form $f(L)$, we have $P \in \mathcal{U}_L \cap \mathcal{K}(X)$. It follows easily that $P + (\lambda - L)(I - P)$ is a one-to-one mapping of X onto X, and its inverse certainly belongs to $\mathcal{U}_L$. Therefore the operator

$$\pi(P + (\lambda - L)(I - P)) = \pi(\lambda - L) = \lambda E - \pi(L),$$

where E is the identity in $\mathcal{U}_L/(\mathcal{U}_L \cap \mathcal{K}(x))$, is invertible. It follows that $\lambda \notin \sigma(\pi(L))$, which shows that

$$\sigma(\pi(L)) \subset \sigma_{ess}(L).$$

Now suppose that $\lambda \notin \sigma(\pi(L))$. Then there exist operators $B \in \mathcal{U}_L$ and $C \in \mathcal{U}_L \cap \mathcal{K}(X)$ such that

$$B(\lambda - L) = (\lambda - L)B = I + C.$$

This implies that for $k = 0, 1, 2, \cdots$

$$\mathcal{N}((\lambda - L)^k) \subset \mathcal{N}((I + C)^k)$$
$$\mathcal{R}((\lambda - L)^k) \supseteq \mathcal{R}((I + C)^k).$$

It follows from the Riesz theory of compact operators that the conditions $(C_1') - (C_3')$ are satisfied, and now Proposition A.3.4 implies that $\lambda \notin \sigma_{ess}(L)$ which takes care of the inclusion

$$\sigma_{ess}(L) \subseteq \sigma(\pi(L)). \quad \square$$

THEOREM A.3.6 (SPECTRAL MAPPING THEOREM).

a) *Let L be a bounded linear operator, and f a complex-valued function which is locally holomophic on $\sigma(L)$. Then*

$$\sigma_{ess}(f(L)) = f(\sigma_{ess}(L)).$$

b) *Let L be a closed unbounded linear operator with nonempty resolvent set, and let f be a complex-valued function which is locally holomorphic on the extended spectrum $\sigma(L) \cup \{\infty\}$. Then*

$$\sigma_{ess}(f(L)) \cup \{\infty\} = f(\sigma_{ess}(L) \cup \{\infty\}).$$

PROOF.

a) Let L be a bounded linear operator, and let D be a bounded open neighbourhood of $\sigma(L)$ with boundary ∂D (positively oriented) such that $\overline{D}$ is contained in the domain of f. Since $\sigma(\pi(L)) \subseteq \sigma(L)$, we can define $f(\pi(L))$ in the usual manner. Since π is a continuous homomorphism, we get

$$\begin{aligned} f(\pi(L)) &= \frac{1}{2\pi i} \oint_{\partial D} f(\lambda)[\lambda E - \pi(L)]^{-1} d\lambda \\ &= \frac{1}{2\pi i} \int_{\partial D} f(\lambda) \pi((\lambda I - L)^{-1}) d\lambda \\ &= \pi\Big(\frac{1}{2\pi i} \int_{\partial D} f(\lambda)(\lambda I - L)^{-1} d\lambda\Big) \\ &= \pi(f(L)). \end{aligned}$$

Obviously, $f(L) \in \mathcal{U}$ and by the previous lemma we get

$$\sigma_{ess}(f(L)) = \sigma(\pi(f(L))) = \sigma(f(\pi(L)))$$
$$= f(\sigma(\pi(L))) = f(\sigma_{ess}(L)).$$

b) Now let L be an arbitrary closed unbounded operator with $\rho(L) \neq \varnothing$. Let $\alpha \in \rho(L)$, and define $\Phi : \mathbb{C} \cup \{\infty\} \to \mathbb{C} \cup \{\infty\}$ by

$$\Phi(\lambda) := (\alpha - \lambda)^{-1}.$$

Let $T \in \mathcal{L}(X)$ be defined by:

$$T = \Phi(L) = (\alpha I - L)^{-1}.$$

It is easy to show that for every $\lambda \in \mathbb{C}$, $\lambda \neq \alpha$ and $k = 1, 2, \cdots$ we have

$$\mathcal{N}((\Phi(\lambda)I - T)^k) = \mathcal{N}((\lambda I - L)^k),$$
$$\mathcal{R}((\Phi(\lambda)I - T)^k) = \mathcal{R}((\lambda I - L)^k),$$

and from Proposition A.3.4 we obtain that

$$\sigma_{ess}(T) = \Phi(\sigma_{ess}(L) \cup \{\infty\}).$$

Note that $0 \in \sigma_{ess}(T)$ because $\mathcal{R}(T) = \mathcal{D}(L)$ is not closed.

$$\sigma_{ess}(f(L)) \cup \{\infty\} = \sigma_{ess}(f \circ \Phi^{-1}(T)) = f \circ \Phi^{-1}(\sigma_{ess}(T)) =$$
$$f \circ \Phi^{-1}(\Phi(\sigma_{ess}(L) \cup \{\infty\})) = f(\sigma_{ess}(L) \cup \{\infty\}). \quad \square$$

If $L \in \mathcal{L}(K)$, and X is infinite-dimensional, then $\mathcal{U}_L \cap \mathcal{K}(X) \neq \mathcal{U}_L$, and we get from Lemma A.3.5 that

$$\sigma_{ess}(L) = \sigma(\pi(L)) \neq \varnothing.$$

Hence we can define the essential spectral radius by

$$\sigma_{ess}(L) = \sup \{|\lambda| : \lambda \in \sigma_{ess}(L)\}.$$

There exists another characterization of $r_{ess}(L)$ analogous to (A.3.1). Before presenting this characterization, we have to introduce the so-called *measure-of-noncompactness.*

Let Ω be a bounded subset of X, then the diameter diam(Ω) of Ω is given by

$$\text{diam}(\Omega) = \sup \{\|x - y\| : x, y \in \Omega\}.$$

DEFINITION. For a bounded subset Ω of X we define its measure-of-noncompactness $\alpha(\Omega)$ by:

$$\alpha(\Omega) = \inf \{d > 0 : \text{ there exist finitely many subsets } \Omega_1, \ldots, \Omega_n \text{ of } X \text{ such that diam}(\Omega_i) \leq d \text{ and } \Omega \subseteq \textstyle\bigcup_{i=1}^n \Omega_i\}.$$

α satisfies the following elementary properties.

LEMMA A.3.7. *Let Ω,Ω_1,Ω_2 be bounded subsets of X:*

a) $\alpha(\Omega)=\alpha(\overline{\Omega})$

b) $\alpha(\Omega)=0$ *if and only if* $\overline{\Omega}$ *is compact.*

c) $\alpha(\lambda\Omega)=|\lambda|\alpha(\Omega),\ \lambda\in\mathbb{C}$.

d) $\alpha(\mathrm{co}(\Omega))=\alpha(\Omega)$. *Here* $\mathrm{co}(\Omega)$ *denotes the convex hull of* Ω.

e) $\alpha(\Omega_1\cup\Omega_2)=\max\{\alpha(\Omega_1),\alpha(\Omega_2)\}$,
$\alpha(\Omega_1\cap\Omega_2)\leqslant\min\{\alpha(\Omega_1),\alpha(\Omega_2)\}$,
$\alpha(\Omega_1+\Omega_2)\leqslant\alpha(\Omega_1)+\alpha(\Omega_2)$.

For an element $L\in\mathcal{L}(X)$ we define its measure-of-non-compactness $|L|_\alpha$ by:

$$|L|_\alpha = \inf\{k>0 : \alpha(L\Omega)\leqslant k\alpha(\Omega) \text{ for every bounded subset } \Omega \text{ of } X\}.$$

From Lemma A.3.7 one easily deduces the following properties of $|\cdot|_\alpha$.

LEMMA A.3.8. *Let $L,K,L_1,L_2\in\mathcal{L}(X)$.*

a) $|L|_\alpha\leqslant\|L\|$.

b) $|\lambda L|_\alpha\leqslant|\lambda|\,|L|_\alpha,\ \lambda\in\mathbb{C}$.

c) $|L|_\alpha=0$ *if and only if L is compact.*

d) $|L+K|_\alpha=|L|_\alpha$ *if K is compact.*

e) $|L_1+L_2|_\alpha\leqslant|L_1|_\alpha+|L_2|_\alpha$.

f) $|L_1L_2|_\alpha\leqslant|L_1|_\alpha\cdot|L_2|_\alpha$.

For every $L\in\mathcal{L}(X)$ the essential spectral radius is given by

$$r_{ess}(L) = \lim_{n\to\infty}|L^n|_\alpha^{1/n} \qquad \text{(A.3.6)}$$

See NUSSBAUM [1970] for a prove of this result. Note that $r_{ess}(L)=0$ whenever L is a compact operator. This also follows from the Riesz theory for compact operators.

2. PSEUDO RESOLVENTS

If L is a closed linear operator on X, then its resolvent operator $R(\lambda,L)$ is an analytic operator-valued function on $\rho(L)$ and satisfies

$$R(\lambda,L) - R(\mu,L) = (\mu-\lambda)R(\lambda,L)R(\mu,L),$$

for $\lambda,\mu\in\rho(L)$. This equation is usually called the resolvent equation. If $D\subseteq\mathbb{C}$ is an open set and $R:D\to\mathcal{L}(X)$ is an analytic function obeying

$$R(\lambda) - R(\mu) = (\mu-\lambda)R(\lambda)R(\mu), \quad \lambda,\mu\in D, \qquad \text{(A.3.7)}$$

then we call $R(\lambda)$ a *pseudo resolvent.*

PROPOSITION A.3.9. *Let $R:D\to\mathcal{L}(X)$ be a pseudo resolvent. The following statements hold.*

a) $R(\lambda)R(\mu)=R(\mu)R(\lambda)$, $\lambda,\mu\in D$.
b) $\mathfrak{N}(R(\lambda))$ *and* $\mathfrak{R}(R(\lambda))$ *are independent of* λ, $\lambda\in D$.
c) *Let* $\mu\in\mathbb{C}$, $x\in X$ *and* $(\lambda-\mu)R(\lambda)x=x$ *for some* $\lambda\in D$, *then this relation holds for every* $\lambda\in D$.
d) $R(\lambda)$ *is the resolvent operator of a densely defined closed linear operator* L *if and only if* $\mathfrak{N}(R(\lambda))=\{0\}$ *and* $\mathfrak{R}(R(\lambda))$ *is dense in* X. *In this case* L *is uniquely determined by* $R(\lambda)$.

If $\lambda\in\mathbb{C}$ and $\lambda_0\in D$ then there exists at most one element $S\in\mathcal{L}(X)$ such that

$$R(\lambda_0)-S=(\lambda-\lambda_0)R(\lambda_0)S=(\lambda-\lambda_0)SR(\lambda_0).$$

If such an S exists, then it satisfies this equation for every $\lambda_0\in D$, and in that case we say that R can be extended to λ. If $\lambda_0\in D$, then

$$R(\lambda)=\sum_{k=0}^{\infty}(\lambda_0-\lambda)^k R(\lambda_0)^{k+1},$$

for every λ which satisfies $|\lambda-\lambda_0|\leq\|R(\lambda_0)\|^{-1}$. On account of these properties, every pseudo resolvent has a unique maximal extension, defined on an open subset of $\mathbb{C}$.

Let X_0 be a closed subspace of X which is invariant under $R(\lambda)$, i.e. $R(\lambda)X_0\subseteq X_0$, $\lambda\in D$. Let $R_0(\lambda)$ be the restriction of $R(\lambda)$ to X_0, and let $\tilde{R}_0(\lambda)$ be the operators induced by $R(\lambda)$ on the quotient space X/X_0. Then both $R_0(\lambda)$ and $\tilde{R}_0(\lambda)$ define pseudo resolvents.

PROPOSITION A.3.10. *Let* $\lambda\in\overline{D}\setminus D$.
a) *The following assertions are equivalent:*
(i) *R can be extended to* λ.
(ii) R_0 *and* $\tilde{R}_0$ *can both be extended to* λ.
(iii) *For some (every)* $\lambda_0\in D$, $(\lambda_0-\lambda)^{-1}\in\rho(R(\lambda_0))$.
b) R *has a pole in* λ *if both* R_0 *and* $\tilde{R}_0$ *have a pole in* λ. *If* p,p_0 *and* $\tilde{p}_0$ *are the respective orders, then*

$$\max\{p_0,\tilde{p}_0\}\leq p\leq p_0+\tilde{p}_0.$$

PROOF.
a) (i)⇒(ii): If the extension S of $R(\cdot)$ to λ exists, then, by continuity $SX_0\subseteq X_0$. Now $S|_{X_0}$ and the operator induced by S on X/X_0 are the extensions of R_0 and $\tilde{R}_0$ to λ.
(ii)⇒(iii): Let $\lambda_0\in D$. There exist operators $S_0\in\mathcal{L}(X_0)$ and $\tilde{S}_0\in\mathcal{L}(X/X_0)$ such that

$$(I-(\lambda_0-\lambda)R_0(\lambda_0))(I-(\lambda-\lambda_0)S_0)=(I-(\lambda-\lambda_0)S_0)(I-(\lambda_0-\lambda)R_0(\lambda_0))=I,$$

$$(I-(\lambda_0-\lambda)\tilde{R}_0(\lambda_0))(I-(\lambda-\lambda_0)\tilde{S}_0)=(I-(\lambda-\lambda_0)\tilde{S}_0)(I-(\lambda_0-\lambda)\tilde{R}_0(\lambda_0))=I.$$

which shows that $\rho(L)\subseteq\rho(\hat{L})$. Now suppose that $\lambda\in\sigma(L)$. Then $\lambda\in\sigma_{ap}(L)$ or $\lambda\in\sigma_r(L)$. If $\lambda\in\sigma_{ap}(L)$, then there exists a sequence $\{x_n\}_{n\in\mathbf{Z}^+}$ in X with $\|x_n\|=1$ such that $\lim_{n\to\infty}(\lambda x_n - Lx_n)=0$. Now $\hat{x}=\{x_n\}_{n\in\mathbf{Z}^+}+c_0(X)$ satisfies $\|\hat{x}\|=1$ and $\hat{L}\hat{x}=\lambda\hat{x}$, and therefore $\lambda\in\sigma_p(\hat{L})$. On the other hand, $\lambda\in\sigma_{ap}(\hat{L})$ implies that $\lambda\in\sigma_{ap}(L)$ as well, and this proves b). If $\lambda\in\sigma_r(L)$, then $\mathfrak{R}(\lambda-L)^{\perp}\neq\{0\}$. Choose $x_0^*\in\mathfrak{R}(\lambda-L)^{\perp}\setminus\{0\}$ and $x_0\in X\setminus\{0\}$. Let $U\in\mathfrak{L}(X)$ be defined by $Ux=<x,x_0^*>x_0$, then $U(\lambda-L)=0$, hence $\hat{U}(\lambda-\hat{L})=0$, and since $\hat{U}\neq 0$ we get $\lambda\in\sigma(\hat{L})$, which proves a).
c) follows from (*).
Now λ_0 is an isolated point of $\sigma(L)$ if and only if λ_0 is an isolated point of $\sigma(\hat{L})$. It follows readily that for such an element λ_0 the corresponding coefficients in the Laurent expansions of $R(\lambda,L)$ and $R(\lambda,\hat{L})$ near λ_0 are related by the canonical mapping. This proves d). □

We end with an illustration.

PROPOSITION A.3.12. *Let X be a Banach lattice and $L\in\mathfrak{L}(X)$ a positive operator. Then $r(L)\in\sigma(L)$.*

PROOF. If $L=0$ then the result is trivial, so assume that $L>0$. Let $r:=r(L)$ and $\lambda\in\sigma(L)\cap\{\lambda:|\lambda|=r\}$. Then $\lambda\in\partial\sigma(L)\subseteq\sigma_{ap}(L)$. Let $\hat{L}$ be the extension of L to $\hat{X}$, then $\lambda\in\sigma_p(\hat{L})$. Hence there exists an element $\hat{x}\in\hat{X}$, $\|\hat{x}\|=1$, such that $\hat{L}\hat{x}=\lambda\hat{x}$. Since $\hat{L}\geqslant 0$ we get, $\hat{L}|\hat{x}|\geqslant|\hat{L}\hat{x}|=r|\hat{x}|$. If $\mu>r=r(L)$, then

$$R(\mu,\hat{L})|\hat{x}| = \sum_{j=0}^{\infty}\mu^{-j-1}\hat{L}^j|\hat{x}|\geqslant\frac{|\hat{x}|}{\mu-r}.$$

Therefore $\|R(\mu,\hat{L})|\hat{x}|\|\geqslant(\mu-r)^{-1}$. Letting $\mu\downarrow r$ we find $\|R(\mu,\hat{L})|\hat{x}|\|\to\infty$ yielding that $r\in\sigma(\hat{L})=\sigma(L)$. □

Bibliographical Notes

CHAPTER 2

The main reference on the subject of nonlinear evolution governed by dissipative (accretive) operators in general Banach spaces should be the book by BENILAN, CRANDALL and PAZY, which is to appear (already for many years!). The first books written on this topic are those of BARBU [1976], BROWDER [1976], DA PRATO [1976], MARTIN [1976] and PAVEL [1982]. The most recent review article is CRANDALL [1986], where the basic references can be found. As review articles we also mention CRANDALL [1976] and EVANS [1978].

The theory of "nonlinear semigroups" in Banach spaces is a fascinating theory but it is by no means "the" solution to the problem of well-posedness of nonlinear evolution equations in Banach spaces. For this general problem we refer the reader to the monumental work of Kato. For a recent review article on this subject see KATO [1986]. The theory of linear semigroups plays a very important role in the study of nonlinear evolution equations in Banach space, in particular for quasilinear and semilinear equations. Concerning semilinear equations we mention the recent book of VON WAHL [1985]. The theory developed in this chapter is not yet representative for the theory of evolution governed by accretive operators since it only treats the homogeneous initial value problem. The inhomogeneous problem is essentially more difficult. However there are already interesting examples of nonlinear contraction semigroups, which can be obtained by Crandall-Ligett's theorem. One of them is treated in Chapter 4. The results of Sections 2.3 and 2.4 are due to BRÉZIS and PAZY [1970], [1972], and were extended to the inhomogeneous case by BÉNILAN [1972].

CHAPTER 3

References in book form on semigroup theory include BELLENI - MORANTE [1979], BUTZER and BERENS [1967], VAN CASTEREN [1985], DAVIES [1980], DUNFORD and SCHWARTZ [1958], DYNKIN [1965], FATTORINI [1983], FRIEDMAN [1969], GOLDSTEIN [1985], HILLE and PHILLIPS [1957], T. KATO [1976], S. KREIN [1971], LADAS and LAKSHMIKANTHAM [1972], NAGEL [1986], PAZY [1983], REED and SIMON [1978], RIESZ and NAGY [1955], SHOWALTER [1977], STEIN [1970], TANABE [1979], WALKER [1980], YOSIDA [1980] and others.

The content of this chapter is fairly standard, except the theorem of Da Prato - Grisvard of Section 3.6 which does not appear in the references cited above. For further extensions to the noncommutative case we refer the reader to the original paper of DA PRATO and GRISVARD [1975]. The results in Section 3.5 are due to CLEMENT, DIEKMANN, GYLLENBERG, HEIJMANS and THIEME [to appear].

For a survey of the basic notions of functional analysis which are freely used in this chapter we refer the reader to the Appendix of BUTZER and BERENS [1967] and to the Appendix of BREZIS [1973]. One remark about the notation E_h^A. This notation is not standard. In the context of nonlinear semigroups one usually considers accretive operators and $(I+\lambda A)^{-1}$ is called the resolvent of A. In the linear theory one usually considers dissipative operators and the corresponding resolvent of A is given by $R(\lambda,A):=(\lambda I-A)^{-1}$. Since the main part of this notes is concerned with the linear theory, we adopted the "dissipative" notation, but we still use the operator $E_\lambda^A=(I-\lambda A)^{-1}$ instead of $R(\lambda,A)$, which has the nice interpretation of being the operator obtained by performing one "Euler-implicit" step. Hence E_h^A denotes the "Euler operator" associated with A.

CHAPTER 4

Classical references for nonlinear parabolic equations are the book by FRIEDMAN [1964] and the book by LADYŽENSKAJA, SOLONNIKOV and URAL'CEVA [1968]. The equations treated there are nondegenerate. For Problem (I) this means that $\phi'\in(\mu,\mu^{-1})$ on $\mathfrak{D}(\phi)$ for some $\mu\in(0,1)$. Degenerate equations for which ϕ' vanishes at a single point of $\mathfrak{D}(\phi)$ were first considered by OLEINIK, KALASHNIKOV and CHZHOU [1958]. Later much attention was devoted to properties of nonnegative solutions of Problem (I) with $\phi(s)=|s|^m\,\mathrm{sign}(s)$, $m>0$. Observe that for $0<m<1$, $\phi'(r)\to\infty$ as $r\to\infty$ and for $m>1$, $\phi'(0)=0$. For this special ϕ, the partial differential equation is called the porous media equation. For a survey on this equation up to '81, see PELETIER [1981]. For later results, see for instance ARONSON, CAFFARELLI and KAMIN [1983] or VAZQUEZ [1984]. Further references are given there. An example of an equation where ϕ' vanishes on an interval in $\mathfrak{D}(\phi)$ can be found in BERTSCH, DE MOTTONI and PELETIER [1984].

The existence and uniqueness of a solution w of the perturbed problem (P_ϵ) in the case $0\in\beta(0)$ and $0\in\mathrm{int}\,\mathfrak{R}(\beta)$ is from BREZIS and STRAUSS [1973], who studied this problem in a much more general setting. Also the proofs of the important estimates, Propositions 4.9 and 4.10, are given there. The results when passing to the limit for $\epsilon\downarrow 0$ are from BENILAN, BREZIS and CRANDALL [1975], who also considered the higher dimensional case. The results for the case $\mathcal{R}(\beta)\subseteq(0,\infty)$ are from CRANDALL and EVANS [1975]. The continuous dependence on ϕ of bounded weak solutions of Problem (I) was treated by BENILAN and CRANDALL [1981a]. They also characterized the set $\overline{\mathcal{D}(A_\phi)}$. The regularity property given in Section 4.6, for semigroup solutions of Problem (I′) is from BENILAN and CRANDALL [1981b]. In fact, due to a result of VERON [1979], this semigroup solution $u(t,x)$ is bounded on $[\delta,\infty)\times\mathbb{R}$ for every $\delta>0$. Therefore it coincides with the weak solution of Problem (I′). ARONSON and BENILAN [1979] proved for nonnegative weak solutions of Problem (I′)

$$u_t \geqslant -\frac{u}{(m+1)t} \quad \text{in } \mathcal{D}'((0,\infty)\times\mathbb{R})$$

by using a comparison function argument. This result is slightly better than (4.30). It is also sharp, as follows by comparison with the explicit BARENBLATT-solution [1952]. A similar result for a more general ϕ is given by CRANDALL and PIERRE [1982].
Finally we remark that all results concerning the case $0\in\phi(0)$ and $0\in \operatorname{int}\mathcal{D}(\phi)$ can be generalized to higher space dimensions.

CHAPTER 5
The basic theory of analytic semigroups is classical. The main results can be found in the book of HILLE and PHILLIPS [1957]. More recent expositions are to be found in the text books of DAVIES [1980], PAZY [1983], and KRASNOSELSKII, ZABREĬKO, PUSTYLNIK and SOBOLEVSKII [1976].

CHAPTER 6
This chapter discusses the link between regularity theory for inhomogeneous initial value problems on the one hand, and the interpolation theory of Banach spaces on the other.

The main result of this chapter is Theorem 6.10, which is due to DA PRATO and GRISVARD [1979]. This theorem is a special case of an earlier result of theirs, namely Theorem 3.11 of DA PRATO and GRISVARD [1975]. The proof we give here is the one of their later paper DA PRATO and GRISVARD [1979].

In order to formulate Theorem 6.10 we give a brief discussion of the "continuous interpolation method". The definition of this method we give is taken from DA PRATO and GRISVARD [1979], just as the alternative descriptions given in Theorems 6.5 and 6.6. It can be shown that the continuous interpolation method I_θ is closely related to the usual real interpolation method. In fact $I_\theta(E)$ is the closure of E_1 in $(E_1,E_0)_{\theta,\infty}$ (see BERGH & LÖFSTRÖM [1976, Lemma 3.1.3], TRIEBEL [1978, Section 1.7, page 40] and our Theorem 6.3).

For more information on interpolation theory in general, and on function spaces we refer the reader to the textbooks of BERGH & LÖFSTRÖM [1976], and of TRIEBEL [1978]. These books also contain many historical notes.

The discussion of interpolation theory is followed by a short account of some of the standard results on inhomogeneous initial value problems. These results, and some generalizations and related results can be found in PAZY'S book [1983].

CHAPTER 7

During the last decade the literature on positive semigroups has grown considerably. The main reference is the recent book edited by NAGEL [1986]. In this chapter we do not attempt to present a complete survey of the existing literature, but try to give some of the relevant references. We mention in particular the survey paper of BATTY and ROBINSON [1984] on positive semigroups in ordered Banach spaces.

Theorem 7.4 and Proposition 7.6 are due to GREINER, VOIGT and WOLFF [1981], and Proposition 7.7 and Corollary 7.8 are taken from Section 2.5 in BATTY and ROBINSON [1984].

Section 7.2 is mainly based on the paper of ARENDT, CHERNOFF and KATO [1982]. See also BATTY and ROBINSON [1984], Sections 1.6 and 2.1. Proposition 7.16 and Corollary 7.17 can be found in Section 2.2 of the latter paper. Corollary 7.15 goes back to PHILLIPS [1962].

The first part of Section 7.3, in particular Theorems 7.27 and 7.29, is based on ARENDT, CHERNOFF and KATO [1982] and also on BATTY and ROBINSON [1984]. The positive off-diagonal property has been considered originally by EVANS and HANCHE-OLSEN [1979] on norm bounded generators. The application of Theorem 7.29 to $C(K)$ spaces is taken from ARENDT [1984c]. The second part of Example 7.31 and Example 7.32-(ii) are due to BATTY and DAVIES [1983]. Example 7.32-(i) appears in ARENDT [1984a].

The results on the automatic uniform continuity of C_0-semigroups in Grothendieck spaces with the Dunford-Pettis property, in particular in L^∞, are taken from LOTZ [1984], [1985]. It was conjectured by Baillon that every C_0-semigroup in L_∞ is uniformly continuous. KISHIMOTO and ROBINSON [1981] showed that every positive C_0-semigroup in L_∞ is uniformly continuous. LOTZ [1985], and independently COULHON [1984], proved that any positive C_0-semigroup in a Grothendieck $C(K)$-space is uniformly continuous.

The abstract Kato inequality originates from the distributional inequality $\mathrm{Re}[\mathrm{sign}\, f \cdot \Delta f] \leqslant \Delta |f|$, $f \in L_1^{loc}(\mathbb{R}^n)$, for the Laplacian Δ on $\mathbb{R}^n$, which was proved by KATO [1973]. A few years later, B. Simon realized that the Kato inequality for Δ is related to positivity of the semigroup generated by Δ in $L_2(\mathbb{R}^n)$. In fact, it was shown by SIMON [1977] that positivity of a contraction semigroup generated by a form-positive self-adjoint operator A in $L_2(\mu)$ is equivalent to an abstract Kato inequality for A. In this situation, the Kato inequality is closely related to the so-called Beurling-Deny criterion for positivity (see e.g. REED and SIMON [1978], Theorem XIII. 50). It was conjectured by NAGEL and UHLIG [1981] that the abstract Kato inequality (when properly formulated) characterizes positivity of C_0-semigroups in Banach lattices. This conjecture was established, independently by ARENDT [1984a], [1984b] and by SCHEP [1985]. Our exposition in Section 7.4 follows ARENDT [1984b].

CHAPTER 8

PERRON [1907] discovered that for a square matrix $A=(a_{ij})$, all of whose elements a_{ij} are strictly positive, the spectral radius $r(A)$ is a simple eigenvalue with a strictly positive eigenvector. FROBENIUS [1912] showed among other things that the spectrum of such a matrix is fully cyclic. The book of SCHAEFER [1974] contains a rather detailed description of the spectral theory of positive matrices.

Example 8.1 is due to ZABCZYK [1975], but can also be found in DAVIES' book [1980] and in the lecture notes of SCHAPPACHER [1983]. Lemma 8.2 is due to HILLE and PHILLIPS [1957]. The proof we have given here, closely follows that of SCHAPPACHER [1983]. Proposition 8.3 (or parts of it) can be found in most textbooks on semigroup theory, for example HILLE and PHILLIPS [1957] and PAZY [1983]. To our knowledge, Proposition 8.4 is due to WEBB [1985]. Proposition 8.5 was proved by HILLE and PHILLIPS [1957]. A much shorter proof, using the Gelfand transform, was developed by DAVIES [1980]. The essential type $\omega_{ess}(A)$ was introduced by VOIGT [1980] and independently by PRÜ'SS [1981]. However, the first author was seemingly unaware of the characterization of the essential spectral radius of a bounded linear operator using the measure-of-noncompactness (see Appendix 3.1), and he used (8.5) to define $\omega_{ess}(A)$. Our definition of $\omega_{ess}(A)$ (see (8.6)) is that of Prüss. Proposition 8.6 was originally formulated by PRÜ'SS [1981], however he did not give a proof. This was first done by WEBB [1985].

Most of the spectral theory of positive semigroups was developed by the Functional Analysis group in Tübingen, in particular by GREINER [1981, 1982], and our exposition in Sections 8.2 and 8.3 is extracted from GREINER [1981]. Of course, most of the results of this chapter can also be found in the excellent Lecture Notes edited by NAGEL [1986]. Theorem 8.7 is due to GREINER, VOIGT and WOLFF [1981]. We note that in GREINER [1981], Theorem 8.14 is proved under the weaker assumptions which are mentioned in Remark 8.15. The exposition on Markov semigroups in Section 8.3 is taken from DAVIES [1980] (we note that some authors use the terminology Feller-Markov semigroup in stead of just Markov semigroup; in view of the contributions of FELLER [1971] to the theory of stochastic processes this is certainly justified). For a more thorough exposition on Markov semigroups we refer to DYNKIN's book [1965]. GOLDSTEIN [1985] also pays relatively much attention to this subject. Finally the monograph of VAN CASTEREN [1985], which discusses the application of semigroups to probability theory, deserves mentioning. Theorem 8.17 is again due to GREINER [1981]. To conclude with, we mention DERNDINGER [1980] and DERNDINGER and NAGEL [1979], where some related results can be found.

CHAPTER 9

The exposition in Section 9.1 is mainly based on NAGEL [1982], [1984]. Example 9.3 is due to GREINER, VOIGT and WOLFF [1981]. Proposition 9.4 originally goes back to DATKO [1970] and, in the present form, can be found in the book of PAZY [1983]. Theorem 9.5 is due to Greiner, and can be found in DERNDINGER [1980]. Theorem 9.6 is essentially due to GEARHART [1978], but the proof is taken from NAGEL [1982]. The result of Theorem 9.7 can be found in NAGEL [1982] and in GREINER and NAGEL [1983]. As shown by Example 9.4, the equality $s(A)=\omega_0$ does not hold in general for a positive C_0-semigroup on a Banach lattice X. However, if $X=L_p(\mu)$ with $p=1,2$ or ∞, then $s(A)=\omega_0$ does hold for positive C_0-semigroups ($p=1$ is Theorem 9.5; $p=2$ is Theorem 9.7; if $p=\infty$, then, by Corollary 7.37, a C_0-semigroup is uniformly continuous, and so the equality trivially holds). As far as we know, the problem whether $s(A)=\omega_0$ holds is still open for the remaining values of p. In this connection we mention the following result of VOIGT [1985]. If $\{T(t)\}_{t\geqslant 0}$ is a positive C_0-semigroup defined in $L_p(\mu)$ for all $p\in[p_1,p_2]$ with $p_1<2<p_2$, and if the type is independent of $p\in[p_1,p_2]$, then the equality $\omega_0=s(A)$ holds for all $p\in[p_1,p_2]$. The proof of this result uses interpolation techniques.

The last part of Section 9.1, in particular Proposition 9.8 and Corollary 9.9, is taken from NEUBRANDER [to appear]. Results concerning the asymptotic behaviour of linear C_0-semigroups are scattered all over the literature. Many results can be found in the lecture notes by SCHAPPACHER [1983]. Two important fields of applications are *transport theory* and *structured populations dynamics.* References concerned with the first issue are the book by KAPER, LEKKERKERKER and HEJTMANEK [1983], and the articles by VIDAV [1967, 1970], GREINER [1984a], and VOIGT [1984]. Of historical interest is the older paper by BIRKHOFF [1959]. Structured population dynamics is getting relatively much attention presently. We mention the book by WEBB [1985], the lecture notes edited by METZ and DIEKMANN [1986], and the work by GREINER [1984b] and HEIJMANS [1985, 1986].

Many of the results of Section 9.3 were known for bounded perturbations $B:X\to X$. In this respect we mention PHILLIPS [1953], VIDAV [1968, 1970], and VOIGT [1980, 1984]. The more general theory in Section 9.3 is due to CLEMENT, DIEKMANN, GYLLENBERG, HEIJMANS and THIEME [to appear].

CHAPTER 10

The results of Section 10.1 are due to HEIJMANS [1985, 1986]. In DIEKMANN, HEIJMANS and THIEME [1984], a similar model is studied from a different point of view. The results in Section 10.2 are due to CLEMENT, DIEKMANN, GYLLENBERG, HEIJMANS and THIEME [to appear]. Other applications of semigroup theory to structured population models can be found in WEBB [1985], and in particular in METZ and DIEKMANN [1986].

APPENDIX 1

There is an extensive literature on convex analysis, and we have not tried to be complete in this appendix. Much more information on this and related subjects can be found, for example, in the books of AUBIN and EKELAND [1984], BREZIS [1973], DEIMLING [1985], EKELAND and TEMAM [1976] and HOLMES [1975], from which most of the material in this section is taken.

APPENDIX 2

For an exposition of the theory of ordered Banach spaces we refer to BATTY and ROBINSON [1984]. For the theory of Banach lattices and positive operators we refer to SCHAEFER [1974] and ZAANEN [1983].

APPENDIX 3

A smooth introduction into the spectral theory of linear operators is provided by the book of TAYLOR and LAY [1979] and the book of KATO [1976]. A third reference is the monograph of DOWSON [1978], which is completely concerned with spectral theory, with the emphasis on spectral operators.

There are many definitions of the essential spectrum in the literature, and inevitably it is attended with a lot of confusion. The definition we give here is due to BROWDER [1961]. Other definitions of essential spectrum occurring in the literature are those of the Weyl-, the Wolf- (or Fredholm-), and the Kato essential spectrum (see SCHAPPACHER [1983] and KATO [1976]). The Browder essential spectrum is the largest of them (we refer the interested reader to a paper of NAGY [1976] where eleven different essential spectra are defined, together with their inclusion scheme.

Theorem A.3.1 is taken from TAYLOR and LAY [1979]. Theorem A.3.3 is due to BROWDER [1961]. Proposition A.3.4 is an immediate consequence of Theorem V.10.2 of TAYLOR and LAY [1979]. The Spectral Mapping Theorem A.3.6, and the preparatory Lemma A.3.5 are due to GRAMSCH and LAY [1971]. The measure-of-noncompactness $\alpha(\cdot)$ introduced in this appendix is called the *Kuratowski measure-of-noncompactness.* A very similar notion is the so-called *ball measure-of-noncompactness* $\beta(\cdot)$ defined as follows. Let Ω be a bounded subset of X, then

$$\beta(\Omega) = \inf\{r>0 : \Omega \text{ can be covered by a finite number of balls having radii less than or equal to } r\}.$$

$\alpha(\cdot)$ and $\beta(\cdot)$ are equivalent in the sense that

$$\frac{1}{2}\beta(\Omega) \leqslant \alpha(\Omega) \leqslant 2\beta(\Omega).$$

A number of elementary results on $\alpha(\cdot)$, $\beta(\cdot)$, $|\cdot|_\alpha$ and $|\cdot|_\beta$ (the definition of which is similar to that of $|\cdot|_\alpha$), including Lemma A.3.7 and Lemma A.3.8 can be found in NUSSBAUM [1970], MARTIN [1976] and SCHAPPACHER [1983]. The characterization of the essential spectral radius of a bounded linear operator given by (A.3.6) is due to NUSSBAUM [1970]. We note that for most other definitions of the essential spectrum, the essential spectral radius is the same. If X is a so-called Π_m-space (see GOLDENSTEIN and MARKUS [1965]), then $|L|_\alpha$ and $|L|_\pi := \|\pi(L)\|$ define equivalent seminorms on $\mathfrak{L}(X)$ (here $\pi(L)$ is given by Lemma A.3.5) and in this case (A.3.6) follows immediately from Lemma A.3.5. To our knowledge it is still an open question whether or not $|\cdot|_\alpha$ and $|\cdot|_\pi$ are equivalent for general Banach spaces X.

Some straightforward results on pseudo resolvents like Proposition A.3.9 can be found in the books of PAZY [1983] and KATO [1976]. Proposition A.3.10 is due to GREINER [1981].

The idea to associate a sort of generalized eigenvector (i.e. an element of the extended Banach space $\hat{X}$) with an element in the approximate point spectrum originates from BERBERIAN [1962]. The results in Section A.3.3 where first stated in this form by LOTZ [1968], but can also be found in §V.1 of the book of SCHAEFER [1974].

Bibliography

S. ANGENENT [1986]. Abstract parabolic initial value problems. *Report no. 22, Mathematical Institute, University of Leiden.*

S. ANGENENT [to appear]. Analyticity of the interface of the porous media equation. *Proc. AMS.*

W. ARENDT [1984a]. Kato's inequality; a characterization of generators of positive semigroups. *Proc. Roy. Irish Acad. Sect. A 84,* 155-174.

W. ARENDT [1984b]. *Generators of Positive Semigroups and Resolvent Positive Operators.* Habilitationsschrift, Tübingen, 1984.

W. ARENDT [1984c]. Generators of positive semigroups. In: *Infinite-Dimensional Systems.* Proc. Conf. on Operator Semigroups and Applications, 1-15, Lecture Notes in Mathematics 1067, Springer-Verlag, 1984.

W. ARENDT, P.R. CHERNOFF, T. KATO [1982]. A generalization of dissipativity and positive semigroups. *J. Operator Theory 8,* 167-180.

D.G. ARONSON, PH. BÉNILAN [1979]. Régularité des solutions de l'équation des milieux poreux dans $\mathbb{R}^N$. *C.R. Acad. Sci. Paris Ser. A-B, 288,* 103-105.

D.G. ARONSON, L.A. CAFFARELLI, S. KAMIN [1983]. How an initially stationary interface begins to move in porous medium flow. *SIAM J. Math. Anal. 14,* 639-658.

J.-P. AUBIN, I. EKELAND [1984]. *Applied Nonlinear Analysis.* John Wiley and Sons, New York, 1984.

J.B. BAILLON [1980]. Caractère borné de certains genérateurs de semigroupes linéaires dans les espaces de Banach. *C.R. Acad. Sc. Paris t. 290,* 757-760.

J.M. BALL [1977]. Strongly continuous semigroups, weak solutions and the variation of constants formula. *Proc. AMS 63,* 370-373.

V. BARBU [1976]. *Nonlinear Semigroups and Differential Equations in Banach Spaces.* Noordhoff, Leyden, 1976.

G.I. Barenblatt [1952]. On self-similar motions of compressible fluids in porous media [Russian]. *Prikl. Mat. Mek. 16,* 679-698.

C.J.K. Batty, E.B. Davies [1983]. Positive semigroups and resolvents. *J. Operator Theory 10,* 357-363.

C.J.K. Batty, D.W. Robinson [1984]. Positive one-parameter semigroups on ordered Banach spaces. *Acta Appl. Math. 1,* 221-296.

A. Belleni - Morante [1979]. *Applied Semigroups and Evolution Equations.* Oxford University Press, Oxford, 1979.

Ph. Bénilan [1972]. *Equation d'évolution dans un Espace de Banach Quelconque et Applications.* Thesis, Orsay, 1972.

Ph. Benilan, H. Brezis, M.G. Crandall [1975]. A semilinear elliptic equation in $L^1(\mathbb{R}^N)$. *Ann. Scuola Norm. Sup. Pisa, Serie IV, Vol. II,* 523-555.

Ph. Benilan, M.G. Crandall [1981a]. The continuous dependence on ϕ of the solutions of $u_t = \Delta\phi(u)$. *Indiana Univ. Math. J. 30,* 161-177.

Ph. Benilan, M.G. Crandall [1981b]. Regularizing effects of homogeneous evolution equations. *In: Contributions to Analysis and Geometry.* D.N. Clark, C. Pecelli, R. Sacksteder (eds.), 23-39, Baltimore, 1981.

Ph. Benilan, M.G. Crandall, A. Pazy [to appear]. *Nonlinear Evolution Governed by Accretive Operators.*

S.K. Berberian [1962]. Approximate proper vectors. *Proc. AMS 13,* 111-114.

J. Bergh, J. Löfström [1976]. *Interpolation Spaces, An Introduction.* Springer-Verlag, Berlin, 1976.

M. Bertsch, P. de Mottoni, L.A. Peletier [1984]. The Stefan Problem with heating: appearance and disappearance of a mushy region. *Nonl. Anal. 8,* 1311-1336.

G. Birkhoff [1959]. Positivity and criticality. *Proc. Symp. Appl. Math. 10,* 116-126, AMS, Providence.

H. Brezis [1973]. *Operateurs Maximaux Monotone et Semigroups de Contractions dans les Espaces de Hilbert,* Math. Studies 5, North-Holland, Amsterdam, 1973.

H. Brezis, M.G. Crandall [1979]. Uniqueness of solutions of the initial value problem for $u_t - \Delta\phi(u) = 0$. *J. Math. Pures et Appl. 58,* 153-162.

H. Brézis, A. Pazy [1970]. Accretive sets and differential equations in Banach spaces. *Israel J. Math. 8,* 367-383.

H. Brézis, A. Pazy [1972]. Convergence and approximation of semigroups of nonlinear operators in Banach spaces. *J. Funct. Anal. 9,* 63-74.

H. Brezis, W. Strauss [1973]. Semi-linear second-order elliptic equations in L^1. *J. Math. Soc. Japan, 25* 565-590.

F.E. Browder [1961]. On the spectral theory of elliptic differential operators. *Math. Ann. 142,* 22-130.

F.E. Browder [1976]. *Nonlinear Operators and Nonlinear Equations of Evolution in Banach Spaces.* Proc. Symp. Pure Math. 18, AMS, Providence, 1976.

P.L. Butzer, H. Berens [1967]. *Semigroups of Operators and Approximation.* Springer-Verlag, Berlin, 1967.

J. van Casteren [1985]. *Generators of Strongly Continuous Semigroups.* Pitman, Boston, 1985.

P. Chernoff [1968]. Note on product formulas for operator semigroups. *J. Funct. Anal. 2,* 238-242.

Ph. Clement, O. Diekmann, M. Gyllenberg, H.J.A.M. Heijmans, H.R. Thieme [to appear]. Perturbation theory for dual semigroups. I. The sun-reflexive case. To appear in: *Math. Ann.* Parts II - IV in preparation.

T. Coulhon [1984]. Suites d'operateurs sur un espace $C(K)$ de Grothendieck, *C.R. Acad. Sc. Paris, t. 298, série I, no. 1,* 13-15.

M.G. Crandall [1973]. A generalized domain for semigroup generators. *Proc. AMS 37,* 434-439.

M.G. Crandall [1976]. An introduction to evolution governed by accretive operators. In: *Dynamical Systems: An International Symposium.* L. Cesari, J. Hale and J. La Salle (eds.), 131-165, Academic Press, New York, 1976.

M.G. Crandall [1986]. Nonlinear semigroups and evolution governed by accretive operators. In: *Proc. Symp. Pure Math. 45, Part I,* 305-337, AMS, Providence, 1986.

M.G. Crandall, L.C. Evans [1975]. A singular semilinear equation in $L^1(\mathbb{R})$. *MRC Technical Summary Report # 1566.*

M.G. Crandall, T. Liggett [1971a]. A theorem and a counterexample in the theory of semigroups of nonlinear transformations. *Trans. AMS 160,* 263-278.

M.G. Crandall, T. Liggett [1971b]. Generation of semigroups of nonlinear transformations on general Banach spaces. *Amer. J. Math. 93,* 265-298.

M.G. Crandall, M. Pierre [1982]. Regularizing Effects for $u_t + A_\phi(u) = 0$ in L^1. *J. Funct. Anal. 45,* 194-212.

M.G. Crandall, L. Tartar [1980]. Some relations between nonexpansive and order preserving mappings. *Proc. AMS 78,* 385-390.

G. da Prato [1976]. Applications croissantes et équations d'évolution dans les espaces de Banach. *Inst. Math. Vol. II.* Academic Press, London, 1976.

G. da Prato, P. Grisvard [1975]. Sommes d'opérateurs non linéaires et équations différentielles opérationelles. *J. Math. Pures Appl. 54,* 305-387.

G. Da Prato, P. Grisvard [1979]. Equations d'évolution abstraites non linéaires de type parabolique. *Ann. Mat. Pura. Appl. IV, 120,* 329-396.

R. Datko [1970]. Extending a theorem of A.M. Liapunov to Hilbert space. *J. Math. Anal. Appl. 32,* 610-616.

E.B. Davies [1980]. *One-parameter Semigroups.* Academic Press, London, 1980.

K. Deimling [1985]. *Nonlinear Functional Analysis.* Springer-Verlag, Berlin, 1985.

R. Derndinger [1980]. Uber das Spektrum positiver Generatoren. *Math. Z. 172,* 281-293.

R. Derndinger, R. Nagel [1979]. Der Generator stark stetiger Verbandshalbgruppen auf $C(X)$ und dessen Spektrum. *Math. Ann. 245,* 159-177.

O. Diekmann, H.J.A.M. Heijmans, H.R. Thieme [1984]. On the stability of the cell size distribution. *J. Math. Biol. 19,* 227-248.

H.R. DOWSON [1978]. *Spectral Theory of Linear Operators.* Academic Press, London, 1978.

N. DUNFORD, J.T. SCHWARTZ [1958]. *Linear Operators. Part I: General Theory.* Interscience, New York, 1958.

E.B. DYNKIN [1965]. *Markov Processes I.* Springer-Verlag, Berlin, 1965.

I. EKELAND, R. TEMAM [1976]. *Convex Analysis and Variational Problems.* North-Holland, Amsterdam 1976.

D.E. EVANS, H. HANCHE-OLSEN [1979]. The generators of positive semigroups. *J. Funct. Anal. 32,* 207-212.

L.C. EVANS [1977]. Nonlinear evolution equations in an arbitrary Banach space. *Israel J. Math. 26,* 1-42.

L.C. EVANS [1978]. Application of nonlinear semigroup theory to certain partial differential equations. In: *Nonlinear Evolution Equations.* M.G. Crandall (ed.), 163-188, Academic Press, New York, 1978.

H.O. FATTORINI [1983]. *The Cauchy Problem.* Addison-Wesley, London, 1983.

W. FELLER [1971]. *An Introduction to Probability Theory and its Applications II (2nd ed.).* John Wiley & Sons, New York, 1971.

A. FRIEDMAN [1964]. *Partial Differential Equations of Parabolic Type.* Prentice-Hall, Englewood Cliffs, NJ, 1964.

A. FRIEDMAN [1969]. *Partial Differential Equations.* Holt, Reinhart and Winston, New York, 1969.

G. FROBENIUS [1912]. Uber Matrizen aus nicht-negativen Elementen. *Sitz-Berichte Kgl. Preuss. Akad. Wiss. Berlin (1912),* 456-477.

L. GEARHART [1978]. Spectral theory for contraction semigroups on Hilbert spaces. *Trans. AMS 236,* 385-394.

L.S. GOLDENSTEIN, A.S. MARKUS [1965]. On a measure of noncompactness of bounded sets and linear operators. Studies in Algebra and Math. Anal. [Russian], *Izdat, Karta Moldovenjaski, Kishinev,* 45-54.

J.A. GOLDSTEIN [1985]. *Semigroups of Linear Operators and Applications.* Oxford University Press, New York, 1985.

B. GRAMSCH, D. LAY [1971]. Spectral mapping theorems for essential spectra. *Math. Ann. 192,* 17-32.

G. GREINER [1981]. Zur Perron-Frobenius-Theorie stark stetiger Halbgruppen. *Math. Z. 177,* 401-423.

G. GREINER [1982]. Spektrum und Asymptotik stark stetiger Halbgruppen positiver Operatoren. *Sitz. -ber. Heidelb. Akad. Wiss., Math. - Nat. Kl., 1982, 3 Abh.*

G. GREINER [1984a]. Spectral properties and asymptotic behaviour of the linear transport equation. *Math. Z. 185,* 167-177.

G. GREINER [1984b]. A typical Perron-Frobenius theorem with applications to an age-dependent population equation. In: *Infinite Dimensional Systems.* Proc. Conf. on Operator Semigroups and Applications, 86-100, Lecture

Notes Math. 1067, Springer-Verlag, 1984.

G. GREINER, R. NAGEL [1983]. On the stability of strongly continuous semigroups of positive operators on $L^2(\mu)$. *Annali Scuola Normale Sup. Pisa 10,* 257-262.

G. GREINER, J. VOIGT, M. WOLFF [1981]. On the spectral bound of the generators of semigroups of positive operators. *J. Operator Theory 5,* 245-256.

H.J.A.M. HEIJMANS [1985]. *Dynamics of Structured Populations.* PhD Thesis, Amsterdam, 1985.

H.J.A.M. HEIJMANS [1986]. Structured populations, linear semigroups, and positivity. *Math. Z. 191,* 599-617.

D. HENRY [1981]. *Geometric Theory of Semilinear Parabolic Equations.* Lecture Notes in Mathematics 840, Springer-Verlag, 1981.

E. HILLE, R.S. PHILLIPS [1957]. *Functional Analysis and Semigroups.* AMS, Providence, 1957.

R.B. HOLMES [1975]. *Geometric Functional Analysis and its Applications.* Springer-Verlag, New York, 1975.

H.G. KAPER, C.G. LEKKERKERKER, J. HEJTMANEK [1982]. *Spectral Methods in Linear Transport Theory.* Birkhäuser, Basel, 1982.

T. KATO [1973]. Schrödinger operators with singular potentials. *Israel J. Math. 13,* 135-148.

T. KATO [1975]. Quasi-linear equations of evolution, with applications to partial differential equations. In: *Lecture Notes in Mathematics 448,* 25-70, Springer-Verlag, 1975.

T. KATO [1976]. *Perturbation Theory for Linear Operators (2nd ed.).* Springer-Verlag, Berlin, 1976.

T. KATO [1978]. Trotter's product formula for some nonlinear semigroups. In: *Nonlinear Evolution Equations.* M.G. Crandall (ed.), 155-162, Academic Press, New York, 1978.

T. KATO [1985]. Nonlinear equations of evolution in Banach spaces. In: *Proc. Symp. Pure Math. 45, Part II,* 9-24, AMS, Providence, 1986.

W. KERSCHER, R. NAGEL [1984]. Asymptotic behaviour of one-parameter semigroups of positive operators. *Acta Appl. Math. 2,* 323-329.

A. KISHIMOTO, D.W. ROBINSON [1981]. Subordinate semigroups and order properties. *J. Austral. Math. Soc. (series A) 31,* 59-76.

Y. KOBAYASHI [1974]. On approximation of nonlinear semigroups. *Proc. Japan. Acad. 50,* 729-734.

M.A. KRASNOSELSKII, P.P. ZABREIKO, E.I. PUSTYLNIK P.E. SBOLEVSKII [1976]. *Integral Operators in Spaces of Summable Functions.* Noordhoff, Leyden, 1976.

S.G. KREIN [1971]. *Linear Differential Equations in Banach Spaces.* Translations AMS 29, Providence, 1971.

G.E. LADAS, V. LAKSHMIKANTHAM [1972]. *Differential Equations in Abstract Spaces.* Academic Press, New York, 1972.

O.A. LADYŽENSKAJA, V.A. SOLONNIKOV N.N. URAL'CEVA [1968] *Linear and Quasilinear Equations of Parabolic Type.* Translations of Mathematical Monographs, Volume 23, AMS, Providence, 1968.

H.P. Lotz [1968]. Uber das Spektrum positiver Operatoren. *Math. Z. 108,* 15-32.

H.P. Lotz [1984]. Tauberian theorems for operators on L^∞ and similar spaces. In: *Functional Analysis, Surveys and Recent Results III.* K.-D. Bierstedt and B. Fuchssteiner (eds.), 117-133, North-Holland, Amsterdam, 1984.

H.P. Lotz [1985]. Uniform convergence of operators on L^∞ and similar spaces. *Math. Z. 190,* 207-220.

G. Lumer, R.S. Phillips [1961]. Dissipative operators in a Banach space. *Pacific J. Math. 11,* 679-698.

R.H. Martin [1976]. *Nonlinear Operators and Differential Equations in Banach Spaces.* John Wiley & Sons, New York, 1976.

J.A.J. Metz, O. Diekmann (eds) [1986]. *Dynamics of Physiologically Structured Populations.* Lecture Notes in Biomathematics 68, Springer-Verlag, 1986.

I. Miyadera, S. Ôharu [1970]. Approximation of semigroups of nonlinear operators. *Tôhoku Math. J. 22,* 24-27.

R. Nagel [1982]. Zur Charakterisierung stabiler Operatorhalbgruppen. *Semesterbericht Funktionalanalysis Tübingen, Wintersemester 1981/82.*

R. Nagel [1984]. What can positivity do for stability?. In: *Functional Analysis, Surveys and Recent Results III,* K.-D. Bierstedt and B. Fuchssteiner (eds.), 145-154, North-Holland, Amsterdam, 1984.

R. Nagel (ed) [1986]. *One-parameter Semigroups of Positive Operators.* Lecture Notes in Mathematics 1184, Springer-Verlag, 1986.

R. Nagel, H. Uhlig [1981]. An abstract Kato inequality for generators of positive semigroups on Banach lattices. *J. Operator Theory 6,* 113-123.

B. Nagy [1976]. Spectral mapping theorems for semigroups of operators. *Acta Sci. Math. 38,* 343-351.

F. Neubrander [to appear]. Laplace transform and asymptotic behaviour of strongly continuous semigroups. *Houston J. Math.*

R.D. Nussbaum [1970]. The radius of the essential spectrum. *Duke Math. J. 38,* 473-478.

O.A. Oleinik, A.S. Kalashnikov, Y.-L. Chzhou [1958]. The Cauchy problem and boundary problems for equations of the type of non-stationary filtration [Russian]. *Izv. Akad. Nauk SSSR Ser. Mat. 22,* 667-704.

N.H. Pavel [1982]. *Analysis of Some Nonlinear Problems in Banach Spaces and Applications.* Mimeographed Notes, University of Iasi, 1982.

A. Pazy [1983]. *Semigroups of Linear Operators and Applications to Partial Differential Equations.* Springer-Verlag, New York, 1983.

L.A. Peletier [1981]. The porous media equation. In: *Applications of Nonlinear Analysis in the Physical Sciences.* H. Amann et al. (eds), 229-241, Pitman, London, 1981.

O. Perron [1907]. Zur Theorie der Matrices. *Math. Ann. 64,* 248-263.

R.S. Phillips [1953]. Perturbation theory for semi-groups of linear operators. *Trans. AMS 74,* 199-221.

R.S. Phillips [1962]. Semigroups of positive contraction operators. *Czech. Math. J. 12,* 294-313.

J. PRÜSS [1981]. Equilibrium solutions of age-specific population dynamics of several species. *J. Math. Biol. 11,* 65-84.

M. REED, B. SIMON [1978]. *Methods of Modern Mathematical Physics IV: Analysis of Operators.* Academic Press, New York, 1978.

S. REICH [1986]. Nonlinear semigroups, holomorphic mappings and integral equations. In: *Proc. Symp. Pure Math 45, Part 2,* 307-324, AMS, Providence, 1986.

F. RIESZ, B. SZ-NAGY [1955]. *Functional Analysis.* Ungar, New York, 1955.

H.H. SCHAEFER [1974]. *Banach Lattices and Positive Operators.* Springer-Verlag, Berlin, 1974.

W. SCHAPPACHER [1983]. *Asymptotic Behaviour of Linear C_0-semigroups.* Lecture Notes, Quaderni, Bari, 1983.

A.R. SCHEP [1985]. Weak Kato-inequalities and positive semigroups. *Math. Z. 190,* 305-314.

R.E. SHOWALTER [1977]. *Hilbert Space Methods for Partial Differential Equations.* Pitman, London, 1977.

B. SIMON [1977]. An abstract Kato's inequality for generators of positivity preserving semigroups. *Indiana Univ. Math. J. 26,* 1067-1073.

E.M. STEIN [1970]. *Topics in Harmonic Analysis Related to the Littlewood-Paley Theory.* Annals of Math. Study. Princeton University Press, Princeton, 1970.

M.H. STONE [1932]. On one-parameter groups in Hilbert space. *Ann. Math. 33,* 643-648.

H. TANABE [1979]. *Equations of Evolution.* Pitman, London, 1979.

A.E. TAYLOR, D.C. LAY [1979]. *Introduction to Functional Analysis.* John Wiley & Sons, New York, 1979.

H. TRIEBEL [1978]. *Interpolation Theory, Function Spaces, Differential Operators.* North Holland, Amsterdam, 1978.

J.L. VAZQUEZ [1984]. The interface of one-dimensional flows in porous media. *Trans. AMS. 285,* 717-737.

L. VERON [1979]. Effets regularisant de semi-groupes non lineaires dans des espaces de Banach. *Annales Faculti des Sciences Toulouse 1,* 171-200.

I. VIDAV [1968]. Existence and uniqueness of nonnegative eigenfunctions of the Boltzmann operator. *J. Math. Anal. Appl. 22,* 144-155.

I. VIDAV [1970]. Spectra of perturbed semigroups with applications to transport theory. *J. Math. Anal. Appl. 30,* 264-279.

J. VOIGT [1980]. A perturbation theorem for the essential spectral radius of strongly continuous semigroups. *Monatsh. für Math. 90,* 153-161.

J. VOIGT [1984]. Positivity in time dependent linear transport theory. *Acta Appl. Math. 2,* 311-331.

J. VOIGT [1985]. Interpolation for (positive) C_0-semigroups on L_p-spaces. *Math. Z. 188,* 283-286.

W. VON WAHL [1985]. *The equations of Navier-Stokes and Abstract Parabolic Equations.* Vieweg, Braunschweig/Wiesbaden, 1985.

J.A. WALKER [1980]. *Dynamical Systems and Evolution Equations.* Plenum Press, New York, 1980.

G.F. WEBB [1985]. *Theory of Nonlinear Age-dependent Population Dynamics.*

Marcel Dekker, New York, 1985.

M. WOLFF [1978]. On C_0-semigroups of lattice homomorphisms on a Banach lattice. *Math. Z. 164,* 69-80.

K. YOSIDA [1980]. *Functional Analysis (6th ed.).* Springer-Verlag, Berlin, 1980.

A.C. ZAANEN [1983]. *Riesz Spaces II.* North-Holland, Amsterdam, 1983.

J. ZABCZYK [1975]. A note on C_0-semigroups. *Bull. Acad. Polon. Sci. Sér. Math. Astron. Phys. 23,* 895-898.

E.H. ZARANTONELLO [1973]. Dense single-valuedness of monotone operators. *Israel J. Math. 15,* 158-166.

Subject index